T0138274

The Recombinant University

synthesis

A series in the history of chemistry, broadly construed, edited by Angela N. H. Creager, Ann Johnson, John E. Lesch, Lawrence M. Principe, Alan Rocke, E. C. Spary, and Audra J. Wolfe, in partnership with the Chemical Heritage Foundation.

The Recombinant University

Genetic Engineering and the Emergence of Stanford Biotechnology

DOOGAB YI

THE UNIVERSITY OF CHICAGO PRESS   CHICAGO AND LONDON

DOOGAB YI is assistant professor of history and science and technology studies at Seoul National University, where he teaches the history of science as well as science and the law.

The University of Chicago Press, Chicago 60637
The University of Chicago Press, Ltd., London
© 2015 by The University of Chicago
All rights reserved. Published 2015.
Printed in the United States of America

24 23 22 21 20 19 18 17 16 15   1 2 3 4 5

ISBN-13: 978-0-226-14383-5 (cloth)
ISBN-13: 978-0-226-21611-9 (e-book)
DOI: 10.7208/chicago/9780226216119.001.0001

Library of Congress Cataloging-in-Publication Data

Yi, Doogab, 1974– author.
 The recombinant university : genetic engineering and the emergence of Stanford biotechnology / Doogab Yi.
     pages cm — (Synthesis)
 Includes bibliographical references and index.
 ISBN 978-0-226-14383-5 (cloth : alkaline paper) — ISBN 978-0-226-21611-9 (e-book)
 1. Recombinant DNA—Research—California—Stanford—History—20th century.
 2. Genetic engineering—Research—California—Stanford—History—20th century.
 3. Biotechnology—Research—California—Stanford—History—20th century. 4. Stanford University—History—20th century. I. Title. II. Series: Synthesis (University of Chicago. Press)
 QH442.Y54 2015
 572.8'77—dc23

                          2014031905

♾ This paper meets the requirements of ANSI/NISO z39.48-1992 (Permanence of Paper).

FOR JAWON

Contents

Acknowledgments

I have incurred many debts in the course of this book project, and I am very happy to have the chance finally to acknowledge some of the many people to whom I am indebted. First, I would like to thank Angela Creager, Michael Gordin, Graham Burnett, and Bruno Strasser, for their scholarly guidance and help at the initial stage of this book project. Their scholarly erudition, sharp historical analysis, and encouragement are greatly appreciated.

While book writing can be a solitary activity, I have been very fortunate to be surrounded by wonderful individuals in a supportive environment. I especially learned so much from those at the History of Science Program at Princeton. Among them, I would like to thank Dan Bouk, John DiMoia, Nathan Ha, Kevin Kruse, Elizabeth Lunbeck, the late Michael Mahoney, Tania Munz, Joe November, Daniel Rodgers, Alistair Sponsel, Jeris Stueland, Helen Tilley, and Matt Wisnioski, for their valuable feedback and patience. Those at the Office of History, National Institutes of Health (NIH), and National Library of Medicine provided a collegial environment and useful suggestions for revising my manuscript. Among them, I would like to thank Eric Boyle, David Cantor, Brian Casey, Chin Jou, Sharon Ku, Robert Martensen, Todd Olzewski, Sejal Patel, and Laura Stark. A few NIH scientists and former administrators, especially Mel Depamphilis, the late Norman Latker, and Bernard Talbot, provided their recollections and offered valuable suggestions on my book draft. I owe special thanks to Dr. John E. Niederhuber, then director of the National Cancer Institute, for supporting my Stetten Fellowship from 2008 to 2011.

Many historians of science whom I encountered during this project have provided stimulating points of discussions and sources of inspirations. I especially thank Elizabeth Berman, Soraya de Chadarevian, Nathaniel Comfort, Jean-Paul Gaudillière, Sally Hughes, Myles Jackson, Mark Jones, Daniel Kevles, Cyrus Mody, Michel Morange, Buhm Sohn Park, and Nicholas Rasmussen, for providing useful suggestions and insightful comments.

This book project would not be possible without generous help from Stanford scientists and administrators, who have supported and kept genuine interest in my progress. Current and former Stanford scientists, especially Paul Berg, Stanley Cohen, Ronald Davis, David Hogness, A. Dale Kaiser, the late Arthur Kornberg, Peter Lobban, Janet Mertz, and Charles Yanofsky, offered their recollections and valuable guidance for my research and even provided access to materials still in their possession. Katherine Ku, director of Stanford's Office of Technology Licensing (OTL), generously granted access to its archive. She was kind enough to offer me a desk and a chance to observe Stanford's OTL's operations and meetings. Those at the Stanford University's History and Philosophy of Science and Technology Program, especially Jessica Riskin, provided a valuable institutional affiliation. Most of my research was conducted in the collections of the Stanford University Archives. I would like to thank the archivist Margaret Kimball and library staff members for their excellent services. I also thank Daniel Rakove for many tips for life at Palo Alto.

This book project has been generously supported. First, I appreciate Princeton's academic and financial support for my education and research. Its support has been superb and truly world-class, and I remain grateful for the opportunities provided. I gratefully acknowledge the support of Princeton University's History Department and the Graduate School, as well as fellowships from the National Science Foundation (SES-0522502) and the Chemical Heritage Foundation. The Association of Princeton Graduate Alumni, American Institute of Physics, the Caltech Archives, the National Cancer Institute, and the National Science Foundation provided financial support for my research trips and scholarly travels.

Pieces of this book have appeared in print elsewhere, and I acknowledge their publishers for allowing them to appear here. Chapters 2 and 3 is a greatly revised and expanded version of my article "Cancer, Viruses, and Mass Migration: Paul Berg's Venture into Eukaryotic Biology and the Advent of Recombinant DNA Research and Technology, 1967–1974," *Journal of the History of Biology* 41 (2008): 589–636, reprinted with permission from Springer Science+Business Media B.V. A part of chapter 5 was published previously as "Who Owns What? Private Ownership and the Public Interest in Recombinant DNA Technology in the 1970s," *Isis* 102 (2011): 446–74, reprinted with permission from the University of Chicago Press. For permission to quote from documents in their collections, I thank the Archives of the Stanford Office of Technology Licensing, the Department of Special Collections of Stanford University Libraries, the National Library of Medicine, and the Office of History of the NIH. Many individual scientists, such as Paul Berg, Stanley Cohen, Jack Griffith, David Hogness, A. Dale Kaiser, the late Arthur Kornberg, the late Norman Latker, Peter Lobban, Janet Mertz, and Harold Varmus, generously permitted me to use documents and photographs still in their possession.

As the book was coming into final form, Karen Darling at the University of Chicago Press, Audra Wolfe and the board of *Synthesis*, and anonymous referees for the Press made insightful scholarly suggestions and provided calm and warm guidance. I sincerely thank them for their generous and learned readings of my book manuscript. At the initial stage of my book revision, Sheila Ann Dean provided much needed editorial help. I sincerely thank Elissa Park for her meticulous job in editing the manuscript. These readers (and others) saved me from many gaffes. For those that remain I must take full credit.

During my transition back to Korea, my former teachers at Seoul National University, especially Yung Sik Kim, Kiyoon Kim, and Sungook Hong, provided their help and constant encouragement. My colleagues in the Program in History and Philosophy of Science at Seoul National University, especially Jongtae Lim and Hyungsub Choi, welcomed me. I am happy that I can contribute to the thriving community of science and technology studies in Korea as their colleague.

To my family I owe the greatest thanks. My parents and parents-in-law allowed my travel to the other side of the Pacific, patiently supporting my scholarly wandering. The Schlosssteins became my family in the United States, and their welcome made my stay more comfortable. During the course of this book project, I have been blessed with two children, Madison Sun-Hyung and Daniel Sun-Jun Yi. Their presence has brought a constant joy and appreciation of life. My wife, Jawon, has been a constant source of love and support. We have shared all the difficult and joyful moments in life, and she makes my life sweet. Her wisdom and sense of humor and humility have kept me going, and I thank her with deepest gratitude and sincere love. This book is dedicated to her.

Introduction

When we spliced the profit gene into academic culture, we created a new organism—the recombinant university. We reprogrammed the incentives that guide science. The rule in academia used to be "publish or perish." Now bioscientists have an alternative—"patent and profit."[1] — Paul Berg, Stanford biochemist

The author of the above statement, Paul Berg, was well placed to proclaim that biotechnology had transformed academic biology, thereby creating the modern commercial university in the latter half of the twentieth century. He was a key biomedical researcher at Stanford University, where scientific developments and institutional decisions had played an important role in the emergence of biotechnology as a commercial, private enterprise in the 1970s. When biotechnology was new, especially when it was virtually synonymous with recombinant DNA technology, Berg played a crucial role in its scientific development and demonstrated that any foreign sources of DNA sequences could be recombined.[2] Members of the Stanford Biochemistry Department, to which Berg belonged, and the biomedical research community in the San Francisco Bay Area, contributed to the adoption of recombinant DNA molecules in a variety of applications in molecular biology and genetic engineering. Notably, the Stanford geneticist Stanley Cohen and the University of California biochemist Herbert Boyer, who were part of this early recombinant DNA research network, succeeded in propagating and cloning recombinant DNA molecules inside a bacterial host in 1973, thereby devising a biological factory for manufacturing useful biomedical products. Recombinant DNA technology, as it developed in the Bay Area, then emerged as a novel technology for genetic engineering.[3]

As speculations about genetic engineering grew into a reality with the gene cloning experiment by Cohen and Boyer, the mid-1970s became a time of both enthusiasm and fear in the academic life of biomedical researchers. On May 20, 1974, a *New York Times* article, titled "Animal Gene Shifted to Bacteria: Aid Seen to Medicine and Farm," speculated on enormous agricultural, medical, and industrial potentials of Boyer and Cohen's gene-cloning technology.[4] If scientists could "transplant" animal genes for agricultural, medical, and industrial benefits, they could also clone and express these genes inside microbes, which, in turn, would produce useful gene products, such as fertilizers, antibiotics, and hormones, like insulin, in a mass scale. The following month *Newsweek* ran a story on "The Gene Transplanters," who aimed to turn microbes into "factories" for producing an entire array of valuable chemicals and drugs by transplanting the appropriate genes into bacteria.[5] One scientist even claimed that Boyer and Cohen's genetic engineering technology "may completely change the pharmaceutical industry's approach to making biological elements such as insulin and antibiotics."[6] As anticipation about agricultural, medical, and industrial uses of genetic engineering escalated, Stanford University and the University of California filed a patent application for recombinant DNA technology on behalf of Cohen and Boyer at the US Patent and Trademark Office in November 1974.[7]

Although excitement about genetic engineering and its medical breakthroughs increased, so did resentment over genetic engineering experimentation and commercial promotion regarding recombinant DNA technology. Some concerned scientists and the members of the public debated whether the genetic manipulations of life using the technology posed public health risks; they were concerned because some of the initial genecloning experiments involved the combining of tumor viruses or drug-resistant genes with bacterial genes.[8] The mixing of genes from different species, which some regarded as an affront to the sanctity of life, aroused strong ethical objections to genetic-engineering experiments. The patenting and commercialization of recombinant DNA technology by academic scientists exacerbated such concerns. Many wondered whether profit-seeking motives resulting from this commercialization trumped a naive wish of scientists to see recombinant DNA technology get "transferred to private industry so that public benefits come out as soon as possible."[9]

Promoters of biotechnology exuberantly expressed confidence in the possibility that private ownership would accelerate economic and medical innovations. In 1976, when Boyer, as one of the coinventors of recombinant DNA technology, and the venture capitalist Robert Swanson

founded the start-up biotechnology company Genentech, molecular biologists began to envision an alternative and more lucrative scientific life,
much as Berg had observed. Within a few years, entrepreneurial scientists
achieved what initially seemed impossible—they manufactured medically
useful molecules, notably insulin and human growth hormones, on a large
scale by recombining genes and cloning them. Early biotech ventures
such as Genentech, Biogen, and Amgen, demonstrated that an important
drug like insulin could be mass produced by cloning and expressing genes
in bacteria and turning them into biological factories. The new possibilities, both in science and business, were even more easily imagined after
Boyer's initial investment of five hundred dollars grew to be worth about
seventy million when his company offered its public stock at Wall Street
in October 1980.[10] Right after the debut of biotechnology at Wall Street,
Boyer "rushed out to purchase a Porsche Targa, [and] Swanson [became]
'the first boy millionaire of biotech.'"[11]

Early biotech entrepreneurs, however, had to overcome crucial scientific and regulatory uncertainties involving genetic engineering, along
with the moral and cultural ambiguities that their academic counterparts
pointed out; they often accused them as sellouts. Relentless drive to com-
pete and succeed in a race to clone genes indeed pushed these early entrepreneurial scientists dangerously close to transgressing the shifting regulatory, legal, and ethical boundaries. For example, scientists at Genentech
were accused of using a cloning vector (for the transfer of recombinant
DNA) whose safety had yet to be certified by federal biosafety regulation.
They were also suspected of a "midnight raid" of an academic laboratory
for key research materials, which led to a bitter legal fight between Genentech and the University of California, San Francisco (UCSF).[12] Some
academic researchers who witnessed these instances, such as Berg at Stanford and Keith Yamamoto at UCSF, strongly expressed anxiety about the
moral life of scientists, asking how best to accommodate profit-seeking
motives while maintaining their obligations to the academic community
and the public.[13]

In this book I revisit the emergence of biotechnology and the shifting
scientific, institutional, and moral landscapes that attended the commercialization of academic research in the 1970s, focusing on the academic
community in the San Francisco Bay Area. This was where recombinant
DNA technology was developed and adopted as the first major commercial technology for genetic engineering. My account differs from the more
standard narrative of the emergence of biotechnology and the American

entrepreneurial university.[14] The story of private initiative and venture capitalism in commercial biotechnology, told primarily from the perspective of its promoters, has already been well chronicled and has become largely mythologized as an entrepreneurial success.[15] In that narrative, a strike of genius led to the invention of recombinant DNA technology. Subsequently, a group of entrepreneurial scientists and research administrators heroically reformed old-fashioned norms, regulations, and rules in academia to capitalize on this innovative technology. Institutional and legal arrangements enabled the commercialization of science under a new intellectual property regime and provided a novel model for the entrepreneurial university in the age of declining public support. This obliged academic researchers to generate not only knowledge but also profit. Biotech entrepreneurs, through the infusion of venture capital and a relentless pursuit of innovations in science and business, successfully commercialized the technology, thereby bringing both unforeseeable economic profits to shareholders and medical benefits to the public.

This book presents the history of biotechnology not only as accounts of individual creativity, entrepreneurial venture, and private capitalism but also as a story of collective decisions that were shaped, contested, and eventually made in the same academic community that was central to the emergence of biotechnology. I have admittedly included the perspective of the critics of biotechnology who suggested that biotechnology was as much a promise as an achievement.[16] To them, its promoters often confused private profits with the public interest amidst their personal enormous financial gain. Biotechnology's initial market products, such as insulin, often resembled conventional drugs and did not bring immediate medical miracles to the public as initially hoped. The detractors further regarded entrepreneurial scientists at academic laboratories who sought the commercialization of biomedical technologies as the sellouts who undermined academic norms and culture, which had been vital to creative research. My aim here, however, is not to highlight the contentiousness of biotechnology, which tends to simply echo critiques of commercialization. Nor do I aim to idealize science as an "essentially cooperative and communal effort" in the discussion of biotechnology's impact on academic culture and norms.[17]

The commercialization of recombinant DNA technology in the Bay Area established the framework for some of the fairly larger changes in the biomedical enterprise, including transformations in research practice and culture in the life sciences, economic and moral shifts in biomedical research, and changing relationships among academy, government, and industry. Academic researchers and administrators at Stanford played a crucial role in the scientific and political genealogy of biotechnology; this genealogy encompassed the economic and legal transformations that triggered the commercialization of recombinant DNA technology, and the realignment of the public obligations and the moral life of academic scientists. Bay Area scientists, university administrators, and government officials were fascinated by and increasingly engaged in the economic and political opportunities associated with the privatization of academic research; they became preoccupied by the threats and promises of shifting relations between the pursuit of scientific truth and personal profits and reward structures for scientists, as well as by their efforts to develop a new academic identity engendered by commercial contacts. The attempts of Stanford scientists and administrators to demonstrate the relevance of academic research were increasingly mediated by capitalistic conceptions of knowledge, medical innovation, and the public interest, resulting in the filing of the recombinant DNA patent application.

My close examination of changing scientific agendas, legal practices, and moral assumptions about commercialization in the Bay Area academic community intends to tell a much broader story of the reconfiguration of both academic institutions and commercial enterprise in biomedical research. With the decline of American economic productivity in the 1970s, knowledge as a form of intellectual property became a key solution for promoting economic innovation and medical progress.[18] Some academic administrators, scientists, industrialists, and government officials subsequently tried to bring about legal shifts and moral realignments that encouraged the privatization of academic research for public benefit. This local story of the emergence of biotechnology thus illustrates broader developments in academic institutions, government research policies, and the pharmaceutical industry, which all encouraged privatization. Indeed, in this book I argue that biotechnology was initially a hybrid creation of academic and commercial institutions held together by the assumption of a positive relationship between private ownership and the public interest; this hybridity reflected particular cultural, legal, institutional, and economical contexts for biomedical research that were unique to the United States in the 1970s.

Call for Relevance and the Technical Implementation of Life

This book charts the fate of a particular academic biomedical research community, from its post–World War II growth and its political crisis and opportunity in the 1970s to its involvement in genetic engineering, in order to illuminate the contradictions, opportunities, and paradoxes inherent in commercialization of biology. The biomedical research community in the San Francisco Bay Area primarily consisted of scientists at University of California, Berkeley, UCSF, and Stanford University, all of which experienced exponential growth with an unprecedented increase in economic and political support during the post–World War II years. By building on wartime successes of laboratory-based medical breakthroughs (e.g., discovery of penicillin), leaders of voluntary health organizations including the National Foundation for Infantile Paralysis and the American Cancer Society, politicians, and private citizens cultivated cultural and political support for curing diseases through laboratory investigations. This support entailed large-scale federal patronage of biomedical research, especially in areas that underwent rapid growth, such as molecular genetics, biochemistry, and virology.[19] This patronage system that encouraged the molecularization of the life sciences underscored medical aspirations as a critical cultural and political force that shaped the nation's biomedical research policy.[20] Indeed, the rapid rise of federal support enabled the proliferation of biochemistry, biophysics, microbiology, and molecular biology as autonomous scientific disciplines as their academic departments developed under the broad framework of biomedical research.

The focus of this book on the Department of Biochemistry at the Stanford Medical School illustrates this post–World War II burgeoning of biomedical research. Its establishment in 1959, followed by the relocation of the medical school from San Francisco to the main campus at Palo Alto, was the result of an ambitious, conscientious effort by the university's administrators and scientists to capitalize on postwar organizational, scientific, and political trends; they wanted to strengthen the profile of biomedical research.[21] The Stanford Biochemistry Department, chaired by Nobel laureate Arthur Kornberg, reflected the ascendency of a particular biomedical research style that developed in the context of cultural and political enthusiasm for a war against disease. Many post–World War II biomedical researchers insisted that their basic, fundamental pursuit of the molecular understanding of life would eventually result in medical innovations like penicillin, rather than directly resorting to immediate medical outcomes. Until the mid-1960s, some major breakthroughs in molecular biology, such as the elucidation of the DNA helical structure and the operon model of genetic regulation in *Escherichia coli*, provided a solid foundation, if not an immediate medical outcome, for a new kind of biomedical research.

As lay activists, politicians, and academic researchers and administrators thoroughly embraced the prospect of biomedical research actively conquering diseases, a potential for political backlash emerged when the work could not meet the public's fervent demands for medical miracles. From the mid-1960s, when federal support for biomedical research reached one billion dollars for the first time, medical patrons and politicians increasingly called for medical relevance in biomedical research. The growing insistence for pertinent research coincided with the countercultural movement of the late 1960s, which demanded a shift toward more socially useful medical research (as opposed to medical applications for the military, for example). In 1971, the Nixon administration implemented a large-scale national campaign called the "War on Cancer" in order to conquer the disease by 1976, the American Revolution Bicentennial. These cultural and political developments presented a particularly challenging environment for molecular biologists and biochemists, whose research had primarily focused on the molecular understanding of bacterial viruses and bacteria.[22]

This book indicates how biomedical researchers in the Bay Area responded to an increasing demand for medical relevance during the 1970s and provides a nuanced understanding of the changing scientific and political opportunities for biomedical research. What were the implications of the postwar patronage system, based primarily on the anticipation of medical progress, for practitioners and institutions of biomedical research? In what ways did scientists, academic administrators, and government officials respond to this new political shift? Historians of science have just begun to examine the history of molecular biology beyond the cracking of the genetic code in the mid-1960s, an event often touted as the monumental success of the molecular approach to studying life.[23] More problematic has been how commentators, at a time when large-scale research projects, such as the Human Genome Project and commercial biotechnology companies like Genentech, have promised scientific breakthroughs, biomedical innovations, and financial benefits, simply tend to extend this triumphant narrative of molecular biology of the 1960s well into the 1980s.[24] What happened after this supposed climax of molecular biology in the mid-1960s, in fact, amounted to an intellectual and political crisis of medical relevance in the 1970s.[25]

With no tangible medical progress in hand, the response of biomedical

researchers to this crisis had a significant impact on the trajectory of molecular biology and the emergence of a particular form of biomedical enterprise in the 1980s. A "mass migration" of biomedical researchers who
researched simple organisms like bacteria and bacterial viruses toward
more complex eukaryotic organisms, such as animal viruses and human
cells, was one such crucial response to call for relevance.[26] Some Stanford
biochemists, including Paul Berg and David Hogness, began to work on
the biology of higher organisms. Inspired by their experimental success
with prokaryotes in the two preceding decades, Berg and Hogness tried
hard to pioneer a biochemical and genetic approach in eukaryotic biology.
This boundary crossing from prokaryotic to eukaryotic systems allowed
recombinant DNA technology and other significant new research techniques and agendas to emerge.

At another level, in analyzing the scientific and political genealogy of
recombinant DNA technology, I emphasize the epistemic dynamics in
biological experimentation through which an object in a particular research system became transformed into a research technology.[27] This line
of analysis allows me to provide an alternative to the truncated, teleological narrative of the technology's progression toward genetic engineering.[28] I show, in particular, how Stanford biochemists Berg and
Peter Lobban's artificial synthesis of recombinant DNA molecules reconstituted "life" as a technology in the biology of higher organisms.[29]
Subsequently, Stanford biochemists' recombinant DNA research cross-
pollinated with the work of other San Francisco Bay Area researchers
in a large network that opened a new array of possibilities for genetic
manipulations in the early 1970s. Some scientists—notably the Stanford
biochemist Lobban—recognized recombinant DNA research's possible
application to genetic engineering even before its experimental implementation. As early as 1969, he thought that, if useful gene sequences
could be joined together, then the resulting hybrid DNA molecules
would be a source for the mass production of useful medical and scientific molecules, such as insulin.[30] Before long, Cohen and Boyer cloned
recombinant DNA. Their subsequent research, especially their collaborative work with the Stanford biochemist John Morrow using animal DNA,
made the medical application of recombinant DNA technology an experimental reality.[31] Indeed, the technical implementation of life engendered a set of new connections among different research agendas, experimental systems, and funding patronages, which ultimately led to the
emergence of genetic engineering.

Private Ownership and the Public Interest

The emergence of recombinant DNA technology, with its commercial and medical prospects, provided new institutional and political opportunities and challenges. At the outset, transforming economic and legal assumptions about academic research in the 1970s, especially with regard to its ownership and use, prompted some university administrators, scientists, and government officials to claim proprietary rights over biomedical knowledge and practices. By examining how recombinant DNA technology evolved from a research technology shared by molecular biologists to an intellectual property belonging to private institutions and inventors, I ask the following questions. In what economic, legal, and institutional contexts did this new technocultural entity—genetic engineering—become a legal and commercial form or biotechnology? Can an academic institution claim the ownership of a research result funded largely by the taxpayers? How did scientists, university administrators, and government officials begin to reconceive private ownership as a new way to liberate biomedical discoveries for public benefit? And in what ways did this moral rendering of privatization justify a new set of expectations for commercialization about the use of research results arising from government or public support? During this period, the basic tenets of the Bayh-Dole Act of 1980 (also called the Patent and Trademark Law Amendments Act), which gave universities and small businesses control of their inventions resulting from government support, was first proposed, debated, and eventually adopted.[32]

The discussions of patenting recombinant DNA technology with which a number of Bay Area scientists, university administrators, and government officials were involved in turn points to a shifting institutional relationship of academia with government and private industry. Stanford University had charted the postwar realignment of the academy, the government, and industry in the latter half of the twentieth century.[33] In the years during and after World War II, Stanford had capitalized on the expansion of military-related engineering and physics research, aggressively acquiring funds from the Department of Defense. At the same time, the university cultivated relationships with private industry (mostly in electronics), contributing to the rise of Silicon Valley.[34] In other words, Stanford administrators and scientists tried to maintain the university's financial, intellectual, and political independence from its powerful patrons by asserting scientific research as the core mission of the academic university.[35] Similarly, Stanford administrators adopted this strategy of promoting the profile of biomedical research through its cultivation of federal patronage in rebuilding its medical school during the late 1950s. Federal support for biomedical research in universities had grown exponentially, and the National Institutes of Health (NIH) became the largest funding source of scientific research by early 1960, surpassing the Department of Defense.[36] When Stanford relocated its medical school to the main campus in Palo Alto in the late 1950s, it promoted biomedical research by channeling ample federal funds into the new site. Stanford's heavy reliance on federal funding in its institutionalization of biomedical research provided the institutional and scientific backdrop for reconceiving the commercialization of academic research in the 1970s when economic and political shifts diminished public support.

At the heart of the debate over private ownership of recombinant DNA technology was the redefinition of what public knowledge meant in the academy, as well as its relation to private knowledge in industry. Until the early 1970s, patenting had only played a minor role in the transfer of biomedical knowledge from university settings to private industry.[37] Results from academic biomedical research, mainly funded by the federal government and especially by the NIH, remained in the public domain for the promotion of their widest medical use. Critics of unpatentable knowledge, however, pointed to the growing industrialization of medically related sciences and the pharmaceutical industry, which brought innovations through the patenting of critical chemical substances, methods, and practices.[38] Those in the pharmaceutical industry in turn opposed the use of unpatentable knowledge because of concerns about uncertain proprietary interests and undue competition. Moreover, amidst worries about declining economic productivity in the United States during the 1970s, the profitable use of academic research emerged as a crucial national issue among university administrators, government officials, politicians, and the public at large.

At the same time, criticisms of the public character of academic knowledge and its underuse became a subject of a national discussion among conservative scholars in the new field of law and economics.[39] Research universities, seeking alternative resources of support in the context of dwindling financial and political support for academic research, began to mobilize a renewed belief in the marketplace.[40] If universities could generate royalty incomes from academic patenting, much like the commercializing patterns in the twentieth-century pharmaceutical industry, they could enormously benefit financially.[41] Federal agencies that were under budgetary pressures could in turn reduce the amount of research support if universities were provided with legal and institutional means to generate additional income. Thus, government officials and university administrators tried to open a new legal avenue for privatization by implementing broader legal shifts in federal patent policies; these had initially evolved from the NIH's institutional patent agreements (IPAs).[42] By the early 1970s, then, patenting presented a compelling opportunity for universities to demonstrate the economic and social benefits of biomedical research at a time when politicians and the public demanded relevance and tangible benefits from government-supported research.[43]

This book shows how Stanford research administrators, in alliance with some federal officials, promoted the private ownership of recombinant DNA technology as a viable means of disseminating its medical and industrial potential to the public within the changing political and economic landscape for biomedical research in the 1970s. Stanford administrators and some government officials, encouraged by the recent critics of unpatentable public knowledge, claimed that scientific discoveries and inventions left in the public domain would hinder the wider diffusion of their tangible medical and economic benefits. They claimed that private ownership of inventions arising from public support would also offer a powerful means of instigating biomedical discoveries for everyone's benefit. I argue here that the emergence of these capitalistic conceptions of knowledge in the 1970s provided a broad institutional and legal framework for the privatization of recombinant DNA technology.

In some ways, these mid-twentieth-century changes in the privatization of biomedical research in academic institutions reflect the broad historical development of the industrialization of medically related sciences and the pharmaceutical industry, which had a long history preceding biotechnology.[44] By recognizing the extent to which biomedicine had already been commercialized by the mid-twentieth century, I highlight in this book how the privatization of recombinant DNA technology reconfigured the boundary between the public knowledge of the academy and the private knowledge of industry.[45] The post–World War II institutional arrangement among academy, government, and industry, as well as their respective roles in maintaining public and private domains of knowledge, all disintegrated. Consequently, promoters of both biotechnology and privatization articulated that private ownership possibly benefited the public interest provided a moral justification for commercializing academic inventions, thus making it a new public obligation for academic researchers. Indeed, through its institutional and legal rearrangements with regard to the privatization of recombinant DNA technology, Stanford University became the progenitor of the biotechnology industry, setting the stage for the commercialization of biomedical research.[46]

Moral and "Capitalistic" Economies of Science

From the outset, the patenting of recombinant DNA technology by university licensing offices triggered opposition and controversy among the Stanford scientists who had customarily exchanged and shared research ideas, materials, and tools. The ensuing debates involved issues including scientific credit, inventorship, and obligations, and were especially heated between Stanford biochemists and two outsiders to the Stanford Biochemistry Department, Cohen and Boyer, who had been designated as the technology's inventors. Berg and other biochemists and molecular biologists in the Bay Area, some of whom had been collaborators with Cohen and Boyer, expressed serious reservations about commercialization through patenting. Can a scientist "own" a research technology in basic molecular biology? Were Cohen and Boyer the sole co-inventors of recombinant DNA technology, even though others had contributed to the process? What would be the moral implications of this privatization among biomedical researchers, whose traditional assumptions about public knowledge and reciprocal exchange helped foster scientific productivity and regulate their competitive scientific pursuits?

This book examines how members of the biomedical research community in the Bay Area tried to grapple with the consequences for academic institutions and culture resulting from the commercialization of recombinant DNA technology. In order to direct analytic attention to communal and moral issues in laboratory life arising from commercialization, I explore the Stanford biochemists' moral economy of science. This encompassed communally held views about the proper ways of organizing laboratory life; about social norms and obligations in scientific exchange and knowledge production; and about customs and rules in the distribution of resources.[47] Indeed, Stanford biochemists had developed a distinctive research culture that entailed sharing research materials, instruments, and monies; this was partly sustained by their shared interests in DNA. More importantly, by enabling a vibrant circulation of scientific ideas and materials, their communal research culture significantly contributed to the development of recombinant DNA technology. Cohen and Boyer's cloning work relied on the exchange of scientific ideas, techniques, and research materials among the Stanford biochemists, who also provided the moral, social, and material background against which biotechnology emerged. In turn, this moral economy of science implicitly suggested the immorality of secrecy and unfettered competition, which fueled the controversial and lengthy debate over the invention and ownership of recombinant DNA technology.

Thus, the Stanford Biochemistry Department provides an excellent case through which one can observe how customs, rules, and moral obligations—that is, the moral economy of science—came into being in a local context, and how they shifted and collided with a "capitalistic" economy of science based on proprietary claims on knowledge and its products amid commercialization.[48] This moral economy of science approach has one obvious advantage: it aims to reach beyond existing scholarly literature on the impact of the commercialization of science in academe, which has often been built on a set of abstract distinctions between public and private knowledge, academic research and industrial development, and scientists and entrepreneurs.[49] This set of distinctions has provided an underlying analytic framework that deals with issues that include the deregulation of recombinant DNA research, the impact of the commercialization of biology on the academy, and the rise of the biotechnology industry followed by the patenting of recombinant DNA technology. The prevalent historiography of commercialization often takes "the Mertonian norm of open science" somewhat literally in its analysis of the impact of profit seeking on academic culture, focusing on how such motives and behaviors encroached on the traditional norms of scientific research such as free exchange of materials and ideas.[50] Robert K. Merton's normative analysis of academic culture, with its prescription of open science, has a limited analytic advantage for illustrating the shifting norms of academic research in a historical context.

Historians have fruitfully examined the material culture of communities of experimentalists in order to substantiate distinctive cultural and moral assumptions underlying their particular ways of experimental life. In the competitive fields of biochemistry, genetics, and molecular biology, concerns about moral economies, as they were related to the sharing of materials and techniques, secrecy, and credit, were widespread and of paramount importance. Each local community developed its particular customs, especially in terms of material exchange and credit distribution, and its moral economy sustained and regulated a specific form of research community. For example, some research communities, like *Drosophila* geneticists at Caltech in the 1930s or the *Caenorhabditis elegans* research community in the 1970s, fostered their own moral economy centered on the exchange of their model organisms. James Watson's story of the elucidation of DNA structure, as well as controversies regarding the access and ownership of gene databases in the post–World War II years, also underline the increased concerns more recently about moral issues in the biomedical research community.[51]

In the Bay Area, with the growing prospect of broader biomedical and commercial applications of recombinant DNA technology, competition for scientific priority, credit, and financial reward intensified among biomedical researchers, especially after the first patent filing of Cohen and Boyer's recombinant DNA–cloning technology. New opportunities and obligations came into being as knowledge was also capitalized through a different sort of privatization in academia, bringing about changes in social rules, customs, moral assumptions, and obligations (as in scientific exchange over research materials) among early recombinant DNA researchers. The shifting moral economy at Stanford was not just critical to understanding alliance and conflict among scientists. More importantly, it presaged the larger economic, political, and moral changes in biomedical research in the 1980s. With the enactment of the Bayh-Dole Act of 1980, a new legal regime for academic institutions and researchers, which allowed for the privatization of academic research for the public benefit, provided what some saw as both a moral obligation and a justification for venturing into commercial biotechnology.

This opening of a legal avenue for academic patenting, with its articulation of a causal link between private ownership and public interest, reshaped the scientific life of biomedical researchers.[52] Just as the field of biotechnology emerged, some of those who were trying to place a monetary value on biomedical research faced mounting legal challenges and moral criticism. In the early 1970s, from the point of view of Stanford biochemists, there was neither certainty nor inevitability—in fact, there was quite a lot of skepticism and opposition—about the emergence of biotechnology as a private enterprise serving the public in a better, more efficient way. To some promoters of biotechnology, however, Stanford biochemists were just "those who left behind" the biotechnology revolution in the 1970s. To a new generation of entrepreneurial scientists, a scientist's pursuit of money, like the pursuit of knowledge, would become a calling and still be thought to play a role in fulfilling public obligations.

Stanford biochemists' venture into commercial biotechnology in the late 1970s and early 1980s—especially their establishment of a commercial biotech venture, DNAX—provides an apt lens through which I examine the shifting perception of commercialization in academia and the emergence of new obligations for academic researchers. In a competitive global economy where the production and management of knowledge increasingly became the key to prosperity, the obligations of academic institutions and researchers to the public were increasingly reimagined in terms of economic output. Indeed, Stanford biochemists, along with like-minded academic administrators, private investors, and scientists, tried to experiment with alternative institutional forms of the biomedical enterprise, which would fulfill obligations toward both shareholders and the public. Commercial biotechnology in the end brought a new financial regime and capitalistic economy in science in academia, linking the academy and industry together in the name of public interest.

Summary of Contents

I begin chapter 1 by examining the establishment of Stanford Biochemistry Department in the context of the post–World War II rise of biomedicine as a major research enterprise. I describe its formation in 1959 as a research department focused on the genetics and biochemistry of DNA; this was the focus of most post–World War II biomedical research. This "DNA Department" developed its own style of research management and collaboration involving arrangements of finances and materials among faculty members. Stanford biochemists' moral and political economies of sharing, along with their concentration on DNA, provide an important background through which I examine the dynamics between experimental developments and research environments, which were changing during the 1970s, especially with the emergence of genetic engineering and commercial biotechnology.

I begin to examine these transforming dynamics, as well as the scientific and political genealogy of recombinant DNA technology, in chapter 2 by considering Stanford biochemist Paul Berg's new research venture into the molecular biology of cancer. I argue that the advent of recombi-

nant DNA technology derived from particular experimental opportunities, as well as from scientific commitments to the biology of higher organisms; this occurred with tumor virus model systems in the late 1960s and
early 1970s, when molecular biology was experiencing an intellectual and
political crisis because of its inability to expand its experimental horizon
beyond simple organisms like bacteria. Berg's use of tumor viruses for
studying eukaryotic biology eventually led to the advent of recombinant
DNA technology as a research tool for gene mapping. I explain how this
artificial synthesis of recombinant genes brought about new possibilities
for reinventing nature.

I investigate how a wide range of experimental hybridizations occurred
as concepts and materials were circulated through a system of exchange
and collaboration in the laboratories of Stanford biochemists where recombinant DNA research was centered. The first experimental success of
cloning recombinant DNA molecules in 1972, by Cohen and Boyer, was
followed by a series of molecular-cloning experiments, and Cohen's laboratory emerged as a central node of the research network. At the same
time, the Stanford biochemists grew embroiled in the public debate on the
possible biohazards of recombinant DNA experiments. In chapter 3, by
analyzing the shifting material, intellectual, and social contexts in which
scientists raced to clone recombinant DNA molecules, I illustrate how the
system of exchange in recombinant DNA research transformed, along
with material and moral consequences.

In chapter 4, I describe the shifting moral economy of science in the research network that formed around Stanford biochemists. While the Stanford biochemists' communal research culture initially contributed to the
series of experimental hybridizations that led to the advent of recombinant DNA technology, I show how Cohen and Boyer's work also eventually deprived the other biochemists of an opportunity to claim their own
priority in gene cloning. Frustration among these biochemists heightened
when Cohen began to take advantage of his possession of a cloning vector. This chapter further demonstrates how the pursuit of patenting by scientists and university patent officers subverted the culture of sharing and
collaboration that had earlier sustained the Stanford biochemist research
network.

In chapter 5, I analyze how academic institutions, government agencies, and the nascent biotechnology industry argued about the legal ownership of recombinant DNA technology in the name of the public interest.
In doing so, I reconstruct how a small but influential group of govern-

ment officials and university research administrators introduced a new framework for the commercialization of academic research by linking private ownership and the public interest. These contested developments at Stanford provide a valuable background for analyzing changes in the academic culture that attended commercialization of academic research.

This new support for academic patenting, promoted by the suggested causal link between private ownership and the public interest, further challenged the moral economy of science at the laboratories of Stanford biochemists. In chapter 6, I examine the diverse motives, rationales, and frustrations of scientists who participated in the biotechnology industry during the transforming academic environment of the 1980s.[53] By evaluating the seemingly impossible shift from Stanford biochemists' opposition to the Cohen-Boyer patent to their own later involvement in the biotechnology industry, I discuss how these scientists accommodated increasing commercialization within the universities. I examine what they conceived of as problematic patterns in the commercialization of biomedicine and investigate how these biochemists tried to build what they regarded as an academic research institute, DNAX, in a corporate setting.[54] I also discuss the fate of DNAX.

In the conclusion, I reflect on the coproduction of both late-twentieth-century biotechnology and the American entrepreneurial university, underlining how both promoters and critics attempted to rethink the relationship of the academy and its products, to industry, government, and the public. The history of recombinant DNA technology shows how scientific, institutional, legal, and moral transformations that attended the commercialization of academic research were inextricably linked to the rise of market-oriented ideas about knowledge and its relation to the public at large. The gospel of private ownership indeed provided a new intellectual regime for publicly supported research and shifted the moral landscape for academic research. The institutional and moral reconfiguration of the research university in turn made commercialization a new public obligation of academic researchers. The result was a new hybrid entity—biotechnology—that binds the academy and industry together in the name of public interest, through the circulation of capital.

CHAPTER ONE

Communal Form of DNA Research

In the summer of 1959, Arthur Kornberg, along with five of his former colleagues at Washington University, St. Louis, arrived at Stanford University in California. They became faculty members of the newly established Department of Biochemistry at the Stanford Medical School, with Kornberg as its chair. Kornberg had accepted the chairmanship of the new biochemistry department two years earlier; since then, he had the unique opportunity to assemble his faculty members, organize research programs, and help design the department's laboratory space in the new building at the medical school on the Palo Alto campus. Kornberg had complete freedom to choose new faculty members, and in a short letter to Robert Alway, acting dean of the Stanford Medical School, he recommended six—Melvin Cohn at the rank of professor; Paul Berg and Robert L. Baldwin as associate professors; and David Hogness, A. Dale Kaiser, and I. Robert Lehman as assistant professors. There were no other explanations about his recommendations, except his admission that "the [application] forms are rather short on detail, but I think it should be realized that each of these men is being sought after now by excellent universities throughout the country."[1]

The Stanford Biochemistry Department, assembled de novo under the strong leadership of Kornberg, was literally a "Kornberg" department. The entire faculty, except for Baldwin who came from the University of Wisconsin, had previously been recruited to Kornberg's Microbiology Department at Washington University as his postdoctoral fellows or junior faculty. As Kornberg characterized, they were an "extended family."[2] Moreover, they had been, according to Kornberg, "working as a 'team,' in the loosest sense of this term, in trying to understand heredity and differentiation at a chemical and molecular level."[3] Kornberg, who was awarded the Nobel Prize in 1959 for his research on the synthesis of DNA, was at the forefront of exploring the biochemical basis of heredity as DNA became widely understood as the cell's most important component, its "master plan" that would direct production of its other, various molecular structures. Kornberg had already isolated an enzyme— DNA polymerase—that was believed to direct the chemical synthesis of DNA; this was the work for which he was awarded the Nobel Prize. The group of scientists he had assembled at Washington University in the early 1950s had been examining the chemical and biological properties of DNA. When most of them accompanied Kornberg to Stanford, he formed a "DNA" department as well.

The establishment of the Stanford Biochemistry Department, chaired by the Nobel laureate Kornberg, exemplified the university's commitment to a new vision for biomedicine. The importation of a strong group of scientists in biochemistry and molecular biology to the Stanford Medical School was a key part of its ambitious effort to establish a center for biomedical and clinical research. In July 1953, Stanford president John E. Wallace Sterling, along with members of the board of trustees, had reached a conclusion that the medical school should be an integral part of the university, both geographically and intellectually. The decision to relocate the medical school from San Francisco to the main campus at Palo Alto, in addition to Stanford's vision of a research-oriented medical school, reflected the board's recognition of the significance of biomedical research and its rising cultural, scientific, and financial status in post–World War II research universities.[4] According to Sterling's plan, Stanford's new medical school would establish its basic biomedical departments along with the reorganized clinical departments, and the proximity of both to the main campus would facilitate a new style of biomedicine in which a broad array of conceptual and technical tools in the experimental life sciences could contribute to medical education and research. Sterling noted also that the faculty of the Stanford Medical School underlined the significance of research:

> We place great emphasis on the creation within the Medical School of the stimulating and exciting environment which stems from maximum productivity and diversity of research. Although patient care is the backbone of the practice of medicine, the great advances in medicine must necessarily come from experimental and clinical investigation. It is an integral part of the responsibility of the medical school to the Nation to expand the horizons of scientific medicine and to break new ground in the conquest and prevention of disease.[5]

The rise of biomedical research at Stanford was orchestrated by administrators, including the university's first provost, Frederick Terman, and
the dean of the medical school, Windsor Cutting; the dean shared the university's strategic pursuit for "steeples of excellence" in a few key areas
of research with the highest growth potentials.[6] At a time when the federal government, especially the National Institutes of Health (NIH), dras-
tically expanded its support for biomedical research, Stanford administrators opportunistically took advantage of this post–World War II trend
to build up their profile in that area.[7] At one level, Stanford's integration of its medical school with the main university reflected a broader
post–World War II realignment in the relationship between medicine and
biology, namely the rise of biomedicine as a "hybrid form of research and
therapy that combines the normal and pathological."[8] At another level,
Stanford's new biochemistry department, with its obstinate focus on basic
research as opposed to clinical care, reflected the post–World War II disciplinary consolidation of biochemistry, biophysics, microbiology, genetics, and molecular biology into the broad framework of basic biomedical
research. For example, biochemistry, or a minor medical specialty called
"medical chemistry," had been a service discipline inside the medical
school, providing a basic biological and biochemical training necessary for
medical students. However, after World War II, biochemistry emerged as
an autonomous and powerful subject area of biomedical research.[9] With
the postwar expansion of the biomedical research enterprise, other included disciplines, such as genetics and molecular biology, began to prolif-
erate as autonomous fields in research universities, attracting ample funding to support their laboratory operations.[10]

In this chapter, I examine the development of Stanford biochemists'
communal form of laboratory life.[11] I first show how Stanford's strategic
appropriation of the expansion of federal patronage for biomedical research led to the establishment of the new biochemistry department, one
that strongly focused on DNA as its research subject. Under Kornberg's
influence, Stanford biochemists developed their shared research interests
in DNA, especially in its biochemical replication and biological activities
like genetic expression and regulation—the central problems in molecular biology in the 1960s and 1970s. Stanford biochemists in turn formed
a particular style of research community at the local level by cultivating
distinctive communal practices among its faculty members. I analyze how
Stanford biochemists tried to foster a research community with distinctive moral and political economies of science by sharing laboratory space,

research instruments and materials, and even monies.[12] At one level, their sharing practices were embedded in their distinctive moral economy of science—communal views about proper ways of organizing their laboratory life; about social norms and obligations in scientific exchange and knowledge production; and about customs and rules in the distribution of resources in the community life. At another level, their communal mode of the department's financial and managerial operations was reflected in their particular, local political economy of science—a small, tight-knit political economic sphere devised through pooling its resources communally while maintaining its broader, rational economic relationship with federal funding for biomedical research. As I show, the moral and political economies of science embedded in the Biochemistry Department evolved from their efforts to sustain a vibrant flow of ideas, materials, and technologies that could sustain the productivity and independence of their research in the increasingly competitive world of biomedical research in the 1960s.

Toward a Biomedical School

In his presentation to Stanford's Board of Trustees in June 1953, President Sterling argued that the medical school should undertake a bold move that would benefit both itself and the university:

> It was argued that the future of medical education is dependent on the course of medical science, and that, in turn, medical science has become increasingly dependent upon the basic physical sciences and upon the social sciences. This key relationship of medical education and science to other scientific fields can best be strengthened and advanced by bringing the Medical School into the closest possible physical and intellectual relationship to the whole University. This is a view to which I subscribe.[13]

The move to the main campus at Palo Alto would be both "physical and spiritual" (figure 1.1). When concluding his study of the Stanford medical school in San Francisco in 1952, Sterling pointed out that the deteriorating educational and clinical facilities were in major need of replacement and refurbishment (figure 1.2). In addition to the "hopeless jumble" of the medical school and Lane Hospital, the former's financial woes were growing worse; it had been losing four hundred thousand dollars annually by 1950.

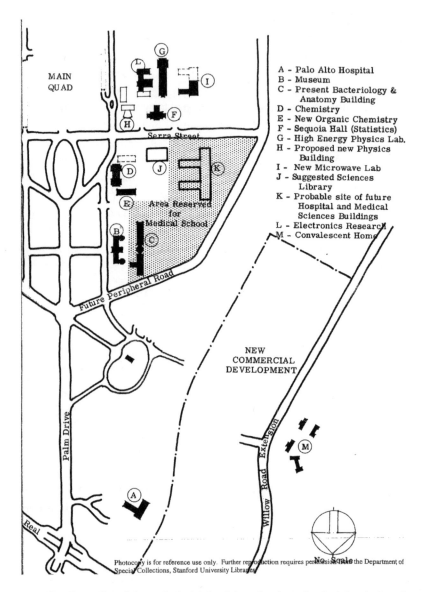

A - Palo Alto Hospital
B - Museum
C - Present Bacteriology &
    Anatomy Building
D - Chemistry
E - New Organic Chemistry
F - Sequoia Hall (Statistics)
G - High Energy Physics Lab.
H - Proposed new Physics
    Building
I - New Microwave Lab
J - Suggested Sciences
    Library
K - Probable site of future
    Hospital and Medical
    Sciences Buildings
L - Electronics Research
M - Convalescent Home

MAIN
QUAD

Serra Street

Area Reserved
for
Medical School

Future Peripheral Road

NEW
COMMERCIAL
DEVELOPMENT

Palm Drive

Willow Road Extension

Real

FIGURE 1.1. Integration of the medical school and the university at Stanford. Stanford president Wallace Sterling presented his plan to relocate the medical school to the main campus at Palo Alto in 1953. The integration was meant to be both geographical and intellectual in nature, heralding Stanford's commitment to a new biomedical school. Image from "Statement to the Stanford Trustees," John E. Wallace Sterling Papers, Stanford University Archives, SC 216, box 63. Reproduced with permission, Department of Special Collections, Stanford University Libraries, Stanford University.

FIGURE 1.2. Stanford Medical School and Lane Hospital. Stanford Medical School and Lane Hospital, previously located in San Francisco, were in a financial and intellectual malaise. Their relocation and integration with the larger university aimed to bring biomedicine, both for its scientific development and financial support, to Stanford University. Image from "Stanford Medical School Becomes a True University School," 1956, p. 11, John E. Wallace Sterling Papers, Stanford University Archives, SC 216, box 63. Reproduced with permission, Department of Special Collections, Stanford University Libraries, Stanford University.

More problematically, the relative lack of medical research facilities and professors meant that the medical school was in danger of failing to take part in the postwar development of biomedical research and the expansion of its federal support. At Stanford, a memo circulated in the late 1950s stressed the urgent need to emphasize research in the newly relocated medical school. Pointing to the fact that the NIH's grants for training and research had increased by 5 times to 7.5 times in eight years following 1950, it was suggested in the memo that the new medical school should "put an emphasis on education, on attracting more doctors (M.D. + Ph.D.) into academic and investigative training careers" as a way to finance the new medical school (figure 1.3).[14] Sterling, along with the dean of the medical school, Windsor Cutting, emphasized that the intellectual and geographic division between preclinical disciplines of the university, such as biology, biochemistry, chemistry, and genetics, and clinical departments of the medical school, such as anatomy, pathology, and physiology, was no longer tenable. Pointing to the "essential unity of biology and the basic medical sciences," and its implications for both medical practice and medical education, Sterling and Cutting further asserted that the progress of medicine increasingly depended on advances in the basic biochemical and biophysical sciences. The geographic and academic integration of the medical school with the rest of the university, they concluded, could provide an unparalleled opportunity for Stanford to institute biomedical research in a truly academic medical school.[15]

The relocation of the medical school to Stanford's main campus also provided a chance to empower dispersed medicine-related faculty members in the biology and chemistry departments at the university. At Stanford, biochemistry had played a traditional service role to medical education, reflecting the state of the discipline since the early twentieth century.[16] For example, most Stanford biochemists, such as J. Murray Luck and Laurence Pilgeram, resided in the chemistry department and were more oriented toward chemistry rather than biochemistry. Moreover, immediately after World War II, Stanford lost one of the pioneers of biochemical genetics when George Beadle moved to Caltech.[17] He was partly attracted to Caltech on account of the pervasive cooperative research among biologists, chemists, and physicists, who were supported by a flow of grant funds to biochemistry and molecular biology.[18] Beadle was further disappointed by Stanford biochemist Hubert Loring's lack of appreciation for the novel approach based in molecular genetics that Beadle had developed in his *Neurospora* experimental system.[19] Loring, a student of Wendell M.

USPHS GRANTS

The growth of National Institutes of Health appropriations for grants (extra-mural) has been as follows:

Fiscal Year	Research Fellowships	Training awards	Research Grants
1945	$ 28,000	$ 29,000	$ 85,000
1950	1,448,057	6,415,000	13,065,000
1955	2,562,000	11,051,000	33,918,000
1956	2,800,000	14,502,000	38,038,000
1957	5,397,000	28,075,000	89,697,000
1958	6,812,000	32,560,000	99,345,000

Training awards: 5 fold increase in 8 years
Fellowships: 5 fold increase in 8 years
Research grants: 7½ fold increase in 8 years

Suggestion: it may be time to put emphasis on
education, on attracting more doctors
~~attracting~~ more
(M.D. + Ph.D.) into academic + investigative
academic + investigative training
careers, through providing facilities + training support.
Relate to unfilled faculty posts, and to estimated
need for 20-25 new medical schools. [Lippard briefed on this]

FIGURE 1.3. Stanford's memo on the postwar growth of National Institutes of Health (NIH) grants. This memo noted the postwar growth of the NIH's grants to research organizations and scientists, which had increased by 5 times in training awards and by 7.5 times in research grants in the eight years since 1950. The memo then suggested that expanding biomedical research in the medical school could attract NIH grants. Image from "USPHS Grants," John E. Wallace Sterling Papers, Stanford University Archives, SC 216, box 62, folder: 9. Reproduced with permission, Department of Special Collections, Stanford University Libraries, Stanford University.

Stanley, preferred a structural approach to biochemistry based on the crystallization and analytical ultracentrifugation of virus particles.[20] Additionally, Edward Tatum, a biochemist from the Department of Biology at Stanford who collaborated with Beadle on their Nobel prize–winning experiment, had also left to join the Rockefeller Institute in 1956, despite having received Stanford's hasty offer of the chairmanship of a new biochemistry department that was then still a "paper organization."[21]

Frederick Terman, newly appointed in 1955 as academic provost at Stanford, played a key role in establishing two new basic biomedical departments (the Biochemistry and Genetics Departments); he accomplished this, along with Dean Cutting, by coordinating the integration of the medical school with the university.[22] The two men shared Stanford's new vision of a research-oriented medical school, as well as the reform of its medical education curriculum. Terman's administrative experience in rebuilding Stanford's engineering school had provided intellectual and strategic resources for building another "steeple of excellence" in biomedicine. As chair of the Department of Electrical Engineering and later as dean of the School of Engineering at Stanford, Terman had transformed its engineering school into one of the top academic centers of electronics, thus helping lay the foundation for Silicon Valley. Building on the Department of Electrical Engineering's initial strength in radio engineering, Terman had instituted a robust set of electronics-related and microwave research programs in the physics and engineering departments, which in turn attracted government grants and industrial contracts during World War II and the Cold War. Terman's experience in the building of Stanford's School of Engineering, and his contribution to the development of Silicon Valley, convinced him of the centrality of research to a university's intellectual status and financial health.[23] Though already a respectable institution of higher education, Terman claimed that Stanford had become too dependent on political and commercial demands from government and industry. He believed that through its research activities the university could become more independent and gain sufficient intellectual strength to make distinctive contributions to society:

> Universities carry on learning and innovation work in the sciences and engineering because it is necessary to do so in order to provide the best possible education at the higher levels. In addition, because of the freedom and low operating costs of universities, they are ideal institutions to carry on research as a service to society.[24]

FIGURE 1.4. Construction of Edwards Building, Stanford Medical School. The medical school moved to the main campus at Palo Alto, ca. 1958. Image from Arthur Kornberg Papers, Stanford University Archives, SC 359, box 31. Reproduced with permission, Department of Special Collections, Stanford University Libraries, Stanford University.

Terman's emphasis on research was indeed well suited to Stanford's ambition of building a new research-oriented medical school on the main campus (figure 1.4).

At another level, the emphasis on the role of research at both Stanford University as a whole as well as its medical school reflected the post–World War II rise of the "federal research economy" that supported an unprecedented level of government-sponsored academic biomedical research.[25] Above all, the rapid rise of the NIH as a major patron for biological and medical research during the postwar period drew the attention of university administrators. Public enthusiasm for biomedical research not only bolstered federal support but also changed the pattern of private support for medicine after World War II. Lay activists energized voluntary health organizations like the National Foundation for Infantile Paralysis and the American Cancer Society, enthusiastically promoting the importance of laboratory-based research in fighting diseases. Their activism in turn convinced politicians and government officials that support for biomedical research held broad political appeal.[26] Administrators of medical schools were acutely aware of the implications of the changing patronage system for American medical education and research. In

1950, George B. Darling, director of Medical Affairs at Yale University and former vice-chairman of the medical division of the National Research Council during World War II, called attention to the rising share of government-sponsored research in medical schools. As "medical research became big business," Darling asserted, medical schools deserved to benefit from their share of the available funds.[27] Several universities, including the Johns Hopkins and Stanford medical schools, made serious efforts to accommodate the emerging emphasis on biomedical research by integrating their medical and university education in the early 1950s.[28]

Stanford administrators took advantage of ample federal research funds for reinvigorating its biomedical research enterprise. In 1957, Terman solicited various funds from federal government and private foundations, collecting more than $5,000,000 (including $1,500,000 from the NIH's Health Research Facilities program) for the relocation of the Stanford Medical School.[29] The large influx of federal research funding was regarded as a stable resource for the operation and expansion of the medical school. Indeed, to Stanford administrators, the establishment of a biochemistry department at its medical school could thus help meet several goals. First, it could introduce recent advances in biochemistry and molecular biology, contributing to the strengthening of the university's biomedical research profile, which could help Stanford's effort to reform its medical education by focusing more heavily on biomedical research rather than on clinical care and training. Moreover, the infusion of federal grants to biomedical research could provide viable financial resources for the growth of both the medical school and the university as a whole.

De Novo Biochemistry Department at Stanford

Establishing a completely new academic division at Stanford dedicated to biomedical research in the highly competitive context of postwar biomedical research was not an easy feat. Major research universities had been competing with each other since the mid-1950s in order to build a critical mass in the burgeoning fields of biochemistry, genetics, and molecular biology. For example, after Harvard's unsuccessful attempt to reorient biochemistry to be more in line with recent biomedical research in the early 1950s, it hired the young molecular biologist James Watson in 1956.[30] MIT lured Salvador E. Luria away from Caltech in the late 1950s in order to introduce the molecular approach to biology in the biology department and revise its biology curriculum.[31] Caltech, whose biology division had successfully attracted self-proclaimed molecular biologists under the leadership of Max Delbrück, faced increasingly fierce competition with other academic institutions searching for faculty members and research grants in biomedical research.[32] Renato Dulbecco, a member of Caltech's biology faculty, wrote to his colleagues in 1961: "due to the great demand for talents, as soon as a man has revealed his ability he becomes established and is eagerly sought by competing organizations ... Thus we think that there is very little chance of attracting 'young' and at the same time outstanding investigators who may not be sought by other competing organizations."[33]

In this competitive environment, Stanford administrators tried to find a promising group of scientists who might be attracted by Stanford's vision for a medical school in which biomedical research would dictate the future trajectory of medicine. Terman even feared that the life sciences departments and the medical school were "destined for mediocrity" if he could not attract prominent biomedical researchers.[34] On January 22, 1957, with Terman presiding, Dean Cutting (Stanford medical school), biochemist Wendell Stanley (University of California, Berkeley [UCB]), and microbiologist C. B. van Niel (Stanford) gathered to discuss potential candidates to head a new biochemistry department at Stanford. They decided to select someone whose strength was in basic science rather than in clinical biochemistry.[35] A group of candidates emerged through their discussion. Their first choice was Kornberg. Five other prominent scientists, including Severo Ochoa, Seymour Cohen, Konrad Block, Christian Anfinsen, and Fred Sanger, were also considered as alternative choices.

Stanley, chair of the newly established Biochemistry Department at UCB, played a decisive role in identifying key candidates, since he had striven to build a freestanding academic biochemistry department.[36] Stanley strongly recommended Kornberg as the "most desirable man for the position."[37] Though Kornberg had earned an MD (from the University of Rochester in 1941) and had been a resident at Washington University's medical school, he had been completely retrained as a research biochemist during his years at the NIH from 1942 to 1952. At the NIH, he had worked on intermediary metabolism that involved nucleotides and coenzymes; this was research that later evolved into his studies of the enzymatic synthesis of nucleic acids.[38] Kornberg highly prized academic rather than clinical aspects of biochemistry. For example, Stanley and Kornberg shared an admiration of Frederick G. Hopkins who had successfully built an autonomous and rigorous biochemistry institution in Britain after World War I, independent of medical and industrial concerns.[39]

Terman soon invited Kornberg to Stanford for a four-day visit in 1957 to offer him the position as chair of Stanford's new Biochemistry Department. Stanford's offer revealed not only the dedication of its administrators to the new department but also the fierce institutional competition between research universities to recruit top biomedical researchers. The university proposed a generous salary (an increase of sixteen thousand to twenty thousand dollars from his salary at Washington University), promised an eventual increase of the department's annual budget from eighty thousand to two hundred thousand dollars, and provided the prospect of brand new facilities and equipment. More importantly, Stanford would give him the freedom to organize his new department. Terman assured him that he could bring his whole research group to the new biochemistry department, and that he would not be saddled with Stanford's existing biochemistry faculty at the Department of Chemistry.[40] Terman also underlined that the establishment of a new biochemistry department would become a central part of Stanford's plan for a research-oriented medical school. Kornberg was further promised that he would play a major role in the reorganization of the chemistry and biology departments at the university: Kornberg could not only recommend candidates to chair the Department of Chemistry but also exert control over a joint biology-biochemistry appointment that would fill Tatum's vacated position as biochemist in the Department of Biology. Indeed, Kornberg was instrumental in recruiting Joshua Lederberg as chair of the new Genetics Department, and he helped bring Charles Yanofsky to fill Tatum's position. Terman did not forget to mention that Kornberg's new biochemistry department could fruitfully interact with Henry Kaplan's strong biophysics department and Cornelius B. van Niels's microbiology section.[41]

Stanford's emphasis on basic biomedical research was particularly attractive to Kornberg, who had often professed his distress about what he perceived as an emerging clinical and practical orientation at both the NIH (especially with the opening of the NIH Clinical Center in 1952) and Washington University School of Medicine during the late 1950s.[42] Kornberg's skeptical attitude toward clinical research reflected his conviction that there was an advantage to employing simple and well-defined experimental systems. His experience as a medical doctor in clinical experiments strengthened this point of view. The relocation of the Stanford medical school also provided an exciting opportunity for him to build an unusual biochemistry research group that was tightly focused on basic biomedical research regarding nucleic acids.

Stanford's offer to Kornberg indeed provided him with an exceptional opportunity to organize a new biochemistry department from the ground up. Kornberg decided to bring his entire research group that he had assembled after becoming chair, in 1953, of the Microbiology Department at Washington University (see figure 1.5). He had first appointed two of his former postdoctoral fellows, I. Robert Lehman and Paul Berg, as faculty members in 1956. Kornberg then hired three other postdoctoral fellows from the Pasteur Institute in Paris: Melvin Cohn, A. Dale Kaiser, and David Hogness. In France, some biologists had even called Kornberg's Washington University department a "Paris in the Midwest" because of its adoption of the Pasteur Institute's biochemical approach to genetics.

FIGURE 1.5. Microbiology Department at Washington University School of Medicine in St. Louis (ca. 1958–59). From 1952 to the spring of 1959, Arthur Kornberg headed the Microbiology Department at Washington University's School of Medicine in St. Louis, building a research group for enzymology. In the first row (those seated), the fourth person from the left is Arthur Kornberg, the sixth is Paul Berg, and the seventh is David Hogness. After his plan to organize a new biochemistry department at Stanford, Kornberg brought five people from the Microbiology Department. Courtesy of Paul Berg.

The core of Kornberg's research group was built on his expertise in the chemical metabolism of coenzymes and nucleic acids. With the identification of DNA's role as the chemical basis of heredity, and the recent addition of molecular geneticists from the Pasteur Institute, research interest in Kornberg's department developed to investigate key molecular aspects of the gene, such as the replication of DNA and protein synthesis.

Kornberg accepted Stanford's offer in June 1957, bringing five people whom he had hired at Washington University in St. Louis—Berg, Lehman, Cohn, Hogness, and Kaiser—and one new hire, the Oxford-trained physical chemist Robert Baldwin from the University of Wisconsin, Madison. From its inception at Stanford, the Biochemistry Department was called the Kornberg Department, as it was essentially an import of Kornberg's DNA and RNA enzymology research group from the Microbiology Department at Washington University. The importation of Kornberg's core faculty (except Jerard Hurwitz who shared research interests with Berg) from Washington University drew ample attention from other biomedical researchers and university administrators, even more so as Kornberg's move to Stanford coincided with his Nobel Prize in 1959 for his work on DNA polymerase. As Dulbecco wrote to his Caltech colleagues, the migration of so many colleagues from one department could be one of the most productive ways of building a prominent research group in the competitive field of biomedical research:

> [I]f we want to attract the men whom we listed above [such as Drs. H. Gobind Khorana and Paul Berg], we must provide them with something that they do not have at the place where they presently are. This is companionship. To do so, we should not try to attract such men individually, but we should try to attract them all at once. That this is so is proven by examples of other organizations. Departments which were organized *de novo* or were completely reorganized during the last few years have succeeded in securing groups of brilliant investigators mostly with a molecular biology approach. Excellent examples are the newly formed Biology Division at Brandeis, the reorganized Biology Division at M.I.T., besides the Biochemistry Department at Stanford.[43]

Indeed, Stanford biochemists were imbued with a strong sense of companionship, especially in terms of their cultural and intellectual cohesion. When they moved to Stanford, Kornberg, Berg, Kaiser, and Hogness settled around the Santa Fe Avenue, where Stanford University provided faculty housing near the campus. These scientists were part of a newly rising intellectual group in postwar American universities. Moreover, Stanford biochemists as a group had developed tightly connected research interests in nucleic acids since Kornberg assembled his department at Washington University around problems of DNA synthesis and replication. As such, some members of the Stanford biochemistry faculty had worked as Kornberg's postdoctoral fellows. Others had been hired partly because of Kornberg's 1956 isolation of DNA polymerase, which had led him to investigate the biological activities of nucleic acids. Their shared research interests in the chemical nature and biological activity of nucleic acids, combined with their institutional cohesion as a research-oriented department, provided a unique context for Stanford biochemists to develop their own experimental work.

The DNA Department

In the 1950s and 1960s, the material and scientific ascendancy of DNA in molecular biology as a chemical embodiment of hereditary information coincided with Kornberg's transition to Stanford.[44] The same year he moved to Stanford, he was awarded the 1959 Nobel Prize in Physiology or Medicine for his discovery of DNA polymerase, an enzyme that helps to synthesize DNA (figure 1.6). In his Nobel lecture, Kornberg underlined that the knowledge of the biochemical synthesis of DNA could open a new array of investigations into the basis of heredity. According to Kornberg, the elucidation of the double-helix structure of DNA in 1953 by James Watson and Francis Crick only provided a "mechanical model of replication."[45] Like other contemporary molecular biologists, such as Max Delbrück and others, Kornberg understood the limited implications of DNA's double-helix structure for understanding biological heredity.[46] Kornberg suggested that his rigorous DNA synthesis system could provide a biochemical means to explore a sequence of reactions that control the transmission and expression of hereditary information, namely, DNA replication and protein synthesis:

> Five years ago the synthesis of DNA was also regarded as a "vital" process. Some people considered it useful for biochemists to examine the combustion chambers of the cell, but tampering with the very genetic apparatus itself would surely produce nothing but disorder. These gloomy predictions were not justified then, nor are similar pessimistic attitudes justified now with regard to the

FIGURE 1.6. Arthur Kornberg with DNA model (1970). Arthur Kornberg established his Stanford Biochemistry Department with faculty members who focused on the molecular and genetic nature of DNA. Reproduced with permission, ©1970 Estate of Yousuf Karsh.

problems of cellular structure and specialized function which face us. High adventures in enzymology lie ahead and many of the explorers will come from the training fields of carbohydrate, fat, amino acid and nucleic acid enzymology.[47]

The faculty in the biochemistry department initially developed their research interest in the reactions involved in nucleic acids when they worked with Kornberg on DNA polymerase as a postdoctoral fellow or a new faculty member in the Washington University department. As a result, Stanford biochemists initially shared a rather unusual research focus as a coherent research group on DNA.[48] Unlike conventional biochemistry departments, Stanford's department focused its research on several lines of investigation related to nucleic acids, eschewing a more comprehensive portfolio of research that might have included carbohydrates, lipids, vitamins, minerals, and bioenergetics. Stanford biochemists instead approached the synthesis of nucleic acids, their role in protein synthesis, and their wider role in genetic regulation and control, from the very wide-ranging perspective of enzymology, genetics, immunology, and physical chemistry. Their evolving intellectual focus on DNA reveals how traditional biochemical research during the 1950s and 1960s was brought to bear on problems of genetic control and regulation.[49] Indeed, Stanford biochemists developed their mutual research interests in the biological activity of nucleic acids from their previous research trajectories; these interests were also sometimes shaped by the use of shared research materials and techniques.

To best understand Stanford biochemists' early research focus on nucleic acids, it is helpful to first follow Kornberg's research trajectory because their collaboration with him was a key scientific factor in the establishment of the department. During his transition from a medical doctor to a research scientist in the 1940s, Kornberg worked on problems of intermediary metabolism, a field of research dedicated to the investigation of chemical pathways of the degradation of nutrients and biosynthesis of cellular constituents, as well as the transfer of energy involved in these processes. Intermediary metabolism was a subject that had been studied largely in medical schools because of its relevance to physiological and nutritional research, and had become a major preoccupation among biochemists, especially as nutrient research shifted toward enzymology.[50] Enzymology, through its focus on the catalytic role of diverse enzymes in chemical reactions, provided a new means to illuminate a sequence of integrated chemical reactions involved in intermediary metabolism.[51]

Beginning his research career as a nutritionist in the nutrition section of the NIH, Kornberg initially became interested in energy transfer in the respiration of cells and tissues, specifically the generation of adenosine triphosphate (ATP), an energy source for cellular metabolism. His subsequent training under Severo Ochoa at the New York University Medical School in 1946 then equipped him with a set of enzymological techniques for investigating biochemical pathways.[52] During his work with Ochoa, Kornberg mastered essential techniques in enzymology: enzyme purification and the assaying of enzymatic activity through the spectrophotometer.[53] In 1947, Kornberg continued his training in enzymology at Carl and Gerty Cori's laboratory at Washington University in St. Louis. Fascinated by the intermediary pathways and bioenergetics involved in ATP synthesis, Kornberg continued to investigate how the addition of a phosphate group (phosphorylation) provides energy for the synthesis of ATP by using kidney cells.[54] After he returned to the NIH, he established the Enzyme Section at the Industrial Hygiene Division with Bernard Horecker and Leon Heppel. Members of this section had maintained their commitment to basic research, sharing their research materials and ideas through a regular enzyme club meeting.

The Enzyme Section at the NIH, with its expertise on nucleic acid biochemistry, provided important experimental systems and research materials for investigating protein synthesis in the 1960s. Leon Heppel's accessible refrigerators at the NIH became a crucial source of polynucleotides, through which Marshall Nirenberg and J. Heinrich Matthaei experimentally correlated each nucleic acid sequence to the corresponding amino acid. Their experiment was touted as the "cracking" of the genetic code.[55] As the biochemist Maxine Singer recollects, "the biochemistry of phosphate-containing compounds became a central interest of several in the original lunch club group."[56] In his subsequent experiments at the NIH, Kornberg realized that the synthesis of large biological molecules, such as nucleic acids, might involve the reaction of (co)enzymes with ATP as its energy source. Indeed, the synthesis and cleavage of a coenzyme into two nucleotide components inspired Kornberg's new venture into the biosynthesis of DNA and RNA by following the involved, radioactively labeled phosphate or carbon.[57]

In 1953, Kornberg was rewarded for his successful elucidation of a sequence of reactions involving coenzymes, inorganic pyrophosphate, and nucleotides with the chairmanship of the Microbiology Department at the Washington University Medical School. His former teachers Carl and Gerty Cori were members of its Biochemistry Department. Along with the Coris, scientists like Joseph Erlanger and Herbert Gasser were in the medical school, making Washington University arguably one of the top medical schools in basic biomedical science in the 1950s. However, not knowing much about microbiology, Kornberg arranged to visit California to take C. B. van Niel's course in general microbiology at Stanford's Hopkins Marine Station. More importantly, Kornberg's exposure to microbiology provided him with another useful experimental organism, the bacterium *Escherichia coli*; its rapid reproduction cycle could provide an ample supply of research materials for biochemical investigations of nucleic acid synthesis. Although Kornberg had been working with molecular extracts such as coenzymes, orotic acid, and ATP from yeast or the liver, he could only elucidate how nucleotides were synthesized and activated in cells. In teaching microbiology courses at Washington University, he came to know that some bacteria enzymes could degrade DNA; this degradation process then provided him with an ample supply of nucleotides.

Moreover, by tagging a nucleotide with radioactive isotopes (phosphorus or carbon isotopes), Kornberg could follow how nucleotides were incorporated into an RNA or DNA chain.

At Washington University during the 1950s, Kornberg began to expand his work on intermediary metabolism involving the synthesis of nucleic acids. Drawing analogies from the work of Carl and Gerty Cori on the synthesis of carbohydrate chains by the enzyme glycogen phosphorylase, Kornberg searched for enzymes responsible for the synthesis of nucleic acids. His colleague Morris Friedkin, who had earlier isolated an enzyme that makes thymidine (a component of DNA), guided Kornberg's attempt to use radioactive-labeled thymidine to elucidate the biosynthesis of DNA. With Lehman, who joined Kornberg's lab as a postdoctoral fellow in 1955, Kornberg began to adopt a crude cell extract as a source of nucleotides. Since Lehman had previously worked with DNA extracted from the burst of E. coli bacterial cells infected by T2 bacteriophage (bacterial virus), they could isolate ample nucleotides and use them for their work on DNA synthesis. They subsequently purified the enzyme DNA polymerase that helped assemble nucleotides into a DNA chain; this was the work that won Kornberg a Nobel Prize in 1959.[58]

As Kornberg and Lehman's work evolved into the biosynthesis of nucleic acids, another Kornberg postdoctoral fellow, Berg, embarked in 1954 on new research into the synthesis of amino acid chains that comprise proteins. Kornberg initially encouraged Berg to work on the formation of acetyl coenzyme A (CoA) as a way to follow an enzymatic reaction that catalyzes an exchange reaction of phosphate between pyrophosphate and ATP, which provides energy in the fatty acid system. Studying acetic acid activation by CoA, Berg unexpectedly discovered another intriguing metabolic pathway that activated the amino acid methionine. Since the activation of acetic acid was thought to be a key to the understanding of the formation of fatty acid chains, Berg reasoned analogically that the enzymatic reaction of CoA could shed light on the intermediary metabolism of amino acids. With the graduate student E. James Ofengand, Berg soon found that the activated amino acid was transferred to a small RNA molecule (also called soluble RNA), and that this acted as an acceptor for the synthesis of amino acid chains into proteins.[59]

This line of work led Berg to protein synthesis research, exploring how DNA is transcribed into RNA and how the latter is subsequently translated into protein. In the 1950s, some prominent biochemists, such as Paul Zamecnik and Mahlon Hoagland at Harvard Medical School, had elucidated the biochemical pathway of the incorporation of amino acids into proteins through soluble RNA.[60] Berg, adopting his enzyme purification technique, continued to isolate several enzymes that activated different amino acids. Through his biochemical work on amino acid synthesis, his research developed further into problems of protein synthesis. By the late 1950s, Berg became well aware of the implications of his amino-acid synthesis system for the illumination of protein synthesis. After Berg moved to Stanford, his group developed two experimental systems for exploring protein synthesis involving transcription and translation. One of his graduate students, Michael Chamberlain, developed an in vitro transcription system, through which he synthesized RNA from DNA by purifying RNA polymerase, demonstrating that RNA played the "messenger" role that transferred biological information from DNA.[61] In addition to Chamberlain's transcription system, Berg's other graduate student, William Wood, showed how small RNAs (which later turned out to be transfer RNA) functioned as precursors for the incorporation of amino acids into a protein.[62] Through his biochemical analysis, Berg critically contributed to one of the central questions of molecular biology, namely, the biological expression of the genetic materials through the flow of genetic information from DNA to RNA to protein.

Though Kornberg and Berg's interest in nucleic acids developed from their early interest in bioenergetics and intermediary metabolism, some other members of the Biochemistry Department at Washington University approached nucleic acids from a perspective based on genetics and immunology. Melvin Cohn, who was initially interested in immunology, was recruited to the university's Microbiology Department in 1953. During six years of postdoctoral work with Jacques Monod at the Pasteur Institute from 1947, Cohn had been working on enzymatic adaptation, a theory that hypothesized that microbes could be chemically "trained" in response to different growth environments. For example, they investigated how *E. coli* could adapt to grow on different sugar sources like lactose, which required cells to synthesize certain enzymes like β-galactosidase (lactose-digesting enzyme). They suspected that microbes could produce β-galactosidase by using its chemical substances present in the growth medium, lactose. Cohn's expertise in immunological methodology, especially the characterization of antigen-antibody interactions, enabled him to determine the specific activity of β-galactosidase. Through his collaboration with Monod and David Hogness, a new postdoctoral fellow from Caltech, Cohn unexpectedly found out that the production of β-galactosidase resulted from its de novo synthesis, suggesting that the synthesis of the enzyme was controlled genetically as opposed to chemically. This was a major reversal of Monod's enzyme adaptation theory, and this finding helped to reframe the Pasteur Institute group's research on induced enzyme synthesis in terms of genetic control and regulation.[63]

After his appointment as an assistant professor at Kornberg's Microbiology Department in 1954, Cohn also educated traditional enzymologists, such as Kornberg and Berg, about Monod and François Jacob's new exciting work on gene expression and regulation. Cohn also helped bring two faculty members, Hogness and A. Dale Kaiser, with whom Cohn had become acquainted at the Pasteur Institute. Hogness was appointed as an assistant professor in 1956 after his postdoctoral work with Monod, Jacob, and Cohn. Hogness's genetics background, especially his studies of mutation and recombination in *E. coli*, helped him to map out how the synthesis of the β-galactosidase in the bacterium was genetically regulated.[64] More importantly, Kornberg's discovery of DNA polymerase provided another exciting opportunity for Hogness to examine the DNA's biological activity. Hogness subsequently worked with the newly appointed Kaiser, a phage geneticist with a PhD from Caltech, who had also done his postdoctoral work at the Pasteur Institute before he joined Kornberg's Microbiology Department in 1956. The motivation of the collaboration between Hogness and Kaiser was to devise a biological system that could introduce Kornberg's polymerase-synthesized DNA into *E. coli* cells in order to examine its biological activity. With Kornberg's intense interest, they were able to invent an experimental system that could introduce DNA from bacteriophage λ (λ*dg*) into an *E. coli* strain named K-12.[65] This so-called Kaiser-Hogness DNA transformation system demonstrated that Kornberg's DNA polymerase indeed helped synthesize biologically active DNA. More importantly, by enabling scientists to introduce DNA into *E. coli*, the DNA transformation system provided a useful tool for further biochemical analysis of the gene. For example, Kaiser was able to make the first physical maps of genes of the phage DNA by tinkering with the DNA transformation system (see figure 1.7).[66]

Kornberg recruited Robert Baldwin at the University of Wisconsin, Madison, who had earned his PhD in physical chemistry from Oxford University, for the Stanford Biochemistry Department. As the only physical chemist in the new biochemistry department, Baldwin had expertise in several methods for sorting out different proteins and other biological macromolecules; these involved measuring different sedimentation

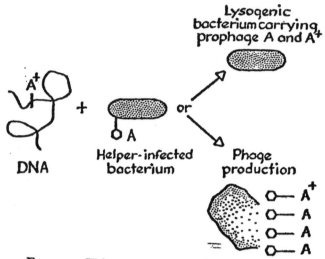

FIGURE 1. This assay system for λDNA utilizes whole bacteria recently infected with complete phage (helper). DNA and helper phage carry different genetic markers so that offspring of the infecting DNA can be recognized. Either a lysogenic or a productive response can be detected by appropriate plating.

FIGURE I.7. A DNA transformation system. A. Dale Kaiser and David S. Hogness demonstrated that one could introduce DNA into the bacterial host when the bacterial host was infected by a helper virus. By adding DNA segments to helper virus–infected bacterial cells, they invented a bacterial transformation system that could examine the biological activity of DNA molecules. Image from A. Dale Kaiser, "Description of Work Leading up to Recombinant DNA," unpublished manuscript, August 1980. Reprinted by permission from A. Dale Kaiser.

behaviors in the velocity ultracentrifuge.[67] When he came to Stanford, he started to work on DNA with Kaiser and helped others who were interested in nucleic acids to investigate their physical characteristics.[68] Baldwin helped other Stanford biochemists to characterize many DNA and RNA molecules used in their experimentation.

By the time the Stanford biochemists set up their new laboratories at the medical school in 1959, their common research interests had thus already gravitated toward DNA, its enzymatic synthesis, and its role in protein synthesis and gene regulation. To summarize, Kornberg and Lehman's research evolved into the replication of DNA, while Baldwin's work on the physical chemistry of the enzymatic synthesis of DNA strengthened these colleagues' work on DNA and RNA polymerases. Starting as Kornberg's postdoctoral fellow before being hired as a faculty member at Washington University, Berg shifted his research away from enzymology and toward the central issues of protein synthesis, exploring the genetic control and regulation in bacteria. Hogness and Kaiser's work on the structure and function of bacteriophage DNA, and Cohn's work on antibody synthesis, also complemented their colleagues' research as Cohn employed immunological methods in order to address the secondary genetic control of protein synthesis. Cohn's, Hogness's, and Kaiser's postdoctoral work at the Pasteur Institute, as well as Monod's and Jacob's frequent interactions with Kornberg's research group, brought a new perspective, which began to weld the connection between biochemistry and genetics during the 1950s and early 1960s, contributing to the development of molecular genetics.

Moral and Political Economies of Stanford Biochemists

As the mechanical model of the DNA double helix gained its chemical and biological relevance due to the biochemical synthesis and replication of DNA in the late 1950s, Stanford biochemists' research interest in DNA presented broader implications for understanding gene expression and regulation.[69] More importantly, the biochemists shared a repertoire of enzymatic and genetic-analysis techniques that enabled them to decipher the biochemical pathways of nucleic acids, especially their roles in the production of proteins; gene expression and regulation—how nucleic acids embody genetic information and how they express and regulate their genetic information inside the cell—emerged as research subjects of intense scientific interest among both biochemists and molecular biologists. Their related projects on the enzymatic synthesis of nucleic acids, the activation of amino acids into proteins, and the introduction of nucleic acids into bacteria to analyze their biological activities, all provided some of the best approaches to understanding the biochemical mechanisms of gene expression and regulation. The Stanford biochemists even organized a monthly meeting, the so-called DNA club at Kornberg's house during this time. In this meeting they discussed their research progress and exchanged information and techniques regarding DNA research.

Stanford biochemists' intellectual and material focus on DNA led to the evolution of a communal form of laboratory life, with its particular assumptions about the moral and political economy of science embedded within their own community.[70] First, Stanford biochemists developed their own moral economy of science, a set of shared beliefs about the proper ways of organizing laboratory life at their department, with particular regard to the production, exchange, and ownership of knowledge and research materials. They, for example, shared most of the department's laboratory space and facilities without exclusively assigning them to individual faculty members and students. Incoming graduate students and postdoctoral fellows were randomly assigned to a particular laboratory space and were strongly encouraged to share laboratory space with members from other research groups. The result was that each laboratory had associations with students and fellows from three or four different groups. The shared space arrangement in turn fostered scientific interactions among different research groups, helping them "extend their experience, knowledge, and skills to related problems."[71] The scientists shared not only laboratory space and major facilities, such as tissue culture rooms and temperature-controlled laboratories, but also equipment like ultracentrifuges and electron microscopes. Those who worked with the genetic regulation of bacteria in Kaiser's group, for example, could obtain enzymes critical to manipulate and investigate DNA's biological activities from the laboratories of Kornberg and Lehman, as well as access their enzymatic techniques and expertise through collegial exchanges. This sharing system also allowed them to achieve economies of scale: the Stanford biochemists found that sharing research instruments, reagents like enzymes, and laboratory space proved to be cost effective because of their common interest in and technical expertise in studying DNA.

At another level, Stanford biochemists' moral economy of science, which emerged initially from and practiced through their sharing customs and rules, postulated key communal obligations and reciprocity supported by the wider consensus of the community. Stanford biochemists' sharing system was meant to facilitate open scientific exchange by encouraging scientists to disclose their novel findings, as well as to gain access to important information, materials, and tools. To be a member of this productive community and to sustain this creative community, one was obliged to reciprocate by disclosing information and granting access to materials. Consequently, their moral economy of science regarded secrecy and unfettered competition among its community members as immoral. Their collegial customs of reciprocity and disclosure thus helped regulate competitive feelings and encourage open exchange of ideas in a situation when they worked on an overlapping research project or collaboration. Indeed, the sharing system was informed and dictated by the communal consensus about how to maintain the productivity and flexibility of their laboratory life. Stanford biochemists insisted that the free exchange of ideas and materials would provide a key cultural and social milieu that would subsequently lead to creative and productive experimental arrangements.

More importantly, Stanford biochemists' sharing practices meant that ideas, tools, and materials were owned in a semicommunal way. They shared their research materials by providing every scientist at the department a common key to its stock rooms. As Berg recollected, he "had complete access to [others' refrigerators]."[72] In a situation where ideas and materials were communally shared and owned, it was crucial for individual scientists to assemble them in a creative way to get scientific recognition for performing original experimental work. Scientific credit would be assigned to those who made an experiment work, rather than to those who vaguely conceived it or provided key materials or tools for the experiment. One the one hand, the experimental imperative generated by the semicommunal ownership of ideas and materials obliged scientists to disclose information and to share materials reciprocally for their own advantage. On the other hand, this particular assumption about scientific credit implied in the semicommunal mode of knowledge production could be "a restraint for someone who is expansive and has lots of ideas," since it would be difficult to get credit for merely suggesting ideas.[73] Their collegial customs of sharing and disclosure, with its particular rule of assigning scientific credit, was embedded in the communal mode of scientific exchange and knowledge production.

Stanford biochemists' moral economy of science was further buttressed by a financial arrangement that pooled all research monies among faculty members. This practice transferred with Kornberg from the Microbiology Department at Washington University. When Kornberg assembled his Microbiology Department at Washington University, he had recruited those who shared his interests in the biochemical synthesis of DNA and its biological activity. As a senior scientist with enviable resources, Kornberg had earlier shared his grant funds with new faculty members he had brought to St. Louis. Because of this common research interest, Kornberg was able to not only use his grant funds to support his postdoctoral fellows or new faculty members but also share his research materials and instruments. As new faculty members developed their own research programs, they also received grants, and they collectively decided to share

their funds and other resources as an act of reciprocity. As Kornberg recollects, "we shared our limited funds, initially supplied as grants to me and
then to the others as their research programs matured; all decisions for
significant expenditures and space allocations were made communally."[74]

In addition, Kornberg used communal resources to entice a new faculty member or to support a more unusual research project that had not
attracted funding. In 1956, he explained this latter practice to Kaiser, a
new faculty member: "As I mentioned in an earlier letter, we pool most of
our resources, which provides additional flexibility. I thought you would
need more for supplies and equipment than you originally figured. If you
should need and find a research assistant, then there will be funds for it."[75]
This pooling practice enabled a new faculty member to "take as many students and postdocs and reagents as anybody else, even though their funds
were grossly inadequate for that." More importantly, for a new faculty
member, this pooling practice called for reciprocity by creating "a legacy
of that kind of indebtedness and ultimately responsibility and sharing."[76]

This sharing practice was continued at Stanford. Kornberg, who already had abundant research funds from the NIH, actively introduced
a system of sharing for the benefit of junior faculty members. When he
moved to Stanford in 1959, for example, Kornberg brought in $139,000 for
research grants and an additional $60,000 for equipment from the NIH,
as well as $85,000 from the National Science Foundation.[77] Other faculty
members, whose funds were rather small compared with that of Kornberg's, obliged to this sharing system. Stanford biochemists maintained
a departmental bookkeeping system for their research grants, recording
only income from each faculty member (e.g., figure 1.8). Since they recorded only total expenditures, each faculty member's individual expenditure was difficult to ascertain.[78] As a result, in Kornberg's recollection,
they had to "periodically . . . declare a moratorium on buying equipment
and expensive reagents and to appeal for initiatives in finding additional
grant support. Without strict accounting, it was natural for a faculty member to believe that he was not overspending. Nevertheless, we resisted the
temptation to adopt the conventional 'every tub on its own bottom' practice."[79] Sustaining their communal mode of operation required dedication
and reciprocity of its members.

Stanford biochemists' sharing customs and their rules of distribut-
ing scientific recognition and communal resources, as well as the obligations of reciprocity in community life, constituted their particular moral
economy of science. At another level, these practices were predicated on

AVAILABLE FUNDS AND PROJECTED INCOME 5/1/69 - 4/30/70

GRANTS:			Equip.	Travel	Other	TOTAL
Baldwin	PHS	12/31/69		1,572	7,834	9,406
		12/31/70			3,048	3,048
	NSF	8/31/69		1,346	1,754	3,100
		8/31/70			12,189	12,189
Berg	PHS	8/31/69		3,889	27,835	31,724
		8/31/70	1,000	2,400	21,512	24,912
	Cal Am Can	6/30/70			12,000	12,000
Hogness	PHS	1/31/70	8,748	787	11,675	21,210
		1/31/71			14,355	14,355
	NSF	8/31/69	2,359	225	7,720	10,304
		8/31/70			23,336	23,336
Kaiser	PHS	12/31/69	1,600		20,814	22,414
		12/31/70			7,724	7,724
	NSF	12/31/69		1,270	3,558	4,828
		12/31/70			6,661	6,661
Kornberg	PHS	8/31/69	801	3,720	65,582	70,103
		8/31/70	11,664	3,464	63,152	78,280
	NSF	2/28/70	2,254	2,770	2,923	7,947
		2/28/71			3,958	3,958
	NASA	6/30/69				-0-
Lehman	PHS	8/31/69		1,382	32,965	34,347
		8/31/70	3,400	1,064	27,000	31,464
Stark	PHS	12/31/69		772	4,384	5,156
		12/31/70			3,892	3,892
	NSF	5/31/70			2,322	2,322
Training Grant PHS		6/30/69			19,546	19,546
		6/30/70			13,270	13,270
			31,826	24,661	421,009	477,496

April Expenditures: $39,383

FIGURE 1.8. Stanford Biochemistry Department budget table (1969). Though members of the Biochemistry Department made each other's income known, they did not count each member's expenses. Rather they pooled research grants, recording only the total expenditure for the department as a whole. This financial pooling practice continued in the early 1980s. Image from Faculty Minutes, Paul Berg Papers, Stanford University Archives, SC 358, box 17, folder: Faculty Meeting, 1969–70. Reproduced with permission, Department of Special Collections, Stanford University Libraries, Stanford University.

their view of the political economy of twentieth-century biomedical research. Kornberg understood that the existence and expansion of biomedical research was increasingly supported by the political willingness
to invest the public's money through government grants. This particular
post–World War II configuration of the political and the economical in
biomedical research laid the foundation for the goal orientation of the
grant system initiated and enormously expanded by the federal government. However, Kornberg thought that the rise of political anticipation
for medical improvements, which provided ample economic resources
to biomedical researchers, could ultimately undermine scientific creativity and intellectual autonomy. As Kornberg later stated, this political
economy of biomedical research was fundamentally at odds with the
nature of scientific research:

> With regard to the support of science, the major flaw is the demand that the sci-
> entist justify a project on the basis of its goals . . . This philosophy is misguided
> in a fundamental way . . . No matter how counter-intuitive it may seem—to the
> scientist as well as to the layman—the most sure and cost-effective route to dis-
> covery is through the creative activity of the scientist or inventor rather than the
> pursuit of a defined goal. The award of a research grant is fundamentally flawed
> when it requires that the applicant chart a path to discoveries that will have
> practical consequences or, at least, will reorient the direction of a discipline.[80]

One key rationale behind the decision to pool research grants among
members of the department stemmed from Kornberg's recognition of the
need for a long-term approach in basic science that could be undermined
by shifting demands from the public and the government. By pooling research monies, Stanford biochemists envisioned a research community
that would not be severely restricted by short-term financial and political
conditions, a small political economic sphere of science where only creative ideas and productive experimentation would matter.[81] The decision
to share research grants among Stanford biochemists reflected more
widespread concerns about the rise of government-funded "directed" research after World War II. Kornberg often worried that the increase of
goal-oriented governmental support for science would hamper scientists'
ability to be creative and flexibly control the direction of their research.
When he left the NIH, Kornberg was distressed with the agency's disease-
oriented research initiatives and its construction of the clinical center in
the early 1950s; he perceived the center as an encroaching clinical and

practical orientation of NIH's research programs. One fellow biochemist at Harvard, Edwin Cohn, shared Kornberg's critical attitude, claiming that scientists "cannot be directed."[82]

This cautious attitude toward federal support in the scientific enterprise was particularly prevalent at Stanford University. Its conservative policies toward governmental intervention in the academy, as historian Rebecca Lowen argues, played a crucial role in promoting Stanford's interaction with private industry.[83] Thus, in the ideological geography of Stanford, Silicon Valley was initially conceived as a means to maintain academic independence by balancing public and private influences on the university. Established in an age of the ascendancy of the federal research economy, Stanford biochemists chose an alternative means to maintain their intellectual autonomy. As Kornberg emphasized, a grant-pooling practice provided a means to achieve intellectual autonomy from external intervention in the laboratory life, offering a solid foundation for creative and productive science.

Sustaining an autonomous and communal group, however, involved a different kind of exclusion. The sharing of grants and instruments made it difficult to accept affiliated faculty members or scientists with somewhat different research interests. For example, Joshua Lederberg in the Genetics Department initially wanted a joint appointment in the Biochemistry Department, but his classical genetics research did not fit well with the Stanford biochemists' evolving direction into molecular genetics. Moreover, not all members of the Biochemistry Department abided by their communal customs, which often demanded shared research interests. After Cohn moved to Stanford, his work shifted toward his main interest in immunology, especially to the problem of antibody synthesis. He felt that his research interests did not fit well with those of the biochemistry department so he eventually left for a position at the Salk Institute after he helped organize it. In addition, because of the communal operation of the department, it was difficult to expand one's own experimental work. Lubert Stryer, who was hired in 1963 after Cohn left, also faced problems with the Stanford biochemists' sharing practices, especially regarding lab space. As his research group expanded, he demanded exclusive lab space. Since the department did not have any room to accommodate his demand, other members of the department suggested that he should not apply for additional grants and place restrictions on the size of his research group. Stryer felt that this communal practice unnecessarily limited the growth of his research group, and he left for Yale.[84]

Nonetheless, in retrospect, Kornberg pointed out that the communal practice of Stanford's DNA department produced a "priceless dividend."[85] Stanford biochemists' moral and political economy of science cultivated lively scientific exchange and productive collaboration. During the first five years after the Biochemistry Department was established in 1959, Stanford biochemists continued to collaborate with one another on a set of DNA research projects, approaching the work with broader perspectives from enzymology, genetics, and physical chemistry. Kornberg, Lehman, and Baldwin succeeded in the discovery and characterization of an enzyme that catalyzes the replication of DNA. Hogness and Kaiser developed a DNA transformation system that enabled scientists to introduce viral DNA into bacterial cells and thus examine its biological activities. Berg's research also involved fundamental discoveries relating to the genetic code and the mechanism of protein synthesis in bacteria.

Divergence in the 1960s

From the late 1960s, the common research interests of this small community of bacterial researchers at Stanford began to diverge into rather disparate directions, ranging from Kornberg's persistent focus on DNA enzymology to Berg's new venture into the genetic expression and regulation of eukaryotic organisms. This shift in intellectual and material condition provided a challenging environment for the maintenance of the moral economy of science at Stanford, which was in part dependent on their common interest on DNA. Some junior faculty members began to shift their research interests into the biology of higher organisms, taking a sabbatical leave in order to prepare for this ambitious new direction. When Berg planned in 1967 to embark on research into the genetic regulation of eukaryotic organisms, he learned about tumor virology in Renato Dulbecco's lab at the Salk Institute as a way to investigate mammalian cell regulation. At the same time, Hogness phased out his bacteriophage work, reducing the size of his group to almost zero. During his sabbatical leave, he traveled to Europe and Australia in order to find the right organism that would allow him to examine the biology of development in higher organisms. Kaiser also developed his interest in developmental biology in the 1960s, investigating how genes are involved with the development of multicellular organisms.

The shifts in research direction among some junior faculty members at the department reflected a larger trend in molecular biology toward both higher organisms and developmental biology during the late 1960s (see chapters 2–4).[86] At the same time, Kornberg's influence over the Biochemistry Department, which had originated in his DNA research group at Washington University, was waning. For example, Stanford administrators were trying to lure Berg away from the Biochemistry Department to become the new head of the Biological Sciences Department on the main campus; they hoped to reinvigorate that department by introducing molecular and cellular approaches in its rather traditional biological research.[87] Berg indicated that he wanted to concentrate on his new cell biology project and suggested that he might be interested in organizing a new cell biology department. Though Stanford's plan did not work in the end, Kornberg had to recognize Berg's new research on mammalian cells, and he handed the chairmanship of the Biochemistry Department over to Berg. Though he remained supportive, Kornberg skeptically called Berg a "Pied Piper" who might ruin the successful department, as well as Berg's own promising scientific career, by employing complex and uncharacterized experimental organisms.[88] However, other departmental members like Hogness also embarked on new research in *Drosophila* developmental biology while eschewing his previous research on the genetic regulation of *E. coli* beginning in the late 1960s.

As Stanford biochemists' research diverged and the Biochemistry Department expanded, their moral economy of science was challenged, leading to negotiations regarding the proper way of organizing a productive laboratory life. For one thing, they collectively monitored the population of each research group as a way to prevent any from growing too large; this way, they could preserve their common resource pools (see, e.g., figure 1.9). In the late 1960s, amidst the expansion and diversification of their research, Stanford biochemists began to discuss whether their laboratory practices, which they had cultivated for the productivity of their experimental investigations, would still be beneficial in the new research environment. In 1969, Hogness raised concerns about their random laboratory space arrangements and sharing of instruments, especially after Stryer left over an issue regarding space allocation. Hogness cautiously observed that, "while some mixing of members from different groups does occur, the general result [by the late 1960s] is segregation according to group."[89] He pointed out that the continuing introduction of new research materials (e.g., *Drosophila*, mammalian cell cultures, chicken

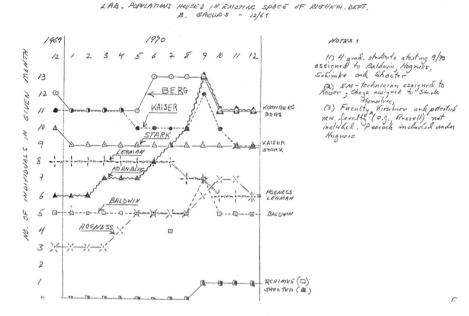

FIGURE 1.9. Laboratory population. In 1969 Stanford biochemists discussed as a group various aspects of their laboratory life. As the Stanford Biochemistry Department expanded in
size, a need emerged to give each faculty member some flexibility in managing their research
space, instead of relying on collective decision making. Image from Faculty Minutes, 1969,
David Hogness, Agenda and Notes about Department: Space, Arthur Kornberg Papers, Stanford University Archives, SC 359, box 11. Reproduced with permission, Department of Special Collections, Stanford University Libraries, Stanford University.

oviducts, and nematodes) would exert "additional pressure for segregation according to group."[90] He wondered "whether it might not be more
useful and efficient in some manner to recognize this condition."[91] Hogness questioned whether the space-sharing practice might no longer be
beneficial to their research because of their diverging interests and changing research materials. He further suggested that a "mild form of decen-
tralization" that would allow individual faculty to organize its own experimental space and materials "would be useful."[92]

Stanford biochemists pragmatically modified their sharing practices, recognizing the diversification of research interests, laboratory instruments, and material demands of each research group. In their 1970
renovation plan, they accommodated each faculty member's need for
exclusive laboratory and equipment space. However, they also tried to
continue some of their communal practices, which they regarded as essen-

tial for the productivity of their research. For example, they maintained some common stock rooms and laboratories and continued to share various research reagents to foster scientific interaction. Despite some disparity among faculty members in their grant sizes, they also continued pooling grant monies in order to preserve their research flexibility and autonomy. Stanford biochemists' moral economy of science, with its particular social practices of sharing and reciprocal obligations in scientific exchange, continued to provide communal ways of organizing the productive laboratory life for Stanford biochemists.

Conclusion

This chapter has examined how, when Stanford University tried to institute biomedical research at its medical school, a group of biochemists came together and developed a particular community of experimental research in the late 1950s. Rather than saddling themselves with the traditional division between the biological and clinical sciences, Stanford administrators tried to integrate the research and clinical endeavor by building a *bio*medical school. In their attempt to establish the Biochemistry and Genetics Departments, Stanford administrators and scientists took advantage of the postwar expansion of the federal research economy, which in turn was buttressed by the post–World War II public and political support for biomedical research. More importantly, Stanford's relocation of its medical school and its articulation of this new vision of biomedicine reflected the postwar emergence of biomedical research and its institutional consequences. Arthur Kornberg and his colleagues thoroughly embraced Stanford's vision of the new medical school and the central role of biomedical research in the progress of medicine.

At the local level, Stanford biochemists developed a communal form of laboratory life by instituting a set of communal practices that would maximize the productivity and flexibility of their research. Developed from their common research interests in DNA, Stanford biochemists' sharing practices illuminated their particular moral economy of science, consisting of shared beliefs about proper ways of organizing their laboratory life, about social norms and obligations in scientific exchange and knowledge production, and about customs and rules in the distribution of resources and scientific credit in the community life. Their sharing customs provided a productive material and political platform for experimentation by giving access to information and materials and by granting intellectual autonomy; their moral economy of science also created a dedicated social and moral context for scientific exchange and knowledge production, obliging them to disclose and share their research results and materials. Their intellectual cohesion and the practice of sharing instruments and research materials (e.g., DNA and enzymes) enabled them to approach DNA's hereditary properties from a more interdisciplinary perspective—biochemical, genetic, and immunological. These emerging patterns of sharing and collaboration in the Biochemistry Department, along with scientists' wariness of goal-directed federal funding, fostered grant pooling among faculty members as a means to sustain their research autonomy during the enormous expansion of federal support for biomedical research. The ethical obligations of reciprocity in community life in turn continued to sustain their moral economy of science as their intellectual and material condition for laboratory life began to be shifted in the mid-1960s.

As we will see in subsequent chapters, Stanford biochemists' culture of vibrant exchange and sharing of ideas and research materials proved to be crucial for their subsequent ventures in the molecular biology of higher organisms in the 1970s, providing a solid platform for productive experimental hybridizations as they began to diversity their field of research. Their experimental culture of sharing underlay creative hybridizations of their experimental systems, helping the scientists to maintain the flexibility of their research when patrons of biomedical research began to call for relevance in research in the 1960s. For example, when Berg tried to develop his experimental system for the exploration of the genetic regulation of eukaryotic organisms, he drew from the particular expertise of Kaiser (on lysogeny), Kornberg (on DNA polymerase), and Lehman (on DNA ligase).[93] More importantly, Stanford biochemists' moral economy of science, particularly their customs of sharing and unstated views about scientific exchange and knowledge production, provides an important backdrop against which we can examine the dynamics between experimental developments and the research environment as they shifted during the 1970s, especially with the advent and commercialization of recombinant DNA technology. Stanford biochemists' communal way of conducting research helps to explain how they later understood and responded to the commercialization of biomedical research when biotechnology emerged from experimental and institutional rearrangements at Stanford in the 1970s.[94]

CHAPTER TWO

"Mass Migration" and Technologies of Gene Manipulation

Stanford biochemists' research in the 1950s and 1960s addressed key questions in molecular biology centered on DNA. Molecular biologists were realizing that DNA was the most important component of the cell, its "master plan" directing the production of a wide variety of included molecular structures. Consequently, Arthur Kornberg and I. Robert Lehman investigated the biochemical synthesis and replication of DNA employing the enzyme polymerase. Molecular biology–oriented junior faculty members, such as Paul Berg, David Hogness, and A. Dale Kaiser, explored the biological activities of DNA as an embodiment of genetic information; they investigated the genetic regulation of bacterial cells and viruses (relatively simple systems) to probe the biochemical and genetic role of DNA. These scientists participated in what Gunter Stent characterized as the golden era of molecular biology when some key questions about the gene were solved. Moreover, Stanford biochemists' focus on DNA as a key material for potential biomedical intervention reflected the ascendency of a particular style of research that developed after World War II; this research developed in the context of cultural and political enthusiasm for a war against disease.[1] The description of the DNA helical structure and its biochemical replication, the "cracking" of the genetic code as a form of linear sequences of DNA, and the elucidation of the operon model of genetic regulation in *Escherichia coli* were all supposed to provide a solid foundation, if not an immediate medical outcome, for a new kind of biomedical research that could intervene in the most fundamental unit of life, the gene.[2]

In the late 1960s, two key members of the Stanford Biochemistry Department, Berg and Hogness, began to seek a new research direc-

tion into the molecular biology of higher organisms. At one level, they
were ambitious enough to venture into the uncharted territory of eukaryotic biology. They understood that the core concepts in molecular
biology, such as the genetic code and the operon model of gene regulation, were developed employing microbes (bacteria and bacterial viruses) as model organisms, whose life processes could be applied to
understanding other higher organisms.[3] After making important breakthroughs in the genetic regulation on bacterial systems, many biochemists
and molecular biologists recognized the need to broaden the investigative horizon of molecular biology. As they recognized, a new intellectual
frontier lay in the biology of higher organisms; there they could examine
whether fundamental biological principles discovered in microbes could
be extended to more complex animals, probing whether "what is true for
E. coli is true for the elephant."[4] Indeed, this period saw a "mass migration" of biomedical researchers from the study of prokaryotic organisms
like bacteria to the cells of eukaryotic organisms, including mammals.[5]
More importantly, the molecular biology of eukaryotes could directly address urgent medical problems by addressing the underlying causes of
disease.

At another level, two Stanford biochemists' foray into eukaryotic
biology resonated well with a rising call in the 1960s for medical relevance in biomedical research. As lay activists and politicians thoroughly
embraced the prospect of biomedical research actively conquering disease, academic researchers began to face political pressure to demonstrate whether their work could meet the public's fervent demands for
medical miracles. In the mid-1960s, when federal support for biomedical
research for the first time reached one billion dollars, patrons and politicians increasingly called for more medically applicable investigations.
The rising call for relevance coincided with the countercultural movement of the late 1960s that demanded a shift toward more socially useful research. In 1971, the Nixon administration implemented a large-scale
national campaign for the "War on Cancer" in order to conquer the disease by 1976, the American Revolution Bicentennial. These cultural and
political shifts presented a particularly challenging environment for molecular biologists and biochemists, whose research had focused primarily
on the understanding of bacterial viruses and bacteria.[6]

Some molecular biologists like Francis Crick, who codiscovered the
double-helical structure of DNA, hesitantly admitted in 1968 that "there
has been nothing as spectacular or as useful as, say, penicillin," despite the

last twenty years of the rapid developments in molecular biology.[7] Those who were at an assemblage of academic disciplines, such as biochemistry, genetics, molecular biology, and virology, which was central to the molecular revolution in the life sciences, however, claimed that the application of concepts and tools of molecular biology to medical problems could radically change the nature of medicine. Crick, for example, suggested that the fundamental understanding of the gene could provide general etiological explanations for certain key diseases like cancer.[8] One Stanford biochemist, Paul Berg, indeed made a conscientious effort to situate his molecular biological work in relation to medicine, pointing out the shifting funding priority toward medically relevant research. In 1970, Berg was chair of the Biochemistry Department at Stanford Medical School. He wrote to his colleague, Arthur Kornberg, about the changing priorities for funding at the National Institutes of Health (NIH), the most beneficent patron of biomedical research:

> The grant situation seems to be deteriorating rapidly. There are all kinds of rumors—many of which are hard to believe—but the gist of the story is that both the level of support and the type of work that will be supported by, for example NIH, is undergoing serious review. One of the rumors is that NIH is taking a very hard look at the question of whether work on *E. coli* and its bacterial phages is within the mission of the NIH—unbelievable![9]

This chapter examines Berg's venture into the molecular biology of higher organisms and its conceptual, medical, and technological consequences for the manipulation of the gene. In particular, I investigate how Berg adopted animal tumor virus experimental system for his explorations into the genetic expression and regulation in mammalian cells, especially the genetic mechanisms of tumorigenesis, garnering enthusiastic support from various government and philanthropic funding agencies for supporting cancer research.[10] He tried to creatively weld tangible connections between gene-regulation studies in prokaryotes, such as bacterial cells and their viruses (bacteriophages), and those in eukaryotes, such as mammalian cells and animal viruses; he did this by drawing conceptual and methodological analogies. As I demonstrate, in his attempt· to find research technologies for the study of the molecular biology of cancer in eukaryotic organisms, Berg's experimental system involving tumor cells and viruses went through several unanticipated shifts that resulted in the advent of recombinant DNA research and technology. Recognizing the dynamics of biological experimentation that brought about recombinant DNA technology, this chapter traces the transformation of Berg's animal virus experimental system from a pursuit of scientific research into a pioneering tool for genetic engineering.[11] Crossing the boundary from prokaryote to eukaryote resulted in the first artificial synthesis of recombinant DNA molecules, leading to the emergence of recombinant DNA technology and other significant new research techniques and agendas.[12]

The construction of artificial recombinant DNA molecules not only provided a technological foundation for gene manipulations but also epistemologically recast genes as a sequence that could be rewritten. By broadening an investigative focus on the development of the material infrastructure and conceptual framework for genetic engineering at the laboratories of Stanford biochemists, this chapter provides an alternative to the canonized history of genetic engineering that tends to focus narrowly on Stanley Cohen and Herbert Boyer's cloning of recombinant DNA molecules and the commercialization of the recombinant DNA–cloning procedures in the mid-1970s.[13] Historians and some scientists have nevertheless pointedly noted that experimental procedures for making recombinant DNA molecules were initially developed by Stanford biochemists Berg and his colleagues, Lobban and Kaiser, in the late 1960s and early 1970s.[14] Even more interestingly, Lobban has often been seen by fellow scientists as one of the scientists who first recognized the potential of recombinant DNA technology for genetic engineering in the late 1960s, several years before Cohen and Boyer filed a patent application for recombinant DNA–cloning technology in 1974.[15] Indeed, Stanford biochemists assembled experimental conditions for genetic engineering not only by constructing recombinant DNA but also by providing conceptual resources for gene manipulations. These material and conceptual resources were in turn actively circulated in a research network formed around the Biochemistry Department, and they enhanced an experimental environment that enabled the development of genetic engineering through the cloning of recombinant DNA molecules.

Animal Viruses and the Molecular Approach to Cancer

Reflecting on the mobilization of scientists during World War II, postwar American political leaders and scientists contentiously debated how to best adapt the wartime scientific enterprise to peacetime conditions.[16] In this early postwar period, lay activists newly energized voluntary health organizations like the National Foundation for Infantile Paralysis (NFIP) and the American Cancer Society (ACS). The organizations therefore played a catalytic role in fostering a new style of biomedical research that mobilized both basic biological researchers and clinical medical researchers for a "war against disease."[17] The subsequent growth of large-scale federal patronage of biomedicine supported a broad array of basic research in biochemistry, genetics, microbiology, and virology; the relevance of this research to conquering dread diseases was in turn exploited by entrepreneurial biochemists and molecular biologists. In the postwar emergence of biomedicine, viruses in particular became one of the prominent model organisms that enabled biomedical researchers to forge links between basic virus research and disease-oriented research, such as polio and cancer.[18]

In the 1950s and 1960s, as the historian Daniel Kevles shows, some biomedical researchers like Renato Dulbecco moved into animal virus research. This influx of scientists like Dulbecco triggered a general shift of focus in animal virology from its clinical to its basic biomedical aspects, including tumorigenesis in mammalian cells.[19] Dulbecco, a former phage geneticist who refashioned himself as an animal virologist with funding from the NFIP, invented a plaque assay for the equine encephalitis virus in 1952; the technical and conceptual implications of this means of determining the virus concentration in an infectious dose were critical for the subsequent development of molecular approaches in animal virology.[20] His plaque assay enabled scientists to isolate the progeny of a single virus particle grown on cloned and cultured mammalian cells, which helped animal virus studies adopt the quantitative style of phage genetics.[21] The 1962 Cold Spring Harbor Symposium on "Basic Mechanisms in the Biology of Animal Viruses" was organized by Dulbecco and signaled the growing interest in animal virology among molecular biologists and biochemists. In a concluding remark at the symposium, Dulbecco asserted that "with the widespread use of techniques for cultivating animal cells *in vitro* ... the full spectrum of the approaches and methodology of molecular biology could be applied to the study of animal viruses."[22]

The illumination of genetic regulation in lysogenic bacteria in the 1950s and early 1960s laid the groundwork for the use of animal viruses as model organisms in cancer research, especially in the viral genesis of cancer.[23] Certain lysogenic (also called temperate) bacteriophages like λ phage did not burst the cells when they infected host bacteria, as a normal viral infection would. Rather, they remained dormant in a special latent form called the prophage. Lysogeny, which involves a bacterium harboring a prophage as hereditary material, captivated molecular biologists at the Pasteur Institute in Paris and at Caltech in southern California, forming a close international network of lysogeny research. André Lwoff, director of the Pasteur Institute, was especially interested in lysogeny since explorations into the conditions under which the prophage perpetuated itself and its DNA without expressing its lytic (pathological) power could provide useful insights into the mechanisms of viral tumorigenesis.[24] In the early 1950s, molecular biologists at Caltech, such as Max Delbrück and Jean Weigle, began to adopt *Escherichia coli* K12, one of the most useful bacterial strains for genetic analysis, for their genetic study of prophage λ, making λ an exemplary virus for the study of lysogeny.[25] Kaiser, a student of Delbrück and Weigle, explored the genetic relation between the prophage λ and its host, and went to the Pasteur Institute as a postdoctoral fellow with François Jacob, who had just started the genetic analysis of lysogenic *E. coli* K12.[26] Through their collaboration, they identified a set of genes that regulated the lysogenization of *E. coli* K12.[27] More importantly, Kaiser's experiment with Jacob also demonstrated that the immunity of lysogenic bacteria was under the genetic control of bacteriophage λ. Before their experiment, the mechanisms of immunity—how "the presence of the prophage confers upon the host bacterium a resistance to infection with the homologous phage and its mutants"—were quite obscure.[28]

Demonstration of the genetic control of lysogeny and viral immunity, as Nadine Peyrieras and Michel Morange show, was a major vindication of the Pasteurian approach for understanding cancer (see, e.g., figure 2.1)[29] Indeed, in his recommendation letter to Kornberg, Jacob underscored the medical implications of Kaiser's λ research: "[Kaiser's] work is of considerable importance in the understanding of the mechanism which allows integration of the viral genetic material and that of the host cell, a phenomenon to which all specialists refer now as a model for the possible action of viruses in cancer genesis."[30] Knowledge of the prophage's genetic association with the host—how the prophage endows both immunity and the capacity to produce phage in bacteria, and under what conditions it expresses its viral characteristics—could provide a basis for developing immunological means of treating diseases caused by viruses.

The genetic regulation of prophage λ in the host cell became a focused research subject in the early 1960s, framing lysogenization and viral cancer genesis in terms of viral gene expression and repression. By drawing analogy from lysogeny in bacteria, many molecular biologists suspected

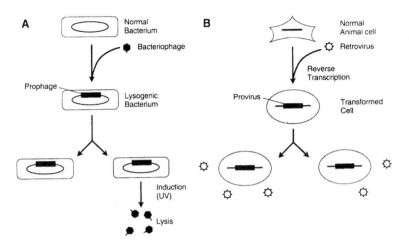

FIGURE 2.1. Analogy between bacteriophage λ and tumor viruses. The analogy between the lysogenic behavior of bacteriophage λ (prophage) and the tumorigenesis of animal viruses (provirus). Images from Harold Varmus, "The Pastorian: A Legacy of Louis Pasteur," in Gorge F. Woude and George Klein, eds., *Advances in Cancer Research*, vol. 69 (New York: Academic Press, 1996), 12 and 14. Reprinted by permission from Harold Varmus.

that tumorigenesis by animal viruses could be understood as a result of the expression of some key viral genes introduced into the host by lysogenic infection. Kaiser's Caltech mentors in virology, Dulbecco and Marguerite Vogt, were at the forefront of tumor virology research. They had embarked on researching viral tumor genesis in mammalian cells using animal DNA viruses like polyoma virus and simian virus 40 (SV40). In 1960, Dulbecco and Vogt showed that the polyoma virus gave rise to two distinctive types of virus-cell interactions, whose "characteristics [were] reminiscent of temperate bacteriophage."[31] They observed that hamster and mouse cells were often transformed into tumor cells by polyoma infection, without the cellular degeneration accompanied by most viral infections. Soon they moved to the newly established Salk Institute in La Jolla, in southern California, where its founder Jonas Salk envisioned the conquest of cancer through virological means, just as he had developed polio vaccine. Dulbecco ambitiously charted his pioneering cancer research of animal tumor viruses in his article in *Science*, suggesting that the structural similarity of animal tumor viruses, such as polyoma and SV40, to lysogenic bacteriophage λ provided a key to understanding the interaction between these viruses and the host cells—tumorigenesis.[32] In his analogical reasoning, Dulbecco relied on Allan Campbell's integration model λ's genetic interaction with its host *E. coli*. Campbell suggested that λ's circular DNA would be cut into linear DNA sequences inside the host cell, and then the linear DNA would be inserted into the host chromosome, which led to lysogeny.[33] Dulbecco called attention to the circular structure of both bacteriophage λ and tumor-producing animal DNA viruses, and further likened the cell transformation by polyoma to the lysogenization of *E. coli* by prophage infection, suggesting that a tumor virus would be integrated into the host chromosome in a form of a provirus, leading to tumorigenesis. He hypothesized: "the absence or low level of virus production and the resistance to superinfection [in the cells transformed virally] are similar to the properties of lysogenic bacterial cultures and suggest that the integrated virus exists as provirus," a noninfectious intracellular form of a virus, like prophage.[34]

Animal viruses thus provided a promising entry point for molecular biologists into cancer research, which was in turn buttressed by political mobilization for medically relevant studies in the 1960s.[35] Howard M. Temin, Dulbecco's graduate student at Caltech in the late 1950s, was another example of a molecular biologist working on cancer. He hypothesized mechanisms of virus-mediated cancer genesis by making analogies between prophage and provirus in the transformed animal cells. Working on the cancer-causing Rous Sarcoma Virus (RSV), Temin boldly claimed that RSV synthesized a DNA provirus from its RNA, suggesting that the genetic information transferred from virus RNA to the host's DNA would cause tumors. He hypothesized the existence of the enzyme reverse transcriptase that would make tumorigenesis possible through viral infection.[36] Both Temin and David Baltimore, a research associate in Dulbecco's laboratory in the mid-1960s, identified reverse transcriptase in 1970, proving Temin's retrovirus hypothesis. Indeed, Dulbecco's laboratory at the Salk Institute became a critical node in the emerging biomedical complex of cancer virus research in the 1960s. Work at the laboratory reconceived cancer in terms of the interaction between cell and virus, whose genetic control was regulated by tumor-generating animal viruses.[37]

Molecular Biology on the Move: "Mass Migration" in Biomedicine

During the early postwar period, biochemists and molecular biologists capitalized on cultural and political enthusiasm for curing dread diseases by adopting virus model systems. Calls for relevance to medical researchers did provoke some ambivalence among scientists who feared increasing governmental involvement in their work. However, the large-scale influx of federal support seemed to provide ample opportunities to flexibly appropriate the mission-oriented research mandate for the scientists' own research interests, at least for a while. In a situation in which lay activists and medical philanthropists, government officials, and scientists thoroughly embraced the prospect of biomedicine generating medical "miracles," there was also potential for an intellectual and political backlash toward biomedical research.[38]

Around the mid-1960s, a perception did indeed emerge among the scientists that basic biomedical research, whose diverse range of research trajectories had consolidated into molecular biology in the late 1950s, was experiencing its own crisis even after its nascent institutional success, emblematized by the founding of the Laboratory of Molecular Biology at Cambridge, England, in 1957.[39] The supposed crisis was two-fold: one was intellectual, and the other was institutional. As for the intellectual predicament, the molecular biologist Gunther Stent proclaimed in 1968 that molecular biology had entered its last phase, or its "academic period." Reflecting on the achievements of molecular biology, such as the structural elucidation of DNA and the formulation of the operon model of gene regulation, he predicted that what remained was to "iron out the details" of the informational paradigm, through which his generation of molecular biologists had explained the key features of the life, such as heredity, in terms of the storage, reading, and transmission of biological information stored in the linear sequences of DNA.[40] He insisted that, by virtue of its very success, molecular biology would no longer present challenging and rewarding problems, and the end of its progress seemed imminent. Stent advised the next generation of researchers to explore uncharted territories in biology, like embryology, the immune response, the origin of life, and the higher nervous system. According to Stent, molecular biology, once an intellectual magnet for creative scientists, was about to decline.

Molecular biology was challenged at the institutional level by those who opposed its "imperialistic" disciplinary politics and its standing as a new postwar discipline.[41] For example, James Watson, a newly appointed molecular biologist at Harvard, devised an expansive plan for the Biology Department that precipitated what the naturalist Edward O. Wilson called the "molecular war" with evolutionary biologists.[42] Often, molecular biologists' scientific and medical claims were bold enough to draw criticism from researchers in a number of other biological fields. One molecular biologist succinctly commented on these criticisms during the controversy over the establishment of the European Molecular Biology Laboratory: "Typical criticisms raised by enemies of molecular biology . . . are that the [genetic] code being solved, the era of molecular biology is over and its proper place is already in the science museum . . . [and] that in spite of its brilliant discoveries, molecular biology has not helped to cure any disease or to improve crops, etc."[43] To some molecular biologists, it seemed that the future of their discipline increasingly depended on its ability to find available intellectual and institutional niches in order to become a constructive part of the expanding biomedical complex.

The publication of commemorative histories and textbooks by the founding generation of self-proclaimed molecular biologists in the mid-1960s reflected the uneasy disciplinary and institutional context of the field. Molecular biologists and biochemists made conscious efforts to articulate their historical roles in order to reassert their intellectual and institutional place in their disciplines.[44] *The Molecular Biology of the Gene*, widely cited as the first textbook of molecular biology, offered a synthesis of universal biological principles operating at the molecular level at the time of its publication in 1965.[45] Its author, Watson, underscored the need for further research into the complexities of higher organisms by adopting the perspectives of molecular genetics and biochemistry.

Watson's call for molecular research on the biology of higher organisms resonated well with the increasing political pressure of the mid-1960s on funding agencies and scientists to conduct research resulting in viable medical applications. Laboratory breakthroughs that dazzled molecular biologists seemed to fail to satisfy the fervent enthusiasm of lay activists and patient groups hoping for substantive medical contributions. The prominent molecular biologist, Crick, optimistically insisted that "because of the very fundamental discoveries which are going to be made," there would be a "change in the nature of medicine."[46] However, physicians and medical patrons were growing impatient with basic biomedical research, whose experimental systems often had been based on rather simple organisms like bacteria and viruses rather than on human cells. For example, in his controversial 1968 book *Cure for Cancer,* the pharmacologist Solomon Garb at the University of Missouri Medical School criticized the inability of the molecular approach in biomedicine to address the complexity of higher organisms. He claimed, "it hardly seems prudent to rely on the molecular-biology approach for practical help in solving the cancer problem during the next few generations."[47] Garb's book influenced powerful medical patrons and activists like Mary Lasker, who began to call for more goal-oriented and applicable biomedical research.[48] Indeed, the subsequent enactment of the National Cancer Act in 1971, which instituted goal-oriented contract research run by government scientists and officials, inaugurated an even more challenging environment for biomedical researchers.[49]

Paul Berg's Venture into Research on Eukaryotes

The heavy reliance of biochemists and molecular biologists on the rhetoric of biomedicine for political and financial patronage backfired in the late-1960s, when attitudes toward science profoundly shifted with growing countercultural critiques of science and technology.[50] This changing cultural context of biomedical research increasingly placed more emphasis on its medical applications rather than on advances in fundamental biological knowledge.[51] Paul Berg, however, had come to recognize that the lack of experimental tools or appropriate organisms for exploring other issues of biology, such as the genetic regulation of higher organisms or their developmental processes, posed a formidable challenge for molecular biologists.

Intrigued by this intellectual challenge, Berg embarked in 1967 on a research project involving the genetic regulation of eukaryotic organisms. In studying the biology of higher organisms, he capitalized on his expertise with bacterial experimental systems, constructively framing the issues of human (cell) biology in terms of the emerging distinction between prokaryotes and eukaryotes. As Susan Spath points out, this distinction was articulated in the early-1960s in order to broaden the relevance of bacterial experimental systems; this was done by underlining the similarities and differences between prokaryotes, like bacteria, and higher organisms.[52] In the 1950s, bacteria gained much of their status as highly productive research organisms for molecular biology and biochemistry. During this period, as Angela Creager has shown, the invention of bacterial sexuality for genetic analysis and the wide availability of bacterial mutants for biochemical analysis, opened new opportunities for representing bacterial genes and their functions through mapping techniques; this had previously been an impossible task for haploid organisms like bacteria.[53] François Jacob, for example, established a genetic mapping system by following the transmission of hereditary factors through bacterial conjugation (gene transfer by direct cell-to-cell contact) and transduction (virus-mediated gene transfer) in mutants of *E. coli*, providing crucial research technologies for elucidating the operon model of genetic regulation.[54]

By the mid-1960s, Berg ventured into research on eukaryotic organisms, recalling that, in "the bacterial field, particularly in the area of gene expression, the outlines were already becoming quite clear. The Jacob-Monod-Lwoff story [of the operon model of gene regulation] was already the dominant paradigm. And yet there was nothing known about gene expression in mammalian cells, in eukaryotes."[55] In his 1968 ACS grant application, Berg situated advances in bacterial experimental systems in relation to their usefulness to new research on higher organisms. He proposed their use as models for investigating the genetic regulation of higher organisms:

> For the past fifteen years or so, my research has been concerned primarily with the biochemistry of genetic processes and more specifically with the enzymic [enzymatic] mechanisms. This approach has uncovered much important information; but equally important, *now we can begin to examine and interpret more complex biologic questions in terms of the concepts developed with the microbial models.* How is the expression of genetic information regulated by intrinsic and extrinsic factors in eukaryotic cells? Are the models of repression induction control of protein synthesis applicable to the genetic organization of higher cells? Or are there new mechanisms and principles to be uncovered? I propose to begin an investigative program to explore these questions using cultured mammalian cells as the experimental system.[56]

Indeed, Berg's move into eukaryotic biology epitomized what later generations of biologists would recall as the "mass migration" of biochemists and molecular biologists into the study of higher organisms from the late 1960s onward. As examples, Sydney Brenner began to work on development using the nematode *Caenorhabditis elegans*; Jacobs moved to the developmental biology of mice; and phage geneticist Seymour Benzer ventured into the behavioral genetics of the fruit fly, *Drosophila*.[57]

Linking Cancer and Eukaryotic Biology through an Animal Virus

In his venture into eukaryotic biology, Paul Berg also made a tangible connection between his new research on the genetic expression and regu-

lation of eukaryotes and cancer research by explicitly drawing the connection between his mammalian cell studies and cancer research through
the use of animal tumor virus experimental systems. His articulation of
medical relevance in the biology of higher organisms resonated well in
the political and medical climate of the 1960s. During the years from 1968
to 1970, with prominent molecular biologists and biochemists, including
Charles Yanofsky, Norton Zinder, and James Watson, Berg also helped
prepare a proposal for the National Program for the Molecular Biology
of the Human Cell. Instituted by the National Science Foundation (NSF),
this program intended to explore a feasible means for studying the expression, organization, and regulation of genes in human cells by using
mammalian cell culture methods and somatic cell genetics. Moreover,
as Toby Appel points out, NSF's human cell biology program "hoped to
link basic research on the human cell to the Nixon administration's war
on cancer."[58] In some respects, the program was also a critical reaction
to the pressure for more medically relevant research. The authors of the
proposal underscored the value of a flexible and basic research system in
dealing with cancer: unlike NASA's Apollo program, "we lack most of the
basic knowledge needed to formulate the technology."[59]

Recognizing all the difficulties of exploring the genetic expression and
regulation of complex eukaryotes, Berg needed to devise a new experimental system that could contain the complexity. Stanford biochemists'
culture of sharing and close interaction in the laboratory was critical for
Berg's move into cancer research. A. Dale Kaiser's work at Stanford on
lysogeny in bacteriophage λ not only inspired Berg's new venture into
eukaryotes but also provided a crucial conceptual resource that enabled
him to open a whole new field of research into the genetic expression
and regulation of mammalian cells (figure 2.2). In other words, bacteriophage λ provided an exemplary model experimental system for Berg's
research on eukaryotes, especially for his cancer research employing an
animal virus. Bacteriophage λ thus mediated prokaryotic and eukaryotic
divisions, and linked biological and medical research.

Berg made his initial acquaintance with bacteriophage λ and its lysogenic behavior at Kaiser's graduate seminar on lysogeny between 1966
and 1967. By the time Berg took Kaiser's seminar, the subject of cell
transformation by polyoma virus and SV40—tumorigenesis—bore particular relevance to lysogeny. Kaiser emphasized the striking similarities
between the two systems: lysogeny with a bacteriophage and tumorigenesis with a tumor virus. In a concluding seminar on "Cell Transformation

FIGURE 2.2. A. Dale Kaiser and the phage lambda (λ) model (1997). A. Dale Kaiser assembling phage lambda (bacteriophage λ) at his seventieth birthday bash at Asilomar Conference Center. Kaiser studied bacteriophage λ whose lysogenic behavior provided an analogical model through which other scientists like Paul Berg considered tumorigenesis. Reprinted by permission from A. Dale Kaiser.

by Polyoma Virus and SV40," Kaiser discussed the possible mechanisms of tumorigenesis in terms of Campbell's integration model of bacterial lysogeny. Kaiser then hypothesized that the viral transformation of mammalian cells might be caused by the integration of an animal virus into the host chromosome.[60]

The analogy made between bacteriophage λ and a mammalian tumor virus inspired Berg to plan his new venture into the genetic regulation of mammalian cells. As noted earlier, it provided tangible connections that mediated prokaryotic and eukaryotic divisions in terms of lysogeny and viral transformation, and it helped Berg to link biological research in gene regulation in eukaryotes with medical research on tumorigenesis. While attending Kaiser's seminar, Berg realized that he could devise a new experimental system to study the genetic regulation of eukaryotic cells that took advantage of the ability of animal viruses, such as SV40 and polyoma, to transform mammalian cells, just as molecular biologists had revealed the genetic regulation of prokaryotes through their research on *E. coli* cells infected by bacteriophages. This mammalian cell experimental system seemed to add another advantage: instead of exploring the expression and regulation of these complex genes, this experimental system could be used to investigate the expression and regulation of viral genes after the virus had infected the mammalian cell. Thus, this system would enable Berg to bypass the difficulties of directly exploring the more complicated genes of the mammalian cell. With this ambitious project in mind, Berg asked Dulbecco whether he could spend a sabbatical leave at his lab at the Salk Institute. In the summer of 1967, notwithstanding Kornberg's reservations, Berg headed to the Salk Institute.

Berg's sabbatical leave at Dulbecco's laboratory and his choice of animal viruses (SV40 and polyoma) not only enabled him to study gene expression and regulation of eukaryotes but also facilitated his participation in the newly emerging subject area of animal tumor viruses. As Berg emphasized in his grant proposal in 1968, the question of viral transformation and cancer research—how animal viruses such as SV40 and polyoma could transform normal cells into tumor cells—was his "immediate intent." As he continued, by "following the entry of the viral chromosome into the infected cell . . . we hope to learn how the entry of the viral chromosome into the host cell chromosomes causes the cancer state."[61]

Berg's meticulous blending of his basic molecular biology research and its potential medical implications was critical for his initial venture into eukaryotic genetic regulation. His investigation of the genetic regulation of tumor viruses in turn helped situate his research in the context of expanding cancer research. By the time Berg arrived at the Salk Institute, Dulbecco's research focused on how animal tumor viruses such as SV40 and polyoma transformed normal cells into tumor cells. In 1967, Dulbecco and J. F. Watkins triggered the "reactivation" of a virus in SV40-transformed cells; by fusing transformed cells with virus-susceptible cells, they produced SV40 virus from SV40-transformed tumor cells.[62] Soon

Heiner Westphal and Dulbecco showed that polyoma and SV40 DNA resided in the transformed cells.[63] With evidence from DNA hybridization, which probed the fidelity of DNA's base-paring between SV40DNA and the DNA from its host chromosome, Joseph Sambrook, Westphal, and Dulbecco further demonstrated in 1968 that SV40 DNA was integrated into the chromosome of host cells.[64]

Dulbecco's path-breaking research on tumorigenesis in mammalian cell systems, especially his explanation of the physical state of the viral genetic information within the transformed cells, provided fertile ground on which Berg built his experimental systems. As Berg emphasized in his 1968 grant application for his cancer research, Dulbecco's SV40 viral induction system "opened up a whole new area of investigation, namely the mechanism of activation of the viral genome and by implication the question of why the viral genome in the transformed cell is not fully expressed"[65] (figure 2.3). Berg was interested in seeing whether "acquisition and maintenance of the transformed phenotype [tumorigenesis] requires the continued presence and expression of at least part of the viral

FIGURE 2.3. Paul Berg discussing his cancer virus research, ca. 1975. Paul Berg explained how his cancer virus research could lead to the discovery of cancer-inducing genes (or oncogenes) using temperature-sensitive mutant strains of tumor viruses. Paul Berg's Special Lecture, Stanford Biochemistry Department Fiftieth Anniversary Photo Collection. Reprinted by permission from Jack Griffith.

genome."[66] If viral integration were sufficient to cause tumorigenesis, he further reasoned, "elimination or inactivation of these essential viral genes would result in reversion to a normal cell growth pattern."[67]

At Dulbecco's lab, Berg and François Cuzin, a postdoctoral fellow from Jacob's lab at the Pasteur Institute, invented a polyoma virus induction system, through which latent polyoma viruses replicate and are released, by adopting Dulbecco's SV40 induction system. Berg traced the synthesis of polyoma DNA in polyoma-transformed tumor cells at low temperature using radioactive labeling and DNA hybridization.[68] He first infected a mouse cell (3T3) with a heat-sensitive mutant polyoma (*Ts-a*), and then examined how the polyoma-transformed cells (*Ts-a*-3T3) were induced to produce virus particles at low temperature. He found that the transformed clones (*Ts-a*-3T3) produced much larger compounds (oligomers) of viral DNA twenty-four hours after low temperature viral induction, which implied that these tumor cells contained the entire polyoma genome. In order to explain the newly synthesized viral DNA in tumor cells, Berg proposed the "excision hypothesis" of viral induction. Adopting the integration model of lysogeny, he hypothesized that "the polyoma genome was [supposed to be] covalently integrated into the host cell chromosomal DNA."[69] The virus integration was believed to be an analogous process to lysogeny in microbial systems: the lysogenic λ phage DNA was integrated into the *E. coli* chromosomes and became dormant. Berg further speculated that when a polyoma virus was induced, it emerged from the integrated host chromosome and started viral replication.[70] In effect, he explained this viral stimulation in terms of the induction model of lysogenic bacteriophage λ.

Beyond the physical model of viral integration, Berg began to adopt a biochemical approach in an attempt to connect the mechanism of viral induction and its tumorigenesis to his initial research interest in the genetic regulation of eukaryotic systems. An understanding of the biochemical dynamics of the viral integration model would provide a means to decipher genetic expression and regulation in mammalian cells. For this, he planned a set of biochemical operations, using his virus induction system, to characterize the molecular pathways of viral infection and its transformation. First, he wanted to trace new protein synthesis in mammalian cells with radioactive isotope labeling after the introduction of an animal virus. Second, Berg suggested that fractionation of the infected cells would provide molecular identification of newly replicating particles and viral specific enzymes. Third, he proposed to employ a technique (analytical radioautography of polyacrylamide gel electropherograms) to label and analyze new virus-specific proteins formed after infection, such as T-antigen and viral capsid proteins. All these biochemical explorations, which he intended to pursue further at Stanford, aimed to provide the means for studying how viral genes were regulated in eukaryotic systems.[71]

Virus as a Vector: Building an Artificial Transduction System

After his sabbatical leave at Dulbecco's lab, Berg phased out his bacterial and enzyme studies and shifted his research entirely to animal cell and tumor virus work. With this change of experimental system, he had to renovate his lab for large-scale virus and cell culture research in a sterile environment. For example, he used ACS grant money to purchase and install a controlled environment incubator for aseptic handling of virus. He also bought a roller apparatus for growing large quantities of mammalian cells. Building on his viral induction system with a large experimental infrastructure, Berg mounted a systematic biochemical exploration of viral oncogenesis, the formation of cancer through a virus.

Berg's experimental research, however, did not progress as initially planned. First, his attempt to trace and identify the proteins induced in cells following viral infection (with radioactive isotopes) did not work because of the background "noise" produced by the highly metabolic and complex mammalian cells. As he explained, "because even resting cells, and serum-starved cells still incorporate sizable levels of radioactive amino acids into their proteins it is extremely difficult to detect new proteins made as a result of virus infection."[72] Instead, he tried to find other, simpler alternatives for biochemical explorations. For example, with graduate student John Morrow, he used an electrophoresis technique to characterize the viral RNA transcripts produced after an SV40 viral infection. By adjusting their repertoire of biochemical tools, they aimed to explore the regulatory mechanisms of viral gene expression in mammalian cells.

Since Berg's initial biochemical approach did not allow direct investigation of viral genes and their functions in mammalian cells, he began to tinker with his tumor-virus experimental system from the perspective of biochemical genetics. In order to deal with difficulties he had faced in his animal virus system, Berg wanted to adopt the productive use of transduction—virus-mediated gene transfer—in a fine-structure mapping of bacterial genes for animal tumor virus research; he hoped this would create a genetic map of tumor viruses, such as polyoma and SV40. In the early 1960s, Berg had become well aware of the power of transduction as a tool for gene mapping through his collaboration with Charles Yanofsky, a colleague at the Department of Biological Sciences at Stanford.[73] Yanofsky played a central role in demonstrating the molecular colinearity between gene and protein, probing whether changes in DNA sequence can produce changes in protein sequence at corresponding positions; he did this by examining a relationship between the linear sequence of mutant sites in a gene and the linear sequence of amino acids in a protein in *E. coli* (by employing P1 transduction).[74] While working on the problem of colinearity between gene and protein, Yanofsky observed that some mutant genes in *E. coli*, such as tryptophan synthetase mutant A36, could be corrected or suppressed. He suspected that the suppression of mutant A36 could be attributed to mutations in transfer RNA (tRNA). In his conversation with Yanofsky at a friendly tennis match, Berg proposed an experimental system involving amino acid incorporation that could determine whether mutations in tRNAs mistranslated a specific mutant codon in a synthetic RNA.[75] Berg and Yanofsky specifically took advantage of transducing bacteriophages in preparing tRNAs that contained a suppressor gene (*su36+*), which in turn misread a mutant gene.

Through the use of transduction, Berg was growing more used to employing genetic tools in analyzing a "subversion of genetic decoding."[76] As transduced bacterial genes were located very close to the genes of their *E. coli* host, transduction was most useful in the high-resolution genetic mapping of bacterial experimental systems. Moreover, as Angela Creager points out, when combined with biochemical operations, "fine-structure mapping was not an end to itself, but rather a tool to investigate the genetic code, metabolism, cellular regulation, and the dynamics of infection, to name but a few topics."[77] Indeed, transduction became a crucial part of research technology that connected genetics to biochemistry, contributing to the development of molecular genetics.

More significantly, Berg's familiarity with transduction provided him with a crucial insight into rebuilding his animal virus experimental system; he was now able to view SV40 as part of a transduction system. Animal viruses, when exhibiting tumorigenic behavior similar to that of bacteriophage λ lysogenic behavior, seemed capable of transducing genes. Berg was particularly intrigued by Dulbecco and Watkins's recent finding that suggested the potential ability of SV40 to transduce genes in the host animal chromosome. In 1967, they had found that SV40 behaved like a prophage, picking up nearby genes when induced from SV40-transformed cells.[78] However, unlike a bacteriophage, whose ability to move genes between bacterial cells through transduction proved to reveal the genetic expression and regulation of bacteria, no animal viruses were known to transduce genes from one cell to another in eukaryotes. In addition, SV40's genome was too small to pick up genes of the host cell, whose chromosomes were much larger than those of bacteria. Furthermore, the coating of the nucleic acid of the virus (encapsidation) would limit the number of genes that could be packed inside since the enclosure (the capsid) is considerably small. Finally, the process of an animal virus picking up genes would be too random to be employed in genetic analysis. To use transduction in mammalian cells, Berg had to find another way to transduce genes through the use of SV40. He recollected the initial conception of his artificial SV40 transduction project:

> Initially, I had serious reservations about the success of such a venture because of the predictably low probability of generating specific recombinants between virus and cell DNA and the limited capacity for selecting or screening animal cells that had acquired specific genetic properties. But it seemed that one possible way out of this difficulty, at least one worth trying, was to produce the desired SV40 transducing genomes synthetically . . . The goal was to propagate such recombinant genomes in suitable animal cells, either as autonomously replicating or integrated DNA molecules.[79]

In his 1970 ACS grant application, Berg put forward his reinvention of the SV40 experimental system as an artificial transduction system, capitalizing on its potential ability to transfer genes into mammalian cells. This led to the experimental construction of recombinant DNA molecules by Berg's research team:

> David Jackson, a postdoc fellow [from Yanofsky's lab], Tom Jovin, a summer visitor from Germany, and I are exploring the feasibility of using SV40 DNA (or PY DNA) to transport non-viral genetic information into the cells they transform. It is our ultimate aim to attach a specific segment of DNA covalently by chemical or enzymatic means to the circular DNA molecule of either virus and then to determine a) if cells can be transformed by such modified DNA's, b) if the new genetic information is integrated with the viral DNA and c) if they express the new information carried by the modified DNA.[80]

What became clear by 1970 was that SV40 was emerging as a research technology for genetic explorations. It marked a critical realignment from Berg's initial use of the animal virus from an object of cancer research to a technology for genetic mapping and subsequently for genetic engineering.[81] At first, Berg wanted to use the ability of animal tumor viruses to integrate into mammalian cells and transform their host into tumor cells as a way to explore the mechanisms of tumorigenesis; he expected to then be able to investigate genetic expression and regulation in mammalian cells. His initial biochemical exploration of SV40 transformation, or animal-virus tumorigenesis, however, became complicated because animal cells synthesized too many proteins to be able to detect viral gene products. Faced with the experimental difficulties noted above, he turned to the potential ability of animal viruses to function as a "vector" that would transfer genes from one cell to another in eukaryotic organisms.[82]

He envisioned that artificial SV40 transduction could be deployed as a mapping tool for studying gene regulation in mammalian cells. The SV40 mapping system would enable scientists to "construct a physical genetic map of SV40 *without requiring a genetic mating system*."[83] Moreover, if a newly introduced DNA segment in SV40 recombinant DNA molecules could then transform the host cell, the usefulness of transduction for gene regulation studies "can be extended to [an] animal cell system."[84] Kenichi Matsubara, a postdoctoral fellow at Kaiser's lab, provided a crucial key to that end. Matsubara had previously derived λ*dv*, a plasmid (an extrachromosomal ring of DNA) that could replicate itself in *E. coli* hosts.[85] Berg thought that if he could derive a plasmid that contained the gal operon (which encodes enzymes for galactose metabolism) from bacteriophage λ, namely λ*dvgal*, its ability to propagate itself as a circular and autonomous entity in *E. coli* would make it appropriate as a genetic vector in a transduction system. If "SV40 DNA can be propagated in *E. coli* as part of the λ*dv* replicon," Berg further contemplated, then perhaps eukaryotic genes could be transduced and their regulatory mechanisms studied. This possibility would take advantage of the well-characterized gal operon system.[86] He cautioned that this novel project still required another technical breakthrough: "if this approach proves useful the challenge will be to expand and increase our access to preparations of specific, homologous (mammalian) genetic DNA."[87]

Berg's first step for constructing an artificial transduction system was to recombine SV40 and λ*dvgal* DNAs, the *gal* operon with the plasmid. Kaiser's research on bacteriophage λ's cohesive ends of DNA provided an important technique for this pursuit, because Berg began by making them for his recombinant DNA experiment. Cohesive ends refer to the "sticky" properties of the end of the molecule of DNA; the ends form base pairs and a complementary pair can bond (anneal) (see figure 2.4). Biochemists learned of them as λ and other temperate phages became useful tools in molecular biology because of their ability to isolate bacterial genes through lysogenization and transduction.[88] In 1963, Al Hershey discovered that the cohesive ends of λ DNAs enabled the temperate λ phages' recombination with the host chromosome.[89] Kaiser soon devised an experiment that probed the molecular nature of the cohesive sites. First, he analyzed the structure of cohesive ends by employing a phage-mediated DNA transformation system, which he designed with Hogness.[90] Kaiser further identified the molecular sequence of the cohesive ends with Ray Wu, a molecular biologist at Cornell.[91]

In fact, Stanford biochemists' distinctive research focus on the biochemistry of DNA metabolism—its synthesis, replication, and genetic expression—and their sharing of research materials, such as the enzymes DNA polymerase and DNA ligase, provided a set of resources that enabled Berg to construct recombinant DNA molecules; these molecules were one of the "priceless dividends" of the department's unique focus and practices.[92] Kornberg and Robert Lehman, Berg's Stanford colleagues, were two prominent figures in DNA enzymology. In 1967 Kornberg had already succeeded in producing the first artificial synthesis of viral DNA.[93] Lehman's laboratory was one of the earliest that simultaneously discovered DNA ligase, an enzyme that helps join DNA fragments together. When Berg tried to synthesize recombinant DNA molecules, he was able to use DNA polymerase and DNA ligase, which were free of contaminating nuclease, from Kornberg and Lehman's refrigerators. A postdoctoral fellow from Yanofsky's lab, David Jackson, undertook an experiment for making SV40-λ*dvgal* recombinant DNA by opening circular SV40 and λ*dvgal* DNAs into linear-form DNAs and joining them. For this artificial synthesis experiment, Jackson had learned one of the critical steps in joining two linear DNA molecules from Peter Lobban; a graduate student of Kaiser's, Lobban had independently discovered an enzymatic DNA joining method around the same time as Berg's work using bacteriophage P22 DNAs (for more on Lobban's research, see next section). Lobban developed a technique to synthesize cohesive ends by adding nucleotides to the end of DNA using the enzyme terminal transferase.[94] By October 1971, Berg's laboratory finally succeeded in making

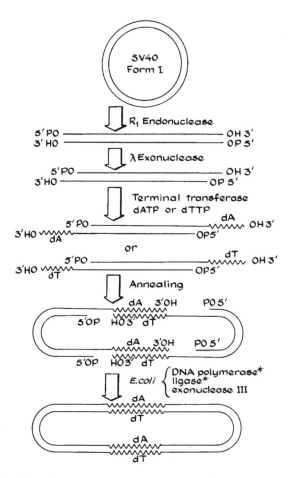

FIGURE 2.4. Making of recombinant DNA: General protocol for producing recombinant DNA. In his American Cancer Society grant application, Paul Berg described the process of making recombinant DNA: "Our approach to synthesizing such potential transducing DNA molecules is as follows: (1) opening the viral DNA rings and preparation of their ends for receiving the new DNA segment; (2) isolation and preparation of the piece of DNA to be inserted; (3) joining of the two DNA's and bringing closure to generate new, expanded circular DNA molecules." Paul Berg, "Viral Oncogenesis and Other Problems of Regulation," 1970, p. 5, Paul Berg Papers, Stanford University Archives, SC 358, box 16, folder: ACS Grant. Image from David A. Jackson, Robert H. Symons, and Paul Berg, "Biochemical Method for Inserting New Genetic Information into DNA of Simian Virus 40: Circular SV40 DNA Molecules Containing Lambda Phage Genes and Galactose Operon of *Escherichia coli*," *Proceedings of the National Academy of Sciences, USA* 69 (1972): 2905. Reprinted by permission from Paul Berg.

recombinant DNA molecules by combining SV40 DNA from a primate tumor virus, and λ*dvgal* DNA from *E. coli* and the bacteriophage λ. The λ*dvgal*-SV40 DNA produced at the laboratory of Berg was a "trivalent biological reagent," which contained most of the genetic information of SV40, the *E. coli* galactose operon, and the λ bacteriophage. Berg aimed to use this recombinant DNA to explore their genetic expression and regulation. Titled "Biochemical Method for Inserting New Genetic Information into DNA of Simian Virus 40," his article heralded a new era for gene manipulation.[95]

Developed from the several investigative shifts following his adoption of animal tumor viruses, Berg's recombinant DNA research had unexpected but profound implications on the technological infrastructure for the manipulation of genes; these implications were especially significant because the research enabled the production of recombinant genes from different organisms. In a way, Berg's reconceptualization of a SV40 virus as a transporting vector reflects a significant epistemic juncture in bacterial genetics that had evolved from the 1950s and 1960s. As Thomas Brock and Creager have argued, the concept of the gene had been extended to include nonnuclear entities, such as fertility factors, episomes and plasmids, which occurred in bacterial conjugation and transduction, suggesting that "all chromosomal genes were potentially mobile."[96] Seen from this perspective, Berg's effort to reconfigure an animal virus as a transducing agent or vector was in essence an attempt to adopt strategies from microbial genetics in order to investigate eukaryotic organisms. Moreover, in an attempt to reconstitute a living organism as a research technology for gene mapping, Berg's construction of recombinant DNA molecules now cast genes as entities that could not only be mobile but also reassembled, a sequence that could be rewritten thorough biochemical operations.

Recombinant DNA as a Technology for Genetic Engineering

As early as 1969, even before the first construction of recombinant DNA molecules, one Stanford biochemist speculated that the ability to amplify and express recombinant DNA could be used for a wide array of agricultural and medical purposes. In his PhD proposal, Peter E. Lobban, a graduate student of Kaiser's in the Biochemistry Department, proposed the potential application of such recombinant DNAs for genetic engineering.[97] By the time Berg was constructing the SV40 transduction system to

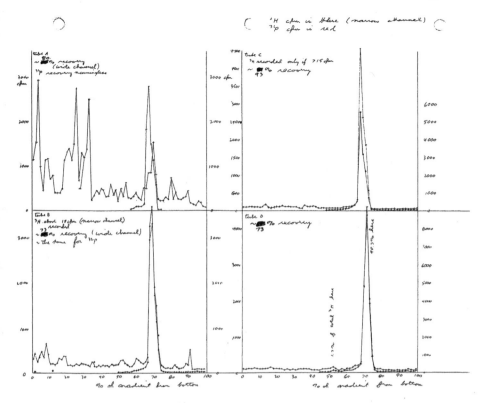

FIGURE 2.5. Synthesis of recombinant DNA. The earliest ethidium bromide–CsCl gradients of DNA samples showed a high-density peak, indicating they were closed, circular DNA recombined by use of Peter Lobban's enzymatic method, A-T litigation. The data were obtained from Lobban's experiment on August 16, 1971 (graph on August 18, 1971). Peter Lobban, Laboratory Notebook, August 1971, Peter E. Lobban Personal Papers, Personal Collection, Los Altos, CA. Reprinted by permission from Peter E. Lobban.

form recombinant DNA molecules from foreign DNA sources like SV40, *E. coli*, and bacteriophage λ, Lobban was independently forming the same from DNAs from a same species, bacteriophage P22. Lobban's project was framed as a study of a general system for joining DNA molecules, and it was technically less demanding as it aimed to chemically combine DNAs of same species origin.[98] By August 1971, Lobban finally succeeded in synthesizing recombinant DNAs by joining P22 DNA molecules (see figure 2.5); subsequent experiments done under improved conditions gave the same results with a higher yield.

Lobban's project on the construction of recombinant DNAs further illustrates another example of remaking life as a research technology. His broader aim in forming recombinant DNAs was to study genetic control and expression in eukaryotic organisms, inspired by analogies between bacterial transduction and viral induction. It was thought difficult to isolate pure genes with no other cellular materials mixed from eukaryotic organisms, in comparison to prokaryotes. In bacteria research, transduction (phage-induced gene transfer) provided a useful tool for producing purified blocks of genes to study gene expression. In higher organisms, there had been no discovery in general transduction phenomenon that enabled molecular biologists to obtain DNA sequences encoding specific sets of genes. Lobban suggested that his recombinant DNA construction method could make it possible to study gene expression in higher organisms; he proposed to do this by infecting recombinant genes via viruses or by whatever means could transfer genes into the host organisms. As he wrote in his dissertation: "If there were a way to join DNA molecules together *in vitro*, then it would be possible in principle to make transducing genomes bearing the genes of any organism by attaching the appropriate DNA to the DNA of a bacteriophage or virus."[99] He mentioned the potential uses of his DNA joining system for studying eukaryotic organisms; these included DNA sources for RNA-DNA hybridization studies and fine-structure mapping of mammalian genes through artificial transduction systems. He also wrote that the joining system could result in a source of medically or agriculturally useful gene products by expressing transducing genes in foreign hosts like bacteria.

However, unlike Berg's group whose main interest was the genetic regulation of eukaryotes and tumor viruses, Lobban clearly foresaw, in both his PhD proposal and dissertation, the potential use of a recombinant DNA transduction system for what was later called genetic engineering. Lobban's suggestion for expressing recombinant DNAs in a foreign host came from his initial interest in problems of antibody synthesis; recombinant DNA–mediated transduction could provide a source for new antibodies if transduced genes could be expressed in the host. In his 1969 PhD proposal, he indeed surmised a potential use of his enzymatic DNA–joining method: if transduced recombinant DNAs could directly express their new genetic information, recombinant DNA–mediated transduction could provide a new source for useful gene products.[100]

With the possibility of recombinant DNA–mediated transformation, Lobban suggested that the ability to insert and express recombinant DNA molecules in bacteria could provide a new kind of technological possibility for a productive use of life. Bacteria could become "factories" to produce medically useful gene products like antibodies. As he put it, "if the bacterial host of the transducing genomes [bearing mammalian genes] is able to transcribe and translate them, it could be used as a source of the gene product that might be far more convenient than the mammalian cells themselves."[101]

Lobban's suggestion for expressing foreign genes in bacteria for the mass production of medically useful molecules was particularly striking to Kaiser and Berg; both recollect that they did not seriously consider such applications before they saw his proposal. As Hans-Jörg Rheinberger has argued, the possibility of creating a transgenic organism—an engineered organism that bears genes deemed culturally and medically useful—with recombinant DNA technology marked the beginning of a significant epistemological shift in biomedical research. The advent of recombinant DNA technology enabled molecular biologists to remake life forms in terms of culturally and medically defined categories. Indeed, to paraphrase Rheinberger, Lobban's proposal marked a radical departure from the earlier intent of molecular biology; its epistemological position had been "the extracellular representation of intracellular processes—i.e., the 'understanding' of life." With the advent of recombinant DNA technology, however, molecular biology entailed the "intracellular representation of an extracellular project—i.e., the deliberate 'rewriting' of life."[102] Molecular biologists then were able to import their culture into cells by modifying life. This shift in the epistemological outlook of the field did not just involve industrial and commercial use of biological organisms.[103] As medical anthropologist Paul Rabinow has suggested, this transition also marked the beginning of a new interaction between nature and culture in the age of biotechnology: "nature will be modeled on culture understood as practice."[104]

Lobban's bold vision of recombinant DNA technology producing useful gene products for medicine, however, was initially regarded as rather elusive, if not dangerous, by other scientists. First, no one had yet succeeded in isolating mammalian genes that could be recombined for genetic engineering experiments. Moreover, making recombinant DNA molecules through enzymatic synthesis was still technically demanding. Lobban himself did not pursue his ideas further after his career shift into medical instrumentation. While the Berg laboratory subsequently set out to experimentally demonstrate whether newly synthesized recombinant genes could be expressed in a biological host, it soon faced controversies regarding the safety of recombinant DNA research (see chapter 3). Berg's recombinant DNA ($\lambda dvgal$-SV40) emerged as a research technology for cancer research, and ironically became an agent of tumorigenesis, provoking unforeseeable public health debates. More importantly, recombinant DNA, a product of the work of Stanford biochemists, soon traveled other trajectories as materials and techniques circulated to the research network formed around the university's Biochemistry Department.

Conclusion

From the mid-1960s, the basic research and rhetoric of medical progress that once fostered the advancement of biomedicine after World War II faced a serious intellectual and political challenge. This challenge was magnified by countercultural critics of science and technology.[105] Berg, a biochemist specializing in bacterial enzymology and nucleic acid metabolism, warily observed the criticisms toward basic biomedical research and strove to demonstrate its relevance for higher organisms. As the intellectual and political context was changing, Berg creatively capitalized on the shift in funding priorities toward medically relevant research with his new work on the genetic regulation of eukaryotic organisms. He did this by adopting tumor-producing animal virus experimental systems. As part of such a system, bacteriophage λ mediated prokaryotic and eukaryotic divisions, enabling the development of a productive analogy between lysogeny and tumorigenesis that linked biological and medical research. The culture of sharing in the Stanford Biochemistry Department provided key resources for Berg's transition to working in eukaryotic biology. This biomedical analogy drawn from Kaiser not only helped Berg navigate the politically charged world of cancer research in the 1970s but also helped him open up the uncharted intellectual territory of molecular investigations in eukaryotic biology.

Through several incremental shifts in the adoption of animal virus experimental systems, Berg's new venture into eukaryotic biology engendered a set of new connections among different research agendas, experimental systems, and funding patronages: first, as an experimental system for examining tumorigenesis; second, as an artificial transduction system for exploring eukaryotes; and later, as a technology for genetic engineering.[106] The advent of recombinant DNA research and technology was thus one of the unexpected outcomes resulting from Berg's adoption of animal virus experimental systems. More consequently, the dynamics of experimentation that transformed Berg's animal virus experimental system from an object for scientific research into a pioneering tool for genetic engineering would be further buttressed by the lively technological and conceptual communication between scientists who formed a research network centered around Stanford biochemists engaged in the study of recombinant DNA (see chapters 3–4).

At the epistemological level, Berg's invention of an artificial transduction system reconstituted particular forms of life as research technologies for gene mapping. This technical implementation of life forms then extended the concept of mobile and modifiable genes in bacteria to the realm of higher organisms, recasting genes as a sequence that could be rewritten thorough biochemical operations.[107] Subsequently, Lobban's articulation of the use of recombinant DNA as a way to possibly produce useful gene products presaged the application of the technology as a tool for genetic engineering. In their attempt to investigate and alter genetic entities—DNA molecules—for higher organisms, Stanford biochemists created experimental conditions for gene manipulations, transforming genes into discreet, mobile, and reassembled entities. Indeed, the prospect of rewriting gene sequences—of engineering life—opened up experimental possibilities for creating new "biotechnological" forms of life through the engineering of nature in culturally and medically defined categories.

In the next chapter we will examine the wide range of experimental hybridizations that occurred as the concepts and materials circulated through the system of exchange and collaboration in recombinant DNA research centered around Stanford biochemists. Their pioneering move into eukaryotic biology and their collaborations, exchange, and sharing of experimental technologies and instruments with scientists in the early network of recombinant DNA research led to the emergence of key technologies for gene manipulations and cloning. I particularly investigate how the race to clone recombinant DNA molecules began to shift the material, intellectual, and social contexts that Stanford biochemists had fostered for the productivity of their research; this race involved Stanford biochemists, such as Berg's graduate students, Janet Mertz and John Morrow, and others who joined the Bay Area network of recombinant DNA research, including Stanley Cohen and Herbert Boyer. While Stanford biochemists' foray into bacterial transformation and gene cloning was put on hold because of fierce public concerns over tumor-virus research, Cohen and Boyer experimentally demonstrated the molecular cloning of recombinant DNAs through a series of three experiments; one of these was completed with Stanford biochemist John Morrow. In the next chapter, we will see how unexpected experimental shifts, regulatory and safety concerns, and opportunistic collaborations reconfigured the system of exchange in recombinant DNA research, with notable material and moral consequences.

CHAPTER THREE

System of Exchange in Recombinant DNA Research and Cloning

In 1971 when Stanford biochemists Paul Berg, Peter Lobban, and their colleagues succeeded in developing new ways to join two DNA molecules together, they revealed a new vista in experimental biology from which key questions in the biology of higher organisms could be explored. Moreover, Berg's group first demonstrated that DNA sequences from foreign sources could be joined by enzymatic means; they synthesized a recombinant DNA molecule by combining λ*dvgal* DNA (from *Escherichia coli* and the bacteriophage λ) with SV40 DNA from a primate tumor virus. As Berg emphasized, his λ*dvgal*-SV40 recombinant DNA could provide an innovative research technology for eukaryotic biology and cancer research:

> The λ*dvgal*-SV40 DNA produced in these experiments is, in effect, a trivalent biological reagent. It contains the genetic information to code for most of the functions of SV40, all of the functions of the *E. coli* galactose operon, and those functions of the λ bacteriophage required for autonomous replication of circular DNA molecules in *E. coli*. Each set of functions has a wide range of potential uses in studying the molecular biology of SV40 and the mammalian cells with which this virus interacts.[1]

If the hybrid DNA molecules were biologically active and could be expressed in the host cell (the *E. coli*), Berg further speculated, he and his colleagues could investigate genetic expression and regulation in animal cells whose complexity otherwise defied genetic and molecular manipulations. While Lobban did not pursue this line of research further after making his career change to medical instrumentation, Berg swiftly moved into the next stage of recombinant DNA research. Beginning in December 1970, his graduate student Janet Mertz undertook a project aimed at replicating or "cloning" λ*dvgal*-SV40 recombinant DNA in *E. coli*; she did this in order to analyze the genes of the cancer-causing virus, SV40.

The construction of artificial recombinant DNA, by providing a research infrastructure amenable for gene manipulations, could have wider material and technological implications, changing how biologists manipulated living materials and how they conceptualized biomedical research problems and applications.[2] Although there was no definite evidence in 1971 that the recombinant DNA molecules that Berg's group had joined together from different biological entities were biologically active and could be replicated in host cells, the potential usefulness of foreign gene cloning for both practical applications and biomedical research drew intense attention from molecular biologists. Beyond basic molecular biology, as Lobban had predicted, introducing and cloning foreign genes at the molecular level could lead to agricultural, industrial, and medical applications.[3] Indeed, Berg's group was not the only group that strived to achieve gene cloning using the λ*dvgal* system. A creative group of visitors appeared in Berg's laboratory as recombinant DNA technology seemed to provide a solid platform for studies in the molecular biology of higher organisms and for a multitude of applications in medical research. There seemed to be no better opportunity for ambitious biologists than the possibilities of gene cloning, and more scientists began to consider it for their own research. Interestingly, Stanford biochemists' customs of sharing and exchange partly contributed to the formation of the network of recombinant DNA research. Those who participated in the network of exchange with Stanford biochemists, especially a group of scientists led by Stanley Cohen in the Stanford Genetics Department and Herbert Boyer at University of California, San Francisco (UCSF), were among the first scientists who tried to clone recombinant DNA molecules for their own research. In the end, it was Cohen and Boyer's collaborative effort that led to the first series of intra- and interspecies gene-cloning experiments between 1973 and 1974, on which they claimed the inventorship of recombinant DNA technology in their patent filing in 1974.[4] The experimental success of gene cloning transformed Lobban's prediction into a reality.[5]

This chapter examines the wide range of hybridizations between different experimental systems that led to the development of recombinant DNA cloning technology as concepts and materials circulated through a system of exchange and collaboration in the early research revolved around Stanford biochemists.[6] The scientists within the network were connected by news and expectations of gene cloning to events in Berg's laboratory; his lab was the center of exchanges of scientific and material innovations regarding gene manipulations. I intend to show how recombinant DNA technology emerged from a dense network of experimental findings and material interchanges among closely interacting and often collaborating groups of scientists. It is not my intention to describe the advent of recombinant DNA cloning technology as a discrete event for which personal and proprietary claims could be staked.[7] I would not go so far as to say that whoever invented the recombinant DNA cloning technology was not mostly a matter of chance. I instead describe how some Stanford biochemists, such as Mertz and John Morrow, as well as Cohen and Boyer, worked with related tools on the same problem of gene replication and cloning; they all did this within the technical infrastructure and expertise suitable for experimental gene manipulations, which were initially created and sustained by Stanford biochemists.

My focus on the system of exchange and collaboration centered around the Stanford biochemists helps to situate the beginning of genetic engineering in a distinctive material and social context, in which scientists interacted with reciprocity and full disclosure, as well as with conceptions about the communal sharing of research materials and tools. This also helps to explain how some scientists and patent examiners came to believe that several Stanford biochemists' work on recombinant DNA research (especially that of Mertz and Ronald Davis) was "a potential infringer [of Cohen-Boyer's patent application] as the closest prior art."[8] My aim here, however, is not to highlight the contentiousness of the dispute over scientific priority. Rather, it is to probe the material and moral background of the scientists' interactions as they competed with each other in a race to clone recombinant DNA molecules.[9] I show how Mertz's experimental strategies, tools, and materials for a cloning experiment were refined and circulated in the network of early recombinant DNA research, where obligatory disclosure and sharing of tools and materials were part of Stanford biochemists' moral economy of science. While she could not proceed with her cloning experiment because of emerging public-health concerns about experimenting with cancer-causing viruses, the circulation of her discoveries and tools continued; because of the communal climate of experimentation at Stanford, these were then easily appropriated by other new participants in recombinant DNA research, such as Cohen and Boyer.

The collaborative work of Cohen and Boyer, through which they assembled significant experimental repertoires for gene cloning, demonstrated that it was possible to propagate foreign genes inside bacteria.[10] Their collaboration with the Stanford biochemist John Morrow, Berg's graduate student, on the cloning of a eukaryotic gene, demonstrated how Cohen began to subvert Stanford biochemists' communal perceptions about research tools and materials. As Cohen and Boyer's plasmid cloning system gained material and technical currency, Cohen began to circulate this system in order to seek collaboration and to advance his scientific priority. His frequent interactions and collaboration with Stanford biochemists particularly made them feel that it was somewhat arbitrary and artificial for Cohen and Boyer to maintain proprietary rights over recombinant DNA–cloning technology. The system of exchange through which early recombinant DNA researchers collaborated and competed had indeed altered, along with its customs and moral assumptions.

Putting Recombinant DNA into Bacteria

By the beginning of fall 1970, Berg's laboratory was at the forefront of recombinant DNA research; its members, including David Jackson and Robert Symons, succeeded by the fall of 1971 in the synthesis of recombinant DNA molecules in vitro by combining two foreign genes. It became the most populated laboratory in the department, and ambitious postdoctoral fellows and graduate students were eager to join to work on the recombinant DNA–cloning experiments that were directed to the biology of higher organisms and cancer research. Janet Mertz was the first graduate student to set her sights on gene-cloning experiments using SV40 (figure 3.1). As a prodigious undergraduate student at MIT, Mertz had performed research on bacterial plasmids and phage λ (bacteriophage λ) in the laboratories of prominent molecular biologists, such as Salvador Luria and Ethan Signer. She had also taken two courses with David Baltimore on animal cells and virology, watching closely when he codiscovered the important enzyme reverse transcriptase in animal tumor viruses in the spring of 1970. During the fall of 1970, Mertz discussed a possible thesis project with Berg that would involve the replication and expression of SV40 using recombinant DNA techniques. As she began to prepare her project on cloning genes of the tumor virus SV40, Berg suggested that Mertz learn advanced techniques in cell culture and animal virology. She

FIGURE 3.1. Janet E. Mertz, October 1981. In the fall of 1970, Mertz joined the Stanford Bio-chemistry Department as a graduate student of Paul Berg. Courtesy of Jonathan M. Kane.

subsequently arranged to attend an annual course on mammalian cells and viruses held at the Cold Spring Harbor Laboratory (CSHL), in New York State, during the summer of 1971.

After the successful artificial construction of λ*dvgal*-SV40 recombinant DNA, those in Berg's Stanford laboratory were considering two potential applications, which involved using the DNA as a research technology for exploring gene expression, replication, and cloning. The first possible application was to introduce known bacterial DNA, such as the galactose operon (a functioning unit of DNA encoding enzymes for galactose metabolism) of *E. coli*, into eukaryotic cells using SV40 as a vector. The second possible application was to introduce animal virus DNA, such as those from SV40, into bacteria after joining them to bacterial replicons (DNA or RNA molecules that have genetic elements necessary for autonomous replication, such as phage genomes and plasmids); attachment to the replicons would allow recombined DNA to replicate and clone its genes inside the host. To achieve the first possible application, Jackson, a postdoctoral fellow in Berg's laboratory, began to develop a method for in vitro attachment of the two DNA molecules. For the second application, and for her dissertation project, Mertz set out to develop a system for introducing SV40 via a bacterial plasmid into *E. coli*. She believed SV40 gene products and functions could be better analyzed in a less complex and well-characterized host, such as *E. coli*, if the genes were replicated and expressed inside that host. Even if they were only replicated in bacteria, it would provide a way to grow mutants of SV40 that could be reintroduced into mammalian cells; she hoped to then observe the effects of the mutations on the life cycle of the virus and the expression of its genes.

Stanford biochemists' techniques and resources, circulated through sharing, collaboration, and informal shoptalk, helped Mertz quickly assemble the crucial experimental tools and materials for her project on recombinant DNA cloning. Initially, researchers in A. Dale Kaiser's lab provided considerable material and technical help. Mertz first had to devise a technique to introduce circular λ*dv* plasmids into *E. coli*. Lobban, a graduate student in Kaiser's laboratory, had tried to improve a calcium chloride method for introducing linear bacteriophage λ DNAs that was originally developed by sabbatical visitors to Kaiser's lab, Morton Mandel and his collaborator, Akiko Higa. In 1970, Mandel and Higa published a paper reporting a technique that could introduce linear, viral DNA into *E. coli*.[11] In December 1970, Lobban taught Mertz his modified version of Mandel and Higa's calcium chloride technique. From Lobban, Mertz learned how to treat the cells with calcium chloride so they would take up linear λ DNAs.[12] Mertz then succeeded in using this technique to introduce the circular λ*dv* plasmid into *E. coli* cells, demonstrating for the first time that a plasmid could be reestablished in *E. coli* by treating the host cells with calcium chloride.

Mertz settled on the bacterial plasmid λ*dvgal* as a potential vector for replicating recombinant DNA molecules and screening for their presence in *E. coli*. For this, she collaborated with phage specialists in Kaiser's lab. First, Kenichi Matsubara, a postdoctoral fellow at Kaiser's lab, provided crucial background knowledge to Mertz's search for a cloning vector. Matsubara had recently isolated a series of deleted variants (*dv*) of the phage λ in which *E. coli* genes involved in use of the sugar galactose (the gal operon) were joined to some of the genes from phage λ, creating autonomously replicating plasmids called λ*dvgal*.[13] Berg suggested to Mertz that she attempt to isolate a λ*dvgal* that contained an entire galactose operon, in other words, all of the genes along with their regulatory signals from

E. coli that are needed for metabolizing this sugar; then Jackson would be able to test whether these well-characterized bacterial genes might be expressed in mammalian cells when linked to SV40. In the winter of 1971, Mertz collaborated with a postdoctoral fellow in Kaiser's lab, Douglas Berg, who already knew how to isolate λ*dv*. Working together, they soon succeeded in isolating their desired plasmid, called λ*dvgal*-120, which carried the entire *E. coli* gal operon.[14] Both were excited about their success, since this λ*dvgal* DNA included the lambda (λ) bacteriophage genes that should enable it to propagate as an autonomously replicating plasmid in *E. coli*. In other words, this λ*dvgal* plasmid could serve as a cloning vector if introduced into *E. coli*.

Mertz next set out to prove that this vector would work for reintroducing and replicating their new λ*dvgal*-120 plasmid DNA in bacteria.[15] She had already shown that the λ*dvgal* DNAs of Matsubara could be reestablished in *E. coli* using Lobban's variation of the calcium chloride method of Mandel and Higa. After a series of laborious experiments in spring 1971, she wrote in her laboratory notebook that she was "beginning to see some larger red [cell] colonies" developing in the *E. coli*—these bacterial colonies were indeed transformed (or genetically modified) by the introduction of the λ*dvgal*-120 plasmid DNA.[16] So, by this time, Mertz had some of the crucial experimental protocols and materials needed to attempt a recombinant DNA–cloning experiment: a bacterial cloning vector, the λ*dvgal*-120 plasmid, and the calcium chloride technique for inserting plasmid or phage DNA into *E. coli* bacteria. She believed that a method being developed in Berg's lab for making recombinant DNAs (SV40-λ*dvgal*-120) in vitro would succeed eventually.[17]

Each experimental protocol and material resource that Mertz developed for the recombinant DNA–cloning experiment was inspired and shaped by her informal interaction and collaboration with Stanford biochemists. Their culture of exchange and sharing contributed to her experimental system, and to a certain extent, it was a product made possible through their communal mode of laboratory life.[18]

In the summer of 1971, Mertz had a chance to present to other prominent scientists her plan for the replication of recombinant λ*dvgal*-SV40 DNA molecules in *E. coli*; she did this while attending a course on animal cells and viruses held at CSHL, where scientists gathered to learn about and exchange new experimental techniques and ideas. She laid out her strategy to use a λ*dvgal*-120 plasmid as a vector to grow individual clones of SV40 and its (insertion) mutants in *E. coli*. Even if the SV40 genes

were not expressed in bacteria, she at least expected to be able to replicate some of the defective SV40 genomes that contain mutations at the
random sites of insertion of the λdvgal-120 DNA. However, her presentation on the cloning of SV40 DNAs in E. coli faced strong reservations
from those who attended her presentation. In particular, one of the in-
structors of the course, Robert Pollack, a biologist at CSHL, expressed serious concerns about the safety of Mertz's proposed experiment. Pollack
worried that inserting the cancer-causing genes of SV40 into E. coli, if successful, might make the bacteria that can reside in the human gut an infectious carrier of human cancers.[19]

Pollack's biosafety concern put the brakes on Mertz's proposed genetic transformation experiment involving λdvgal-SV40 DNA hybrids.
In July 1971, Pollack made a series of phone calls to Berg, who was ad-
vising Mertz on her SV40 experiment, urging him not to implement his
lab's plan for gene replication and the cloning experiment. Berg subsequently consulted Stanford geneticist Joshua Lederberg, who had been
involved in international discussions in the 1970s on the potential use of
biological and chemical agents.[20] Berg inquired about ethical and scientific aspects of biological experiments involving deadly pathogens. Pollack and Berg then agreed to convene a conference where scientists could
address any potential biological hazards of introducing SV40 and other
tumor virus oncogenes as recombinant DNA into bacterial and mammalian cells. Berg also voluntarily imposed a temporary moratorium on
both Jackson's plan to introduce bacterial-SV40 hybrid DNA into mammalian cells and Mertz's plan to introduce λdvgal-SV40 hybrid DNA into
E. coli. In late 1973, Mandel, who had just shifted his research to cancer
biology at University of Hawaii, requested materials necessary for the introduction of λdvgal-SV40 hybrid DNA into E. coli for his investigation
of SV40's genetic regulation and its tumorigenesis. In response, Berg acknowledged the feasibility of such attempts, but noted that he was not
able to provide such materials, pointing to serious safety concerns: "As
you know we have been capable of reintroducing the λdvgal-SV40 hybrid
DNA back into E. coli to ask the same questions you [would like to]. But
we decided more than two years ago to hold off for the same safety considerations you've undoubtedly thought of. For the present I think it's better that way."[21] This self-imposed moratorium on cloning involving tumor
viruses in Berg's laboratory, predating the official government guidelines
for use of recombinant DNAs, continued until the National Institutes of
Health (NIH) guidelines finally permitted it in 1979.[22]

Biohazards and Shifts in Experiment Strategies

While Jackson moved to the University of Michigan for a new faculty position in early 1972, Mertz needed a new direction for her dissertation research, since members of Berg's laboratory had collectively agreed to a moratorium on inserting tumor-inducing SV40 recombinant DNA into *E. coli*. She believed that various forms of SV40 DNA, possibly including her λ*dvgal*-SV40 recombinant DNA, might replicate and express at least some of their genetic properties inside mammalian and *E. coli* cells. Mertz suspected that the host cells might regenerate circular DNA from other forms, making the host carry cancer-inducing genes. Instead, Mertz tried to see whether she could replicate and express recombinant DNA genes across species without using tumor-inducing animal viruses for biosafety. In fact, she performed a series of experiments to see whether she could put bacteriophage λ into human cells; she wondered whether λ with a naturally linked lactose gene from *E. coli*, coding for the utilization of another sugar, could be used to express the lactose gene in a mammalian host cell. For this endeavor, Mertz relied on recently published work by scientists on phage-mediated gene expression in human cells.[23] She wrote in her laboratory notebook that she was interested in finding out whether these bacterial genes carried by the phage λ DNA could be expressed inside human cells. To run this experiment, she used another calcium chloride technique, one that had worked for inserting SV40 DNA into mammalian cells; she hoped to see whether the technique could be used for phage λ DNA as well.[24] In the end, this strategy did not progress rapidly, whereas other feasible approaches were developed.

During their self-imposed moratorium, the Berg group as a whole had to shift to another strategy in order to illuminate the genetic expression and regulation of animal tumor viruses.[25] During the fall of 1971, they devised an approach that combined biochemical analysis of tumor virus mutants in mammalian cell experimental systems with genetic mapping. One class of mutants that Berg had regarded as promising was the naturally arising deletion and substitution mutants of polyoma and SV40 viruses, some of whose DNA sequences were deleted or substituted during rapid viral replication. He had already proposed this molecular genetic approach to SV40 in 1970, suggesting that, if viral mutants expressed their phenotypes in certain conditions, such as high temperatures, they could be employed as a biochemical marker that would illuminate the complex biological functions of viral genes.[26] Moreover, if the various genetic elements of viral mutants could be identified and arranged on the genome of animal viruses, the genetic map could be linked to their biochemical properties, including their functions and RNA products. Thus, the combined use of genetic and biochemical analysis could help decipher the mechanism of viral tumorigenesis and other viral regulatory phenomena.

With Berg's graduate students, Mertz and John Morrow, the deletion and substitution mutant project became the new focus of Berg's laboratory. Morrow, who had already taken a lead in this work, decided to find viral mutants that were deficient in various functions. These mutants were often produced by the deletion of viral DNA during rapid virus multiplication. If deleted portions were not functionally critical, or if they could be grown by complementation with a helper virus, then these viral mutants could be used to infect cells to study their properties, and to see whether they transformed the infected cells into cancerous ones. Using these alternative methods, Morrow and Mertz could analyze viral genes and their functions. Significantly, the position of each deletion in the virus mutants could be located by electron microscopy (the deletion appeared as a loop of single-stranded DNA in the duplex DNA when hybridized with wild-type [normal] virus DNA). Thus, deletion mutants would enable Morrow and Mertz to construct a genetic map of the viral genome, even if they could not clone them in bacteria.

For this deletion mutant–mapping project, Morrow obtained restriction enzymes from several scientists that could cut DNA molecules into pieces. These enzymes became a promising tool for molecular analysis after 1969 when the Johns Hopkins microbiologist Hamilton Smith isolated one that could cut DNA only at specific sites.[27] One of the scientists who provided critical restriction enzymes to Morrow was Herbert Boyer, a biochemist at UCSF. Boyer, who had studied the properties of restriction enzymes in the modification of foreign DNA in bacteria, had isolated *Eco*RI (an endonuclease enzyme coded for the resistance transfer factor, RTFI). From the late 1960s, restriction enzymes became a significant tool for dissecting small genomes of bacteriophages and viruses, such as λ and SV40, because of their exclusive recognition capacities.[28] Morrow wanted to know whether this enzyme could cut SV40 DNA in a specific way. If so, then he could obtain a high-resolution physical and genetic map of SV40 DNA, expanding on the earlier studies of the Johns Hopkins scientists, especially those of Daniel Nathans and his colleagues.[29] Initially, Morrow found some of these restriction enzymes rather useless because they seemed to cut SV40 randomly. However, he soon found that one particular restriction enzyme, *Eco*RI, which had been purified by Boyer, cut SV40 at one unique site in the genome. Morrow undertook a SV40 mapping project by using the *Eco*RI cut on the SV40 chromosome as a reference point from which to measure the relative locations of SV40 genes.[30]

At the same time, Mertz designed a set of experiments to see whether various SV40 DNA segments (topological forms) generated by *Eco*RI could be expressed either in *E. coli* or monkey cells. Mertz's attempt to assay the biological functions of these SV40 DNA segments was inspired by the Caltech molecular biologist Robert L. Sinsheimer; his pathbreaking work had examined the biological infectivity of single-stranded DNAs from φX174 phage.[31] If Mertz found a way to express various forms of SV40 DNA, then their biological functions could be linked to Morrow's physical map, providing a useful guide toward understanding SV40's role in tumorigenesis. In the winter of 1971–72, Mertz performed a series of experiments aimed at examining the infectivity and gene expression of various topological forms of SV40 DNA, including double-stranded, supercoiled DNA (the form present in the SV40 virion particle); double-stranded, nicked, circular DNA; double-stranded, linear DNAs (produced by cleavage with different enzymes including *Eco*RI); and single-stranded, circular DNA.

To her surprise, she repeatedly found that SV40 DNA molecules that had been linearized by cleavage with *Eco*RI restriction endonuclease were quite infectious, producing the virus that contained the usual circular DNA genome with its *Eco*RI site intact. No matter how well she purified this linear DNA to make sure it was not contaminated with circular DNA, which exhibits infectivity, it was still infectious and looked just like wild-type circular DNA when recovered from the monkey cells. *Eco*RI-cut λ*dvgal*-120 behaved similarly when put into *E. coli*.[32] These findings reminded her of phage λ that infects *E. coli* as a double-stranded linear DNA containing sticky (or cohesive) ends; once inside the host cell, these ends anneal and are joined together with an enzyme, called DNA ligase, to form chemically (or covalently) closed, double-stranded, circular DNA molecules. To test whether her *Eco*RI-cut SV40 linear DNA also had sticky ends that were being joined back together once inside the monkey cells, Mertz obtained some *E. coli* DNA ligase from Paul Modrich, another graduate student in the Stanford Biochemistry Department; he had purified DNA ligase for part of his thesis research under his advisor, I. Robert Lehman. She incubated her SV40 DNA that had been linearized with *Eco*RI together with this DNA ligase.

Ronald Davis, a newly hired assistant professor in the Stanford Bio-chemistry Department, suggested to Mertz that he could examine the structure of these linear SV40 DNA molecules treated with DNA ligase with electron microscopy. She examined its structure versus the structure of the same DNA preparation that had not been incubated with the li-gase. She observed that, while less than 0.1 percent of the starting mate-rial was circular DNA, approximately 95 percent of the DNA treated with DNA ligase had become covalently closed circular DNA. Working to-gether, Davis and Mertz showed in May 1972 that *Eco*RI-cut linear SV40 DNA had sticky ends that could be bonded together in vitro by incubat-ing them at refrigerator temperature. They then used this newfound knowledge to demonstrate, by using SV40 and λ*dvgal*-120 DNA, that one could quite easily generate recombinant DNA molecules by simply cut-ting the two different DNAs with *Eco*RI and then incubating them to-gether in the presence of DNA ligase. Mertz and Davis underscored this breakthrough, claiming "any two DNAs with RI endonuclease [*Eco*RI] cleavage sites can be 'recombined' at their restriction sites by the sequen-tial action of RI endonuclease and DNA ligase."[33] Indeed, *Eco*RI's ability to generate sticky ends on DNAs vastly simplified the synthesis of recom-binant DNA molecules.

Aborted Cloning

The discovery that *Eco*RI could generate sticky ends brought about a sig-nificant change by which one could perform genetic manipulations. With this new tool for recombinant DNA experiments, and encouraged by her successful infectivity experiments with *Eco*RI-cut SV40 and λ*dvgal*-120 DNAs, Mertz proceeded with her original plan, but without actually in-serting tumor-inducing genes into *E. coli*, because of the moratorium (see figure 3.2).

Since her aim was to make λ*dvgal*-SV40 hybrid DNA capable of repli-cating inside *E. coli*, Mertz generated rather long polymeric λ*dvgal*-SV40 DNAs; these would insure that the phage λ gene (O gene) needed for replication of the plasmid in *E. coli* would be intact. Thus, she performed the ligation reaction using a high DNA concentration so that most of the DNA would contain both SV40 and two or more head-to-tail cop-ies of λ*dvgal*-120 molecules. She suspected that most of this long chain of λ*dvgal*-SV40 hybrid DNA, with its intact replication gene, could rep-licate inside bacteria.[34] Mertz mixed together the SV40 and λ*dvgal* DNA

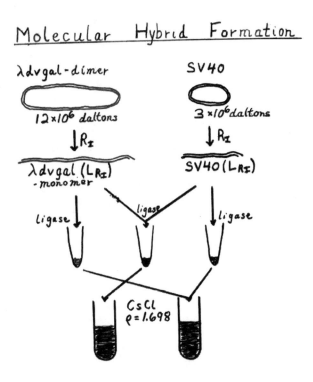

FIGURE 3.2. Janet E. Mertz, "Test for Hybrid Formation," Laboratory Notebook, May 1972. Janet E. Mertz Papers, Personal Collections, University of Wisconsin, Madison. Reprinted by permission from Janet E. Mertz.

cut with *Eco*RI and added DNA ligase to seal the annealed ends. As the data in figure 6 of her published article indicate, "65% of the mass [of the ligated recombinant DNAs] was contained in molecules averaging 7.7 SV40 DNA equivalents in length."[35] Since a recombinant molecule consisting of two copies of λ*dvgal*-120 and one copy of SV40 would be about 5.6 SV40 DNA lengths in size, many of the ligated DNA molecules in her preparation probably contained two or more head-to-tail copies of λ*dvgal*-120, together with one or more copies of SV40 (see figure 3.3). In other words, they probably included recombinant DNAs containing SV40 that would have been able to replicate in *E. coli*. Later, when evaluating Cohen and Boyer's recombinant DNA patent application, US Patent and

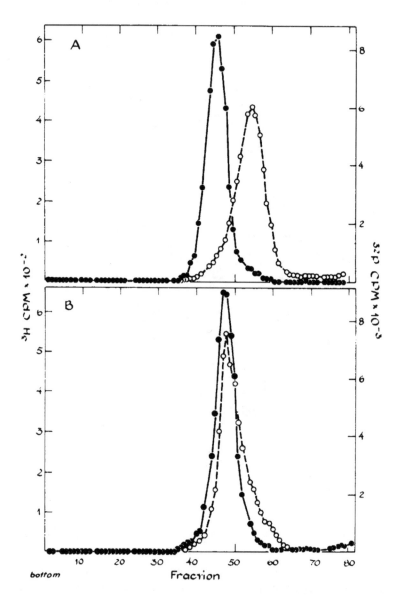

FIGURE 3.3. Janet E. Mertz's hybrid SV40– λ*dvgal*-120 DNA segments. These graphs demonstrate the covalent joining of λ*dvgal*-120 (●–●) and SV40 DNA (○–○) to produce DNA molecules shown in bottom panel *B*; their densities are approximately midway between whose of the SV40 and λ*dvgal*-120 DNA shown in top panel *A*. Examination by electron microscopy of the lengths of the hybrid DNA indicated the likelihood that many contained two or more head-to-tail copies of λ*dvgal*-120, together with one or more copies of SV40. Image from Janet E. Mertz and Ronald W. Davis, "Cleavage of DNA by RI Restriction Endonuclease Generates Cohesive Ends," *Proceedings of the National Academy of Sciences, USA* 69 (1972): 3371. Reprinted by permission from Janet E. Mertz.

Trademark Office examiner Alvin E. Tanenholtz pointed out that the longer chain of λ*dv* polymers could be used for cloning SV40 DNAs.

Mertz presented her data on the construction of polymeric λ*dvgal*-SV40 recombinant DNA in a talk at the Cold Spring Harbor meeting on tumor viruses in August 1972; she suggested that her chimeric DNAs containing SV40 and λ*dvgal* would have been able to replicate in *E. coli*.[36] Although she had successfully created long λ*dvgal*-SV40 hybrids using *Eco*RI, she did not actually introduce this recombinant plasmid DNA into *E. coli* because of the voluntary moratorium in Berg's laboratory. When someone in the audience asked her what she did with this recombinant DNA, Mertz responded that she autoclaved it due to biosafety concerns.[37] Nevertheless, her experiment was a significant step toward gene cloning. Her results provoked more concerns about public health risks involving recombinant DNA research because many scientists could now easily make chimeric DNA, not just the few biochemists such as Lobban and Jackson who had access to the numerous enzymes needed to create recombinant DNAs by synthesizing cohesive ends.

With many scientists alarmed by this series of rapid advances in recombinant DNA research, Berg, along with other prominent scientists and government officials, convened a meeting in January 1973 at the Asilomar Conference Center in California. At the conference, they specifically discussed potential public health risks and laboratory safety procedures of recombinant DNA research involving tumor-inducing viruses.[38] This so-called Asilomar I conference was held about two years before the much publicized Asilomar II (February 1975) conference; the second conference officially called for a research moratorium on recombinant DNA experiments involving tumor viruses.[39] In his concluding remarks at the Asilomar I conference, Berg called for "caution and some serious effort to define the limits of whatever potential hazards exist."[40] He continued: "to do less, it seems to me, is to play Russian roulette, not only with our own health, but also with the welfare of those who are less sophisticated in these matters and who depend on our judgment for their own safety."[41] In the end, Mertz was distressed by her inability to move forward with her recombinant DNA–cloning experiment.[42]

Recombinant DNA in Transit

At the same time, Mertz's discovery of the cohesive nature of DNA ends made by *Eco*RI expanded the possibilities for recombinant DNA experimental procedures. Before *Eco*RI, the process of making recombinant DNA was very technically demanding, and enzymes necessary for recombinant DNA procedures, such as terminal transferase, were difficult to obtain. *Eco*RI did indeed provide a simple and straightforward technology for scientists to combine genes and use them for molecular studies. As Berg wrote to one of his collaborators in England: "We can show experimentally that any two DNA molecules having ends produced by RI endonuclease can be covalently joined. In other words the ability to construct molecular hybrids is enormously extended."[43]

Above all, the discovery of *Eco*RI's property to produce the cohesive ends for combining DNAs enabled other researchers to participate in the use and development of recombinant DNA research, especially those who had formed and sustained the research network around the Stanford Biochemistry Department and beyond. Moreover, by the fall of 1972 and into early 1973, many molecular biologists, such as MIT's Philip A. Sharp who had organized the 1972 tumor virus meeting, learned about Mertz's aborted recombinant DNA–cloning experiment through public discussions regarding recombinant DNA risks.[44] As a young graduate student, Mertz was put into a situation where potential competition on gene cloning could arise quickly: on the one hand, *Eco*RI could be easily employed to produce recombinant DNA beyond Berg's laboratory; on the other hand, the research moratorium on recombinant DNA experiments involving tumor viruses that was initially self-imposed by those in Berg's laboratory provided an opportunity for other scientists to produce recombinant DNA with nonviral materials for gene cloning.

As scientists who participated in the research network surrounding Stanford biochemists became keenly interested in recombinant DNA technologies for their own research, Mertz became increasingly concerned about Berg's frequent disclosure of her experimental protocols. In late 1971, she discovered that Stanley Cohen of the Department of Medicine at Stanford, who frequently interacted with some members of the Biochemistry Department, particularly those of Kaiser's group, was developing a technique for introducing his plasmid DNAs back into bacteria using a modified calcium chloride treatment of the cells.[45] Mertz had succeeded in introducing circular λ*dv* plasmids in early 1971, before Cohen succeeded in inserting plasmids into *E. coli*. At one level, Berg's openness to discuss his group's experimental and technical activities might have been a diplomatic and calculated means to gain access to scientific and technical advances made by other groups. From the perspective of a graduate student who needed scientific priority for securing credit, however, reciprocity might not have been a very affordable strategy. In a letter to Francis Crick in 1973, Berg acknowledged this risky dimension of scientific openness:

> I have had to battle my students and fellows over the issue of whether I talk too freely about ideas and experiments in progress . . . In my view science is not worth doing if secrecy, suspicion and back-biting are rules of the game. Consequently I refuse to accede to the practice of too many others—silence until published; even though on several occasions that degree of openness has come back to haunt and embarrass me.[46]

Though Berg wrote in general terms, scientists in Berg's laboratory, especially Mertz, later came to observe that their openness in scientific exchange required a delicate balance of also asserting one's own accomplishments to insure due scientific credit.

Cohen and the First Cloning of Recombinant DNA

A series of recombinant DNA–cloning experiments by the Stanford geneticist Stanley Cohen and the UCSF biochemist Herbert Boyer emerged from a dense configuration of findings, techniques, and exchanges in the early network of recombinant DNA research revolving around Stanford biochemists. Cohen, who interacted closely with members of the Stanford Biochemistry Department, was one of the new participants in recombinant DNA research. He had been appointed assistant professor in the Department of Medicine at the Medical School in 1968 for his work on drug resistance in bacteria. Cohen's laboratory was located close to the laboratories of biochemists at the Stanford Medical School. Since Cohen's molecular biological approach was rather unfamiliar to other members of the Department of Medicine, he began to interact with Stanford biochemists, participating in their seminars and using their centrifuges and other apparatus.[47] Cohen's close relationship with these biochemists was not only because of his research interest in bacterial plasmids but also partly because of his "family" relationship with Kornberg's early research group. Cohen had been a postdoctoral fellow with Jerald Hurwitz, who had previously worked in Kornberg's Microbiology Department at Washington University in St. Louis in the mid-1950s.[48]

Cohen's physical and intellectual proximity to Stanford biochemists was critical to the transformation of plasmids into experimental tools.[49] As scientists became able to distinguish genes that code for replication from those that code for antibiotic resistance, the search for genes in bacterial plasmids that coded for drug resistance emerged in the 1960s as important to medical and pharmacological research. Following this line of research, Cohen was working since the mid-1960s on the isolation and characterization of plasmid genes that conferred drug resistance to the host microbe. When he first moved to Stanford, he decided to examine R-factors, the intracellular genetic elements of bacteria that transfer drug-resistant markers to other bacteria. His intention was to link specific drug-resistant genes with particular segments of R-factor DNA, thereby probing the molecular nature of R-factors and their replication and transcription mechanisms. Through the isolation of R-factor DNA segments, Cohen hoped to study their genetic properties and find ways to repress their drug-resistant properties.[50]

In his attempt to introduce a circular plasmid into bacterial cells, Cohen frequently interacted with Stanford biochemists, especially Peter Lobban. He knew members of Kaiser's group had developed a calcium chloride technique to introduce linear λ DNA into the host cell. Mertz had already used a modified version of this technique to introduce the circular plasmid (λdvgal-120) into *E. coli* for her research project. Likewise, by adopting a variant of this calcium chloride technique, Cohen intended to introduce physically sheared plasmid DNA into bacteria to locate drug-resistant genes and investigate their properties. By the time Cohen began these experiments, Mertz had also developed her variant of the calcium chloride technique to insert purified plasmid DNA, as well as linear phage DNA, into *E. coli*.[51] Cohen and his undergraduate student, Leslie Hsu, likewise appropriated a variant of the Mandel and Higa calcium chloride technique so that he could introduce nonviral DNA—especially purified plasmid DNA—into *E. coli*.[52]

Cohen was well aware that Stanford biochemists had succeeded in joining DNA molecules, and he wanted to adopt recombinant DNA technology for his plasmid gene-mapping research. More importantly, the implication of recombinant DNA techniques was fairly well known: if these molecules could be replicated and expressed inside the host cell, as Lobban foresaw, useful gene products would be produced in mass quantity. At that time, Mertz wanted to clone recombinant DNA molecules in order to explore the molecular biology of eukaryotic cells and tumorigenesis. Her aborted gene-cloning experiment made scientists that much more aware of the broader potential of gene cloning through the combined use of recombinant DNA molecules and the calcium chloride technique.

Cohen recognized the potential of this recombinant DNA cloning technique for his plasmid research. At a scientific conference in Honolulu, Hawaii, in November 1972, Boyer mentioned to him the Mertz and Davis technique that had just been published for easily making recombinant DNA using *Eco*RI and DNA ligase. Cohen had already determined that *Eco*RI would cleave his large antibiotic-resistance plasmids into fragments that could be analyzed for their drug-resistance properties. Cohen thus reasoned that smaller plasmid DNA segments might contain intact genes (either for conferring drug-resistance properties or for replication) that could be inserted and expressed in calcium chloride–treated *E. coli* bacteria. If he could combine DNA segments containing intact drug-resistance genes with ones containing the genes for plasmid replication, Cohen believed that he could use plasmids as a vector to transport these drug-resistance genes. These recombinant DNA plasmids could be introduced into *E. coli* using his calcium chloride method, whereby the genes propagated inside the bacteria host would express their genetic properties. When the Stanford microbiologist Stanley Falkow later wrote a supporting letter to confirm the conception of recombinant DNA cloning technology by Cohen and Boyer, he testified before the US Patent and Trademark Office that he had been "present when Stanley Cohen and Herbert Boyer discussed their idea for a collaboration." Falkow also wrote: "The idea was to introduce foreign DNA into a plasmid having an antibiotic marker and introduce the resulting hybrid DNA into a bacterial host to see if the hybrid plasmid would be biologically functional."[53]

Cohen and Boyer's close interactions and exchanges with Stanford biochemists regarding recombinant DNA research inspired their collaboration for their gene-cloning experiment. It was at the Hawaii meeting where Cohen seized the opportunity to assemble a repertoire of experimental techniques to build a system for cloning DNA molecules. Cohen discussed with Boyer how he was able to transform the genetic properties of bacteria by inserting plasmid DNA using *E. coli* treated with the calcium chloride technique, a technique developed from his exchanges with Stanford biochemists. As Falkow later recollected, "Boyer then repeated Janet Mertz's observation about 'sticky' DNA ends and some of the work that Peter Lobb[an] was doing with P22 phage in Dale Kaiser's lab to form enzymatic joining of DNA molecules."[54] Boyer also mentioned his

recently published sequencing of the sticky ends produced by cutting
DNA with *Eco*RI enzyme. The usefulness of *Eco*RI in mapping SV40 had
already led to Mertz's unexpected discovery of its property to make sticky
ends for producing recombinant DNA molecules. Cohen then proposed
the collaboration with Boyer, whose lab had become a major source of
*Eco*RI and enzymological expertise.

Cohen and Boyer swiftly cooperated to use *Eco*RI to make recombinant DNA molecules, beginning with Cohen's drug-resistant plasmids.
Cohen isolated the large antibiotic plasmid R6–5, and Annie Chang, a
technician in Cohen's lab who lived in San Francisco, transported it to
the Boyer lab at UCSF (figure 3.4). Boyer then cleaved the purified DNA
of the plasmid R6–5 into multiple DNA fragments with *Eco*RI in order
to generate plasmids like pSC101 and pSC102. *Eco*RI-cut pSC101 and
pSC102 were mixed together and then treated with DNA ligase; since
*Eco*RI cuts leave sticky ends, the DNA fragments from separate plasmids
were fused together to produce recombinant DNA molecules. Boyer returned them to Cohen's lab at Stanford. If DNA segments carrying different drug-resistance genes could be fused, Cohen suspected he could
determine whether the hybrid DNA was biologically functional by treating the *E. coli* host cells with the different antibiotics. Plasmid pSC102,
which carried the gene for resistance to the antibiotic kanamycin, was
spliced to plasmid pSC101, which was resistant to the antibiotic tetracycline. Cohen put this recombinant plasmid (the pSC105 replicon, which
included pSC101 and pSC102 fragments) into bacteria through the calcium chloride treatment. The transformed host bacteria carrying the hybrid plasmid exhibited drug-resistance properties in response to both tetracycline and kanamycin. By March 1973, just a few months after their
collaboration, Cohen and Boyer were able to demonstrate experimentally
that genes from the same prokaryotic species joined together in a test
tube could be propagated and expressed in the host bacteria.[55] This gene-
cloning experiment was the first demonstrating that recombinant DNA
molecules could be replicated and cloned inside a bacterial host.

The striking news of the first intraspecies cloning was first publicly
announced by Boyer at the Nucleic Acids Gordon Conference held
June 11–15, 1973. While the possibility of propagating recombinant DNA
inside bacterial cells had been anticipated, and even feared by some scientists, Cohen and Boyer's 1973 experiment was the first concrete demonstration of the cloning of recombinant genes that had been combined
in vitro. Moreover, the simplicity of their procedures in producing and

FIGURE 3.4. Stanley N. Cohen and Annie Chang, ca. 1975. Reprinted by permission from Stanley N. Cohen.

transferring recombinant DNA, along with their use of drug-resistance genes, made fellow scientists realize that the public health risk associated with recombinant DNA research had to be discussed at an even more comprehensive level than discussed at the previous Asimolar I conference held five months earlier. The Gordon conference's two chairs, Maxine Singer of the NIH and Dieter Söll of Yale University, proposed that the National Academy of Sciences establish a committee to review the risks involving recombinant DNA research using plasmids and animal viruses.[56] Singer and Söll subsequently invited Berg, who had previously led similar biosafety discussions involving recombinant DNA (at Asilomar I) to chair the committee.

Meanwhile, those who were familiar with recent progress in recombinant DNA research in the Bay Area were excited about the availability of a new vector for gene transfer that would be easier to use than λ*dvgal*. Through his demonstration of intraspecies gene cloning, Cohen showed that his small plasmid, that is, pSC101, could serve as a cloning vector that would enable the expression of hybrid genes in the host cell without disrupting its replication gene; it was also more convenient to use for selecting the transformed bacterial host cell, which conferred a drug-resistance property to the host cell when the host took up recombinant DNA containing the drug-resistance gene. Only the transformed host cells could survive (all the other cells would die) after applying an antibiotic, making the selection of transformed colonies simple.[57] Indeed, the pSC101 plasmid provided the first practical cloning vector for use in replicating recombinant DNA molecules in *E. coli*, providing Cohen an enormous advantage in the field of recombinant DNA research and technology. He established his priority on the first intraspecies cloning, and his plasmid system provided a solid experimental platform through which one could clone interspecies and foreign species gene.

Plasmids and the Reshaping of the Recombinant DNA Network

The exchange and collaboration among scientists at Stanford and UCSF on experiments that ultimately led to gene cloning benefitted from a moral economy of science that encouraged local sharing of materials, tools, and ideas at the network of early recombinant DNA research centered on Stanford biochemists. Enlightened self-interest, accompanied by the rhetoric of scientific openness, facilitated scientific and technical exchange and collaboration. At the practical level, this moral economy was based on needs to access research materials, unpublished scientific information, and technical know-how for the productivity of research. Mertz, for example, benefitted from her access to Boyer's *Eco*RI enzyme, and her discovery of *Eco*RI's ability to make sticky ends in turn enabled other researchers to participate in the burgeoning field of recombinant DNA research. Stanford biochemists' help in implementing the calcium chloride method of Mandel and Higa provided a new opportunity in plasmid research for Cohen.

The moratorium in Berg's laboratory on recombinant DNA research with tumor viruses created an uncertain environment, while the local circulation of materials and ideas provided an opening to latecomers in the recombinant DNA field, especially for those who used biological materials not involved with risky tumor viruses. Cohen's subsequent collaboration with Boyer exhibited his opportunistic adoption of recombinant DNA techniques for plasmid gene mapping and cloning. As newcomers in recombinant DNA research, Cohen and Boyer's gene-cloning attempt was yet to be under scrutiny from fellow scientists and the public. In addition, their experiment did not involve cancer-causing viruses, the then focus of biohazard concerns with regard to recombinant DNA research, and they claimed that they did not use too dangerous drug-resistant bacterial strains, such as streptococci and pneumococci, that are resistant to widely used drugs like penicillin They, however, soon cloned a drug-resistance gene from *Staphylococcus aureus* bacteria, whose infections cause difficult-to-treat infections in humans, leading to a major clinical problem later.[58] At a time when the potential benefit and harm of recombinant DNA technology for genetic engineering were being pondered and discussed, Cohen further insisted that it would be ultimately impossible to ensure absolute zero risk in every genetic recombination, and the risk of recombinant DNA research should be assessed with regard to its potential benefit to medicine and agriculture. Cohen characterized those who raised serious biohazard concerns involving recombinant DNA technology as "a horde of publicists—most poorly informed, some well-meaning, some self-serving."[59] Through a remarkably successful collaboration, mediated through the circulation of plasmids and *Eco*RI, Cohen and Boyer quickly assembled a productive repertoire of techniques and materials for their gene-cloning work.

Thus, by early 1973, Stanford biochemists did not have a monopoly on recombinant DNA techniques; shared tools, materials, and ideas also encouraged experimental work elsewhere. The researchers then operated within a competitive environment in which scientists opportunistically forged alliances with each other. Morrow's subsequent collaboration with Cohen and Boyer illustrates how scientists formed a collaborative network to access each other's materials in the race to clone a eukaryotic

gene. After Cohen and Boyer's bacterial gene-cloning experiment had
been proven to be successful, the pSC101 plasmid emerged as the optimum vector to use for cloning genes with recombinant DNA technology.
Cohen now did his best to take advantage of his possession of pSC101,
strategically mobilizing his unique gene transfer vector to gain access to
other valuable research materials and to solicit collaboration. As a result,
the Cohen lab became an almost obligatory rite of passage for researchers
interested in gene cloning; this shifted the central node of recombinant
DNA research away from the Biochemistry Department toward Cohen's
laboratory in the Genetics Department.[60]

One prime example was Cohen and Boyer's collaboration with John
Morrow. Morrow, who frequently communicated with Boyer on his SV40
mapping project with EcoRI in Berg's lab, learned about the first successful intraspecies gene cloning using pSC101 in early 1973. Morrow recognized that Cohen and Boyer's cloning of genes from the same bacterial
species presented an exciting opportunity to attempt the cloning of foreign genes, as his advisor initially had envisioned for SV40 research. Immediately after Boyer's Gordon Conference presentation in June 1973,
where Boyer first publicly disclosed such experimental success, Morrow
discussed a potential collaboration with him. Although no one had yet developed a general technique to isolate genes from eukaryotic organisms,
Morrow had in his possession frog ribosomal DNA. Morrow was finishing
his dissertation on SV40 mapping at Berg's laboratory and had already
decided to work as a postdoctoral fellow in Donald Brown's laboratory.
Brown's lab in the Carnegie Institution's Department of Embryology, located in Baltimore, was at the forefront of investigating eukaryotic gene
expression and was using one of the earliest isolated eukaryotic genes—
ribosomal DNA from the African frog, Xenopus laevis. Because of the ribosomal DNA's unusual buoyancy, Brown was able to isolate the DNA
using centrifugation; he had already investigated the DNA's expression
and developmental processes by inserting the DNA into embryos.[61] Morrow proposed to clone the Xenopus ribosomal DNA, using recombinant
DNA technology, for his investigation of genetic expression in eukaryotes.

Morrow was acutely aware of Mertz's frustration with not being
able to proceed with her SV40 recombinant DNA–cloning experiment;
to avoid a similar restriction, he thought that Xenopus ribosomal DNA
would be a useful eukaryotic gene for molecular cloning since few were
likely to ask questions about biosafety. He also believed that the Cohen-
Boyer plasmid experimental system would provide a more feasible and

efficient platform for the cloning of recombinant DNAs than Mertz's
λdvgal system; the latter lacked the selective power of a drug-resistance
gene and required creation of long-chain DNA to regenerate replication
genes. Thus, instead of collaborating with Mertz in his own laboratory,
Morrow approached Boyer. Morrow suggested that he and Boyer could
go beyond intraspecies gene cloning by introducing ribosomal genes from
Xenopus laevis. Boyer indicated that he would like to try the eukaryotic
gene-cloning experiment, even though at that time nobody knew whether
genes introduced from different species could be replicated or expressed.
Shortly thereafter, Boyer and Morrow invited Cohen to join them in this
collaboration so they could use his pSC101 plasmid and also benefit from
the outstanding help of Cohen's technician, Annie Chang.

With Morrow's *Xenopus* ribosomal DNA, Cohen's pSC101 plasmid,
and Boyer's *Eco*RI enzyme, they had in hand all the reagents needed to
clone eukaryotic genes. Since Cohen and Boyer's first recombinant DNA–
cloning experiment was limited to the replication of DNA fragments from
the same species, the *Xenopus* cloning experiment, if successful, would be
the first demonstration of cloning animal cell genes in bacteria. By October 1973, they showed that recombinant DNA consisting of the plasmid
pSC101 and the frog's ribosomal DNA could be introduced into bacteria
and then replicate (figure 3.5). Importantly, they also demonstrated that
the *Xenopus* DNA was transcribed into RNA, suggesting that it might,
indeed, be possible to clone eukaryotic genes and thereby manufacture
their products through genetic engineering.[62]

Around the same time, Cohen embarked on an interspecies gene-
cloning project, this one involving two different bacterial species. Beginning in late June 1973, Cohen and Chang began to extract plasmid DNA
encoding drug resistance from the bacterium, *Staphylococcus aureus*
(*Staph*).[63] He intended to produce the recombinant *E. coli–Staph* plasmid
to see whether genes from a different bacterium species could be replicated and expressed in *E. coli*.[64] Both of these interspecies gene-cloning
experiments, one using eukaryote genes and another using prokaryote
genes, were performed concurrently from June 1973 to October 1973.
These two experiments occurring in the same time period contributed to
the controversy over the priority for foreign gene cloning. As Morrow
recollects:

> I first learned of the *Staphylococcus aureus–E. coli* chimeric DNAs verbally
> from Stan[ley] Cohen at his laboratory AFTER the success of the *Xenopus*

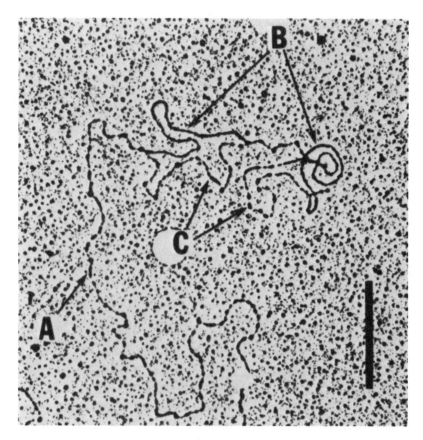

FIGURE 3.5. Transcription of the eukaryotic gene. Electron micrograph of *Xenopus laevis* ribosomal DNA and pSC101 plasmid DNA molecules. Arrow (*B*) shows *X. laevis* ribosomal DNA strand hybridized with molecules of pSC101 DNA. Image from John F. Morrow, Stanley N. Cohen, Annie Chang, Herbert W. Boyer, Howard M. Goodman, and Robert B. Helling, "Replication and Transcription of Eukaryotic DNA in *Escherichia coli*," *Proceedings of the National Academy of Sciences of the USA*, 71 (1974): 1746. Reprinted by permission from Stanley N. Cohen.

frog–*E. coli* cloning was clear, but not submitted for publication. This was long after the June, 1973 Gordon Conference. Stan Cohen said he and Annie Chang needed to complete the *Staph. aureus* paper immediately, because it would not be of general interest after the *Xenopus–E. coli* paper was published. I still remember it, because I was surprised by this parallel effort that Stanley Cohen and Annie Chang were working on. Our joint *Xenopus–E. coli* project had higher priority in my eyes, and I viewed the *Staph. aureus–E. coli* work as an interfering matter that might allow another group to achieve priority. Stan Cohen did not agree with me, and he did not allow Annie Chang to spend more effort on the *Xenopus* project, as I requested.[65]

In the end, Cohen's *Staph* article was published in April 1974, one month prior to the *Xenopus* article, even though Morrow had already presented a formal talk to the Biochemistry Department about the successful cloning of *Xenopus* DNA in the previous fall. By doing so rather than publishing the two articles concurrently, it provided Cohen with the opportunity to claim scientific priority in foreign gene cloning. While Morrow was rather dismayed by this incident, his contribution was highly recognized by the scientific community, earning him a tenure-track position at Harvard University (Mertz also was offered tenure-track positions at prominent universities, such as Harvard, Yale, University of California, Berkeley, and University of Wisconsin, Madison). Morrow's dispute with Cohen later erupted more intensely when Stanford University filed a patent application on recombinant DNA cloning technology identifying Cohen and Boyer as the sole inventors, while including the Morrow et al. paper in the patent application.

Conclusion

This chapter has analyzed the development of recombinant DNA–cloning technology as ideas, tools, and research materials circulated through the early research network revolved around Stanford biochemists. This account offers an alternative narrative to the popular genetic engineering story, one that highlights the close interactions among Stanley Cohen, Herbert Boyer, several Stanford biochemists, and their collaborators. After the initial synthesis of recombinant DNA molecules by Stanford biochemists, including Paul Berg and Peter Lobban, the university's Biochemistry Department had emerged as a center of recombinant DNA research. At a time when scientists began to shift their focus toward more medically relevant subjects, such as the biology of higher organisms, cancer, and other disease-oriented and drug-related research, recombinant DNA technology seemed to provide a breakthrough.

Initially, only Stanford biochemists seemed to have enough technical and material resources to make recombinant DNA molecules; scientists needed access to half a dozen different enzymes and complex and often tacit techniques to combine different strands of DNA by a laborious addition of nucleic acids. When Mertz and Davis discovered that scientists could produce recombinant DNA molecules by simply cutting the DNAs with *Eco*RI and then adding DNA ligase, those who maintained a network of scientific exchange began to adopt this technology. Janet Mertz's λ*dvgal*-120 and Cohen's plasmid systems were the two earliest experimental cloning systems that took advantage of an artificially synthesized recombinant DNA molecule; they each used these to explore questions related to the regulation of products encoded by genes. Their experimental systems transformed their focus of research, including the cancer viruses (Mertz's) and plasmids (Cohen's), into crucial technological tools of gene cloning. Indeed, at an experimental level, both Mertz's λ*dvgal*-120 and Cohen's pSC101 plasmid were reconfigured as components of critical technological systems for cloning a recombinant DNA molecule.

The technical implementation of biological organisms and materials for making hybrid DNA and chimeric clones, however, posed broader public health concerns that reached far beyond scientists' laboratories. First, Mertz's plan to introduce a tumor-causing animal virus into a bacterium that resides in humans raised serious biohazard concerns involving recombinant DNA research. Mertz, as a young graduate student, was strategically unable to advance her experiment amid the public health concerns. However, Cohen and Boyer were not under the intense scrutiny from fellow scientists since they were newcomers to recombinant DNA research. They succeeded in cloning a recombinant DNA molecule using drug-resistance genes before anyone knew enough about the work to object that their cloning experiments might be biohazardous. While this increase in recombinant experimental systems and use of cross-species biological materials proved to be highly productive for scientists, concerns surrounding the tumor-causing virus gave ample advantage to Cohen's plasmid experimental system in being the first to clone recombinant DNA. Cohen's success in introducing drug-resistance genes via plasmids indeed made these biohazard concerns an immediate threat to both scientific work and to the public, forcing scientists at the national level to discuss ways to regulate recombinant DNA research, leading to the much-publicized Asilomar II conference.[66]

As Cohen skillfully exploited the advantages of his plasmid experimental system amid the biohazard controversy involving Mertz's work, the central position that Stanford biochemists had occupied in early recombinant DNA research was undermined. Mertz's λ*dv* experimental system required the rather laborious screening for transformed clones. On the other hand, Cohen's plasmid proved to work as a cloning vector, retaining an intact replication gene when cut with *Eco*RI enzyme; its presence could be easily detected by simply adding an antibiotic to the culture medium that would kill all the other cells. After their first success, Cohen used his plasmid as a strategic tool to build a collaborative network of gene cloning, thereby shifting the center of recombinant DNA research from the laboratories of Stanford biochemists to his own laboratory. Morrow's collaboration with Cohen and Boyer in the first cloning of eukaryotic genes illustrated the reshaping of the recombinant DNA research network.

This reconfiguration of a system of exchange in the early network of recombinant DNA research, both in terms of the conceptual and material circulations, illuminated some central aspects of experimental life of the Bay Area researchers; these included the customs of reciprocal exchange of ideas and research tools and materials, rules of collaboration and competition, and obligations toward fellow scientists and the public.[67] The reshaping of this system of exchange in recombinant DNA research occurred in the highly competitive context of late-twentieth-century American science. On the one hand, Stanford biochemists' earlier customs of sharing and openness, to paraphrase Robert Kohler, served as a "moral code for regulating competitive feelings and privileges," and ensured the efficient production of experimental knowledge. The experimental hybridizations that led to gene cloning occurred as scientists initially felt obliged to share and exchange key ideas, research tools, and materials for their research. Berg's laboratory not only benefitted from this communal mode of scientific life; it eventually helped Cohen and Boyer participate in recombinant DNA research and forge a strategic collaboration with Stanford biochemists.[68] Cohen and Boyer's cloning of recombinant DNA molecules was accelerated and greatly influenced by their interaction with Stanford biochemists.

On the other hand, in a situation where researchers worked intensely to clone genes in hopes of gaining scientific priority and monetary advantage surrounding patents, customs of sharing and exchange were bound to erode. The earlier "communal" ownership of scientific ideas and tools tended to dictate that those who made an experiment work first could claim scientific priority and credit, as opposed to those who first suggested or conceived ideas for experimentation.[69] Morrow's opportunistic alliance with Cohen and Boyer, for example, reflected this imperative of experimental production for scientific priority. Alternatively, Stanford biochemists' inability to complete their cloning experiments, due to their morally conscious, self-imposed moratorium on recombinant DNA work, enabled Cohen and Boyer to assert their priority in gene cloning over Mertz, who significantly contributed to the development of recombinant DNA–cloning technology, including the application of *Eco*RI for making recombinant DNA. The collaboration of Boyer, Cohen, and Morrow on eukaryotic gene cloning then led to a shifting moral economy of scientific collaboration to that of competition within the Bay Area biomedical research community. Cohen's success in gene cloning frustrated Berg and Mertz, stripping them of the opportunity to gain scientific credit for the first cloning of recombinant DNA. Berg, in his recollection, admits that he was "so furious" about Morrow's collaboration with Cohen and Boyer.[70] Stanford biochemists' generally agitated response, in turn, led Cohen to develop an aggressive, proprietary stance regarding the distribution of credit and priority. In the next chapter, we will probe the shifting moral economy of recombinant DNA research as the Bay Area scientists clashed over the dissemination of a cloning vector in the highly competitive context of 1970s biomedical research.

CHAPTER FOUR

Moral and Capitalistic Economies of Gene Cloning

The advent of recombinant DNA technology brought about a multitude of possible applications. These involved both basic molecular biology and genetic-engineering experiments. As ideas, materials, and techniques actively circulated within the network of recombinant DNA research formed around the Stanford Biochemistry Department, scientists began to take advantage of the technology for implementing diverse experimental projects. First, as Stanford biochemists had initially envisioned, recombinant DNA technology emerged as one of the most promising tools for basic molecular biology because of its capacity to produce an artificial hybrid DNA. It offered new experimental possibilities in molecular genetics and developmental biology by enabling scientists to manipulate and clone genes at the molecular level; this was particularly true in eukaryotic biology where it had previously been difficult to acquire genetic crosses for mapping or isolating genes to determine developmental and regulatory effects.

Beyond its use for basic biomedical research, scientists began to envision recombinant DNA technology as a tool for genetic engineering. The possibility of creating a transgenic organism—an engineered organism with genes deemed commercially and medically useful—prompted predictions and speculation that bacteria could turn into factories for useful biological molecules, such as transgenic plants, pharmaceutical therapeutics, and other industrially important chemicals.[1] Those who had sustained their connections with Paul Berg's Stanford laboratory were among those who contemplated a wide array of bold medical applications in genetic engineering. For example, in May 1970, Richard Roblin, a medical researcher at the Salk Institute, wrote to Berg inquiring about the possibility of using recombinant DNA technology for the development of human gene therapy:

> Over the past several months I have been developing an interest in the prospectus for genetic engineering, particularly with reference to ways of treating human genetic defects by gene modification. Several people who have visited Salk recently from Stanford have mentioned a research project going on there involving an attempt to link lac operon DNA and SV40 DNA as a means of constructing a molecule which might integrate foreign DNA into human cells. I thought this sounded interesting and promising and that I would attempt to find out a few more details by writing to you. Without giving away any trade secrets (if you feel there are any) could you tell me more about who is doing the work, and what the system is (i.e. what cells are to be transformed, and how a successful integration would be detected)?[2]

When Stanley Cohen and Herbert Boyer experimentally demonstrated gene cloning, some scientists began to appreciate the prospects of recombinant DNA technology for producing useful molecules. At a more fundamental level of medical intervention, researchers supposed that the technology could make it possible to insert or modify certain genes through a virus, thereby supplementing or correcting defective human genes.

Once the scientific and technological potentials of recombinant DNA technology were widely recognized, competition intensified; the system of technical and material exchanges in early recombinant DNA research in the San Francisco Bay Area became difficult to sustain and was bound to erode. After the success of a series of molecular cloning experiments by Cohen and Boyer, concerns surfaced about both reciprocity and scientific priority in scientific exchange. The Stanford biochemists' collective research culture fostered the sharing of research materials and techniques, which often meant a semicommunal ownership of ideas, materials, and tools. Within their moral economy of science, sharing ideas, research materials, and techniques through informal exchange and casual shoptalk was customary; at the same time, assigning scientific credit to those who conceived ideas, devised research techniques, or constructed experimental materials was difficult. As the historian Robert Kohler notes, in the communal context of limited proprietary rights, the issue of scientific priority would often hinge on those who generated experimental results by assembling a set of ideas, techniques, and materials.[3]

Though the Stanford biochemists applied limited proprietary rights to research tools and materials, it became increasingly difficult to maintain them as communal resources. This growing difficulty occurred as other scientists, especially outsiders to the department, embarked on recombinant DNA research and began to assert personal ownership by limiting access to the involved tools and materials. In Stanford's Department of Medicine, Cohen devised key techniques to use the plasmid pSC101 as a cloning vector in order to transport a recombined DNA molecule. After his collaboration with Boyer, Cohen's laboratory emerged as a center of the network of recombinant DNA research, especially because he possessed the cloning vector. More importantly, Cohen strategically circulated his pSC101, not only to forge collaborations but also to guard his scientific priority by controlling and monitoring its distribution. For example, right after the cloning of *Xenopus* ribosomal DNA, Stanford biochemist David Hogness requested Cohen's cloning vector. Cohen denied Hogness's request, telling him that the plasmid would be distributed only after publication of his first paper with Boyer on intraspecies DNA cloning. This provoked a heated controversy between the Stanford biochemists and Cohen on scientific priority, mutual obligation, and proprietary rights in scientific exchange.

This chapter examines a shifting moral economy of science in the research network that had been centered on Stanford biochemists by analyzing how concerns about the distribution of scientific credit and financial reward diminished the customs of informal scientific exchange, cooperation, and sharing. While the Stanford biochemists' practices of reciprocity, open exchange, and semicommunal ownership of research tools resulted in high productivity, more clarity regarding scientific priority was required in the increasingly competitive field of recombinant DNA technology. Some early researchers became more proprietary regarding their tools and materials, thereby straining the moral economy of science that had earlier been rooted in distinctive communal customs of sharing and free access. By examining the controversy over the dissemination of a cloning vector, I illustrate one of the first instances of the disruption of the moral economy in recombinant DNA research. First, I show how Hogness tried to adopt recombinant DNA technology for his developmental biology project; his efforts illustrate how his understanding of the semicommunal ownership of recombinant DNA technology came into being and then collided with Cohen's claim of proprietary rights. I next analyze how a clash between Cohen and Hogness over the dissemination of a plasmid raised contentious issues of scientific priority, credit, and reciprocal obligations in scientific exchange.

Concerns about the free exchange of ideas and easy access of materials
at Stanford were exacerbated as the prospect of industrial uses of recombinant DNA technology encouraged some scientists and university administrators to assert proprietary interests in the involved research tools
and materials. Cohen and Hogness's dispute over a plasmid, for example,
illustrates the beginning of the collision between a moral economy for
the production of experimental fact and a capitalistic economy for the
generation of profit in recombinant DNA research. A capitalistic conception of knowledge, one that organizes the production, distribution, and
exchange of knowledge centered on issues of private property right and
profit, soon encroached into the field of recombinant DNA amid the rise
of commercial interests in genetic engineering. Stanford University and
the University of California, with commercial aspirations of capitalizing
on their potential ability to produce therapeutically important proteins
like insulin, filed a patent application for recombinant DNA technology
on behalf of Cohen and Boyer. Their assertion of private ownership of the
technology for genetic engineering, along with capitalistic economies of
financial reward, brought a shift in the mode of scientific exchange away
from the ideals of reciprocity and sharing. Shifting moral assumptions regarding reciprocity and exchange in research were exposed by Cohen's
assertion of a certain degree of personal ownership of the tools and materials he had painstakingly constructed. This transition, from limited proprietary rights to personal ownership through patenting of key tools and
materials, illustrates changing conceptions of scientific priority and property rights in biomedical research at the dawn of the age of commercial
biotechnology.

"Double-Helix of Management and Organization"

In the late 1960s, David Hogness, like his colleague Paul Berg, began to
shift his research from bacterial genetics to the biology of higher organisms (figure 4.1). He was interested in what had been rather neglected
by earlier generations of molecular biologists and geneticists, namely, the
problem of development.[4] In this new venture, Hogness drew inspiration
from the operon model of gene regulation in *Escherichia coli* devised
several years earlier by the French molecular biologists Jacques Monod
and François Jacob. The operon model, they had argued, "reveals the genome contains not only a series of blue-prints, but a coordinated program

FIGURE 4.1. David S. Hogness in his Stanford laboratory, ca. 1970. Reprinted by permission from David S. Hogness.

of protein synthesis and the means of controlling its execution."[5] When Monod and Jacob described their operon model at the 1961 Cold Spring Harbor Symposium on Cellular Regulatory Mechanisms, they considered its implications for major issues such as development; they claimed that the "new concepts [of gene regulation] derived from the study of microorganisms will prove of the greatest value" in understanding the biochemical differentiation of cells in higher organisms.[6]

Hogness was part of the younger generation of molecular biologists who were inspired by the value of the operon model for understanding development.[7] In addition to some traditional embryologists, such as Conrad Waddington and Edward Lewis, younger molecular biologists like Sydney Brenner and Walter Gehring also shifted their research after the mid-1960s; they asked how genes might control cellular differentiation and developmental processes.[8] Some, like Brenner and Jacob, adopted the notion of the organism as a machine coordinated by a set of information codes, and they proposed to examine a developmental program encoded inside the gene.[9] Despite initial enthusiasm, perplexing experimental results emerged, which suggested that regulatory mechanisms in higher organisms might operate beyond the level of differential gene transcription.[10] By the early 1970s, the inability to extend the experimental horizon of molecular biology into developmental biology and the biology of higher organisms amounted to what historian Michel Morange has called an "epistemological crisis."[11] While Monod and Jacob famously proclaimed, "what is true of *E. coli* is true of the elephant," in the early 1960s, their belief in the universality of the operon model of gene regulation had waned by the early 1970s.[12] Harvard biologist C. A. Thomas, Jr., conveyed this challenge to the operon model while reviewing current advances in genetic approaches to development:

> If we can extrapolate from these smallest of all biological entities, which display little or no differentiation, to the most interesting eucaryote, man, we see an enormous regulatory problem that confounds description. Certainly the regulation of *E. coli* must be a child's task compared with the regulatory apparatus at work during embryogenesis. When we think of these problems, the double-helix no longer guides us. We must look for another unifying principle. We need a simple way of understanding development and differentiation; *a kind of "double-helix" of management and organization.*[13]

This search for the "double-helix of management and organization" in higher organisms became one of the significant subjects for biologists like Hogness, who pioneered the molecular approach to development. Hogness embedded this emphasis on the organization and management of animal chromosomes with his new project in developmental biology, shifting his research interest from the genome of bacteriophage λ to that of the fruit fly *Drosophila*. He wanted to take advantage of the vast reservoir of genetic data available for *Drosophila*.[14] With its particularly well-studied genetics and cytology, *Drosophila* became an attractive model organism for investigating the organization of animal chromosomes.

Drosophila's cytological data, and especially its chromosomal banding patterns, seemed to provide visual access to the gene on the chromosome (see figure 4.2): each band of densely packed DNA (*dark band*), and sparse interband (*white band*), was believed to form one functional genetic unit called a chromomere.[15] Each dense band, however, consisted of about twenty-five thousand base pairs on average, an enormous number of repetitive DNA base pairs that coded for just one functional gene. What role did repetitive sequences play in gene expression and regulation? More importantly, what did this particular organizational feature of the eukaryotic chromosome say about its replication and regulation?[16] During his 1968 sabbatical leave, Hogness spent his time in the

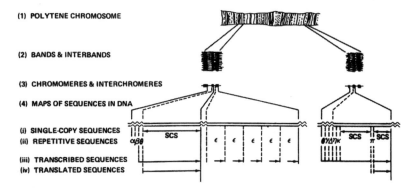

(1) POLYTENE CHROMOSOME

(2) BANDS & INTERBANDS

(3) CHROMOMERES & INTERCHROMERES

(4) MAPS OF SEQUENCES IN DNA

(i) SINGLE-COPY SEQUENCES
(ii) REPETITIVE SEQUENCES

(iii) TRANSCRIBED SEQUENCES
(iv) TRANSLATED SEQUENCES

FIGURE 4.2. Model of *Drosophila* chromosome. The chromosomal banding pattern of *Drosophila* became one of the rare resources that provided direct visual access to the gene and its organizational pattern on the animal chromosome. Each band (densely packed repetitive DNA) and interband were believed to form one functional genetic unit, or a chromomere. Image from David Hogness, Pieter C. Wensink, David M. Glover, Raymond L. White, David J. Finnegan, and John E. Donelson, "The Arrangement of DNA Sequences in the Chromosomes of *Drosophila melanogaster*," in W. J. Peacock and R. D. Brock, eds., *The Eukaryote Chromosome* (Canberra: Australian National University Press, 1975), 3. Reprinted by permission from David S. Hogness.

laboratories of James Peacock in Canberra, Australia; Wolfgang Beerman at the Max-Planck Institute, Berlin; and Edward B. Lewis at Caltech. He absorbed a set of experimental practices and theoretical questions related to *Drosophila* and its chromosomes.

David Hogness and the Cloning of *Drosophila* DNA Sequences

The laboratory culture of Stanford biochemists, with its tradition of collaborative work, encouraged the intermingling of experimental techniques. Stanford biochemists' collaboration, exchange, and sharing of technologies and instruments, as well as their pioneering move into eukaryotic biology, were particularly important for Hogness's venture into the biology of development. With the synthesis of recombinant DNA by his colleagues at Stanford in the summer of 1971, Hogness radically shifted his experimental strategies by adopting the newly developed technology for his analysis of *Drosophila* DNA sequences. Initially, Hogness intended to adopt a chemical mapping strategy, which he had developed for the analysis of λ infection.[17] As he put it in his 1970 grant proposal, Hogness planned to develop a method for determining the topography of transcription over the genome at various stages in the differentiation of specific cell types, asking the question as to whether there are any general rules governing the spatial distribution of active transcriptions specific to that differentiation. The analogy here is the finding that the transcriptions of λ are arranged according to their function and time of expression during development of the infective process.[18]

The technology for the isolation and cloning of recombinant DNA, which initially emerged through Stanford scientists' attempt to explore the molecular biology of higher organisms, provided an innovative tool that enabled the analysis of eukaryotic genome organization. Situated at the center of recombinant DNA research, Hogness was able to take full advantage of the Stanford biochemists' culture of sharing for his developmental biology work, eventually reconfiguring his experimental strategy from chemical mapping to the cloning of *Drosophila* DNA sequences. Recombinant DNA technology allowed Hogness to isolate and clone *Drosophila* DNA segments of developmental significance. In his 1972 grant application to the National Institutes of Health (NIH), entitled "The Arrangement and Function of DNA Sequences in Animal Chromosomes," Hogness planned a systematic study of the organization of the chromosome and its sequences using recombinant DNA.[19] Hogness laid out his plan to determine the "basic laws governing the arrangement of DNA sequences and their replication on eukaryotic chromosomes."[20]

In using recombinant DNA technology, Hogness was indebted to recent advances made by Stanford biochemists, especially those of Lobban and Mertz. For his analysis of *Drosophila* DNA sequences, Hogness intended to use Mertz's λ*dv* plasmid system for constructing a recombinant DNA with *Drosophila* DNA (see figure 4.3). Like Mertz's plan to clone the virus SV40, Hogness would attempt to use the enzyme *Eco*RI to make the cohesive ends of λ*dv*. He also would mechanically shear *Drosophila* chromosomes in order to divide it into all possible linear sequences. He then would cut *Drosophila* sequences and anneal linear λ*dv* to *Drosophila* DNA segments, producing a circular hybrid to be placed into *E. coli* with a calcium chloride method. Then the *Drosophila* DNA segment, he believed, would be cloned inside the bacterial host.

Through Mertz's λ*dv* plasmid system, Hogness wanted to develop a "system for mapping DNA sequences" in eukaryotic chromosomes.[21] He first planned to focus on the *Drosophila* chromomere, a unit of transcription, as a way to investigate the organization of animal chromosomes.

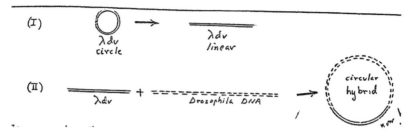

FIGURE 4.3. Construction of *Drosophila* recombinant DNA with plasmid λ*dv*. David Hogness, National Institutes of Health grant proposal, 1972, p. 6, David S. Hogness Papers, Personal Collections, Stanford University. Reprinted by permission from David S. Hogness.

He first intended to isolate and clone *Drosophila* DNA fragments from the chromomere in order to illuminate the puzzling role of repetitive sequences in gene regulation. By hybridizing the isolated repeating sequences onto the *Drosophila* genome, their origins in the genome could be mapped. Through this physical mapping, he claimed that he could identify how the repetitive sequences are organized in the chromosome. More importantly, recombinant DNA technology offered a way to try cloning DNA segments of the *Drosophila* genome. For this, he intended to construct λ*dv-Drosophila* recombinant DNA and to use it for probing its regulatory function. If he could clone repetitive DNA segments, then their regulatory functions, either through their transcription or translation, could be investigated thoroughly.[22]

Tale of Two Plasmids: λdv and pSC101

While Hogness was working on the construction of λ*dv-Drosophila* recombinant DNA, John Morrow told him in October 1973 about the possibility of cloning eukaryotic genes using the pSC101 plasmid. As Hogness later acknowledged,

> I believe I was first informed of the existence of pSC101 as a cloning vector about October, 1973 by Dr. Morrow, who was working with Dr. Cohen. My purpose in obtaining pSC101 was to use it as a vector for cloning eukaryotic DNA. I do not believe that either Drs. Cohen and Boyer were aware of my purpose in using pSC101. My work was independent of Drs. Cohen and Boyer and I was under no obligation to provide any information to either Dr. Cohen or

Dr. Boyer. Dr. Morrow informed me that pSC101 could be used for transformation and had a unique *Eco*RI site.[23]

Both Hogness and Morrow recognized the limitation of Mertz's λ*dv* system; it required dimerization of the cloning vector in the hybrid DNA for
the replication of the λ*dv* sequences (to preserve λ's replication gene).
On the other hand, Cohen's pSC101 vector system allowed scientists to
more easily and efficiently obtain cells with the desired clones: first, one
could replicate vector without being concerned about the destruction of
replication genes in plasmids; second, its use of drug-resistance genes also
enabled efficient killing of the host cells that did not take up the cloning
vector. As early as August 1973, Morrow, as a consequence of his collaboration with Cohen and Boyer, knew one could clone a foreign eukaryotic gene by recombining the pSC101 plasmid with frog ribosomal DNA
(see chapter 3). He presented his *Xenopus* cloning experiment at the
Biochemistry Department's annual Asilomar meeting in October 1973,
and many Stanford biochemists were excited about its scientific implications. For example, Mertz wrote to the MIT scientist Mary Lou Pardue
about the possibility of adopting recombinant DNA technology for Pardue's research on the regulation of gene expression and development of
the frog, *Xenopus laevis*: "The J. Morrow, S. Cohen, and H. Boyer experiments of replicating *Xenopus* r[ibosomal] DNA in *E. coli* by attaching it
to an R factor plasmid using R_1 endonuclease [*Eco*RI] and ligase really
work. They even detect transcription of the *Xenopus* rDNA. Think of the
possibilities!"[24]

Hogness was one of the earliest scientists eager to adopt the pSC101
plasmid system for the cloning of eukaryotic DNA. After learning of
the *Xenopus* cloning success from Morrow, he asked Cohen to make
pSC101 available to his own lab as a potential vector for cloning *Drosophila* DNA. Because he had been able to obtain other key materials
and tools in recombinant DNA research from his biochemist colleagues,
such as the λ*dv* plasmid and other DNA-related enzymes, Hogness regarded them as semipublic resources, developed and shared by fellow scientists. Cohen, however, had begun to consider himself an outsider to the
Stanford biochemists' moral economy of science and considered the plasmid to be his personal resource, one that he had developed as a cloning vector. At first, Cohen tried to use his plasmid pSC101 as a strategic
tool to solicit a possible collaboration. After receiving Hogness's request,
Cohen indicated that he would have to be listed as a coauthor on a pub-

lication resulting from Hogness's *Drosophila* DNA–cloning experiment. From Cohen's perspective, his pSC101 cloning system would be a crucial part of Hogness's experimental system. Hogness, however, rejected this proposal; from his perspective, Cohen's role would be limited to a material contribution of a plasmid he had already reported in the scientific literature. Cohen in fact had published an article describing the isolation of pSC101 in May 1973.[25] Hogness stressed that he had already proposed in his 1972 grant application an experimental procedure for cloning *Drosophila* DNA using the λ*dv* plasmid and would simply be replacing λ*dv* with pSC101. When Cohen's attempt to collaborate with Hogness failed, Cohen ultimately responded by saying that he would provide the plasmid only after the official publication of his paper describing the cloning of genes using pSC101 as a carrier for DNA fragments.[26]

In deliberating the dissemination of pSC101, Cohen was deeply concerned about his scientific priority, even though the *Xenopus* DNA–cloning experiment was essentially completed by then. He retrospectively noted: "At that time, we hadn't produced even an outline for a paper on the *Xenopus* work; [we] were relatively early in the *Staph* work [interspecies cloning], and [we] were still a few months away from publication of our *E. coli* plasmid DNA [the first intraspecies cloning]."[27] Cohen believed that his scientific priority in the cloning of recombinant DNA would have been much less recognized if Hogness, who had already started detailed planning for cloning with the λ*dv* plasmid system, were to publish his cloning of *Drosophila* DNA using the pSC101 plasmid system at about the same time.[28] Cohen was then a junior scientist without tenure and was just beginning to build his research program. Hogness, however, was already a renowned scientist at one of the world's best biochemistry departments, where scientists had contributed to the development of recombinant DNA technology to study the biology of higher organisms. He had already established a bold, innovative project to explore the biology of development using several recombinant DNA techniques. More importantly, Hogness had written his 1972 NIH grant application describing detailed ideas and methods on cloning *Drosophila* genes at around the same time that Cohen and Boyer first came up with their idea for cloning bacterial genes at a November 1972 conference in Hawaii. With his resources and technical expertise, Hogness could open up a new field of molecular developmental biology with his gene-cloning experiment, undermining Cohen's priority.

Although Cohen's temporary refusal to distribute an important ve-

hicle for gene transfer and cloning was not uncommon, it was a challenge
to Stanford biochemists' moral economy of science that encouraged the
free exchange of ideas and tools. It was true that scientists often withheld research materials until their papers were published. This norm,
however, could be interpreted variably, and to a degree it was unacceptable to Hogness. Cohen had already described the isolation of pSC101 in
his May 1973 paper, and Hogness insisted that Cohen's pSC101 should
therefore be disseminated to the wider scientific community. On the other
hand, Cohen insisted that, because his first paper on gene cloning using
pSC101 was yet to be published, the distribution of the plasmid should be
restricted. (Cohen's first cloning paper was published in November 1973,
within weeks of Hogness's request.)

For Hogness, Cohen's unwillingness to share his plasmid was a trans-
gression of the norms of scientific exchange based on mutual obligation
and free access. For example, in line with Stanford biochemists' tradition of in-house collaboration and sharing, they forged material ties that
linked a range of experimental projects in eukaryotic biology after the
development of recombinant DNA technology. Hogness, along with Stanford biochemist colleagues Berg, Ronald Davis, Douglas Brutlag, and
George Stark, undertook a systematic study of eukaryotic chromosomes
and their genetic regulation, submitting a collaborative grant proposal in
1973.[29] Through this grant, they sought additional support for their venture into the biology of higher organisms to maintain necessary research
instruments, such as an advanced electron microscope they shared for
their analyses of eukaryotic recombinant DNA. Accustomed to a communal laboratory culture, Hogness and other Stanford biochemists regarded
Cohen's temporary withholding detrimental to their style of work. While
the academic world of molecular biology became increasingly competitive from the 1960s, the tradition of mutual obligation and limited proprietary rights at Stanford's Biochemistry Department made Cohen's temporary retention of pSC101 seem untenable to Hogness.[30]

At another level, the Stanford biochemists saw pSC101 as both a product of nature and a semipublic tool, whose usefulness as a tool for experimental research had been created by the communal efforts of fellow
scientists. The plasmid SC101 was naturally occurring; its value emerged
when it was deployed as part of a recombinant DNA experimental system
where it functioned as a vector, a vehicle that transferred recombinant
DNA into E. coli for replication. The entire experimental system encompassed a repertoire of techniques and materials for making, transport-

ing, and expressing hybrid DNA molecules, and many of its components were devised and shared by early participants in the research, including Cohen. For example, Cohen could not have performed his early recombinant DNA experiments without access to DNA ligase provided to him by the Biochemistry Department. As such, Hogness believed that the plasmid should be communally held as well.

Moreover, when some Stanford biochemists regarded themselves as pioneers in the field of recombinant DNA research and technology, Cohen's competitive stance damaged their relationship with him. Cohen's decision, in his own recollection, marked a "key turning point in [his] relationship to the Department of Biochemistry" at a more personal level.[31] He was well aware of his debt to the Stanford biochemists. Cohen's appointment at the Stanford Medical School in 1968 was in part due to their strong support. Most importantly, they were impressed by Cohen's work on λ phage and RNA polymerase. Cohen's postdoctoral mentor, Jerry Hurwitz, was a former colleague of Arthur Kornberg's when they were at the Department of Microbiology at Washington University in St. Louis during the 1950s; other Stanford biochemists, such as Berg, Hogness, and A. Dale Kaiser, were Hurwitz's colleagues there as well.[32] When Cohen moved to Stanford, Hogness provided him with temporary space until his new laboratory was ready.[33] Cohen was a fairly isolated molecular biologist at the Department of Medicine where most scientists focused on clinical problems, and he had become a regular participant in seminars and meetings at the Biochemistry Department. He was even able to use instruments and exchange ideas, methods, and materials with Stanford biochemists as he continued to interact with some faculty and graduate students, such as Berg, Kaiser, and Lobban.

A moral economy often implies immorality of acts that would violate shared beliefs about ownership and exchange in a community life. When deprived of the plasmid, Hogness was filled with a sense of moral injustice and eventually decided to obtain the pSC101 DNA from Morrow's refrigerator without consent from Cohen.[34] As the dispute over the possession and dissemination of pSC101 intensified, Berg, who was chair of the Biochemistry Department, tried to arrange a solution. While he concurred that Hogness should not use Cohen's plasmid without consent, he reminded Cohen of the norms and obligations of free scientific exchange from which he and other recombinant DNA researchers benefitted.[35] Berg insisted that a repertoire of recombinant DNA techniques and materials, including Cohen's pSC101 plasmid, was one of the results of an active scientific and technical exchange among those who comprised the network around Stanford biochemists. The productivity of this mode of scientific exchange required that the commitment and bond between a giver and recipient could only be maintained through reciprocity.

Cohen decided to give his pSC101 plasmid to Hogness for *Drosophila* gene cloning a few weeks before the publication of his first bacterial gene-cloning paper in November 1973. The *Xenopus* paper was published in May 1974, and Hogness's first *Drosophila* gene-cloning paper was published in December 1974, giving Cohen plenty of advanced time to claim his scientific priority and credit.[36] While not remembered as an inventor of gene cloning, Hogness was able to make significant progress in developmental biology. After obtaining Cohen's plasmid, he constructed hybrid DNA by joining the bacterial plasmid pSC101 to sheared fragments of *Drosophila* embryonic DNA (see figure 4.4). The colonies of these segment clones served as *Drosophila* DNA segment banks, providing new experimental opportunities with which scientists could approach developmental phenomena at the molecular level. In addition to his creation of the *Drosophila* DNA segment bank, Hogness used recombinant DNA technology for further preparations of segments; he developed a chromosomal mapping technique for the isolation of repetitive DNA sequences and developmentally significant genes.[37]

If Cohen's concern about scientific priority and credit temporarily undermined Hogness's venture into developmental biology, the free circulation of cloning plasmids among academic developmental biologists in the 1970s and 1980s confirmed the productivity of giftlike exchange systems. Hogness's mapping technology, developed as an innovative way to approach development in molecular terms, was disseminated to the community of *Drosophila* developmental biologists; these researchers most notably included the laboratories of Walter Gehring at the University of Basel, William McGinnis at Yale, and Edward Lewis at Caltech. Situated in the hotbed of recombinant DNA technology, Hogness's laboratory at Stanford became a focus of attention from biologists whose interests were in the molecular control of development. Moreover, with a set of powerful recombinant technologies, Hogness's laboratory attracted future leaders in *Drosophila* genomics, such as Gerald Rubin. In the early 1980s, Paul Schedl, who had been a graduate student of Hogness, went on to Gehring's laboratory, helping to establish a *Drosophila* gene bank. Christiane Nusslein-Volhard spent her postdoctoral years in Walter Gehring's lab from 1975 to 1977, learning Hogness's *Drosophila* gene–cloning

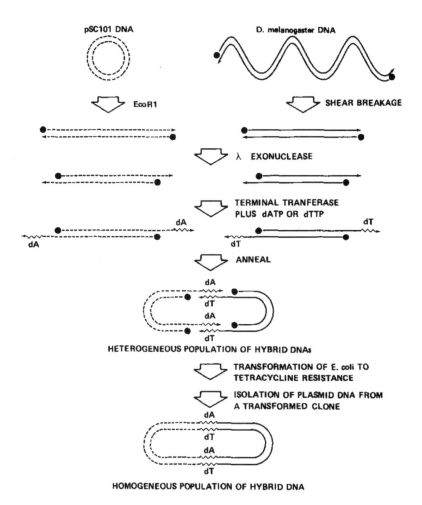

pSC101 DNA

D. melanogaster DNA

EcoR1

SHEAR BREAKAGE

λ EXONUCLEASE

TERMINAL TRANFERASE
PLUS dATP OR dTTP

dA

dT

dA

dT

ANNEAL

dA

dT

dA

dT

HETEROGENEOUS POPULATION OF HYBRID DNAs

TRANSFORMATION OF E. coli TO
TETRACYCLINE RESISTANCE

ISOLATION OF PLASMID DNA FROM
A TRANSFORMED CLONE

dA

dT

dA

dT

HOMOGENEOUS POPULATION OF HYBRID DNA

FIGURE 4.4. Construction of recombinant DNA from *Drosophila* DNA segments. With an enzymatic ligation method developed by Peter Lobban at the Stanford Biochemistry Department, David Hogness constructed recombinant DNA molecules with the pSC101 plasmid, which was eventually given to him by Stanley Cohen. He then was able to clone *Drosophila melanogaster* segments by inserting pSC101 plasmids into *Escherichia coli*. Image from Pieter C. Wensink, David Finnegan, John E. Donelson, and David S. Hogness, "A System for Mapping DNA Sequences in the Chromosome of *Drosophila melanogaster*," *Cell* 3 (1974): 316. Reprinted by permission from David S. Hogness.

technology in order to isolate developmentally interesting genes with Eric Wieschaus, then a graduate student.[38] Indeed, the use of Hogness's genomic analysis through recombinant DNA technology rendered *Drosophila* species the chosen organisms for developmental biology from the 1980s onward.[39]

The following section provides another example of the disruption of the moral economy of science at Stanford. This was prompted by the patent filing of recombinant DNA technology by Stanford and the University of California that listed Cohen and Boyer, outsiders to the Stanford Biochemistry Department, as the sole coinventors. If granted, recombinant DNA patent would provide royalty income to both the universities and its inventors. I investigate how economic considerations further subverted the Stanford biochemists' moral economy as some scientists and university administrators tried to apply full private ownership to those in the field of recombinant DNA research and technology.

Recombinant DNA Technology as Intellectual Commons or Intellectual Property

Peter Lobban's suggestion of the use of recombinant DNA technology for manufacturing useful medical products was discussed more often by the summer of 1973 after Cohen and Boyer successfully cloned foreign genes. The Stanford biochemists talked openly about the possible uses of recombinant DNA technology for genetic engineering, though many of them believed its realization lay in the distant future. Cohen, who was interacting frequently with Lobban during this time to learn how to transfer his plasmids into *E. coli* using a calcium chloride method, might have learned not only about academic uses of recombinant DNA technology but also about its likely industrial applications.[40] The ability to transform an organism into a factory for useful medical molecules via recombinant DNA, as the historian Hans-Jörg Rheinberger puts it, would transform its status to a "*locus technicus*—that is, to the status of a space of representation in which new genotypic and phenotypic patterns are becoming probed and articulated."[41] As biomedical researchers came to grips with the expected economic benefits of recombinant DNA technology, the Stanford biochemists' communal way of doing science—which helped foster productive experimental exchange and collaboration while also regulating competitive feelings and calculating motives—began to be challenged more forcefully. Indeed, the controversy over the patenting of recombinant DNA technology illustrated the collision between the moral and financial interests of science at the coming of age of commercial biotechnology.

As scientists began to grasp the full potential of recombinant DNA technology beyond its role as a research technique, Cohen and Boyer ventured into patenting and commercialization with encouragement from university patent administrators. (For a historical analysis of the privatization of recombinant DNA technology in a broader context, see chapter 5.) News about the industrial application of recombinant DNA technology circulated around Stanford during the summer of 1974.[42] Niels Reimers, an enterprising Stanford patent administrator, began to contact key scientists involved in recombinant DNA research and technology, such as Berg, Davis, and Cohen. Initially, they all wondered whether basic scientific methodologies in molecular biology like recombinant DNA technology, as opposed to particular devices, could be subject to patenting. Technically, it was also impossible for Reimers to file a patent application for recombinant DNA technology based on works by Berg's group; both Berg's recombinant DNA article and Mertz and Davis's paper were published in 1972, and by 1974 both had already passed one year's grace period for filing a patent for publication in the United States. Reimers was able to persuade the reluctant Cohen to consider patent filing for recombinant DNA technology. Both Berg and Davis were not interested in, and were perhaps averse to, patenting. According to Stanford's patent lawyer Bertram Rowland, Berg "adamantly" refused "to file patent applications on his work"; Rowland reflected that consequently "there might not have been a [recombinant DNA] patent."[43] The Stanford biochemists' customs of sharing and exchange of research tools and materials reinforced their sense of semipublic ownership, which made it difficult for them to claim proprietary interest.

In the small, tight-knit community of scientific researchers in the Bay Area, the patenting of what some regarded as communal procedures in molecular biology heightened the tension between the Stanford biochemists and Cohen and Boyer. In their attempt to make recombinant DNA technology proprietary and carve economically viable "intellectual property" out of the research, Reimers and Rowland sought to clearly define what technical procedures constituted its invention. As in the standard practice of maximizing the return on intellectual property, their primary aim was to broaden the scope of the invention of recombinant DNA technology while narrowing down its inventors. Within a few months, university patent administrators at Stanford University and the University of California persuaded Cohen and Boyer to file a patent application. Cohen signed an invention disclosure form on June 24, 1974. For the preparation of the recombinant DNA patent application, he also provided most of the technical information about the invention of the cloning technology, while Boyer was "much less involved."[44] Their plan was to file the application with Rowland representing Cohen and Boyer before the US Patent and Trademark Office (PTO). The patent, if granted, would enable Stanford, the University of California, and their scientists to capitalize on the potential use of recombinant DNA technology for genetic engineering work and to claim royalty shares on any commercial uses of recombinant DNA procedures.[45]

While the initiative for the patenting of recombinant DNA technology came from Stanford's patent administrator, Cohen was intrigued by Reimers's suggestions for several reasons. From Cohen's perspective, a series of three molecular cloning papers seemed to justly establish him as a co-inventor of recombinant DNA technology, along with his major collaborator, Boyer, at University of California, San Francisco (UCSF). This legal determination of the inventors seemed to relieve Cohen of his concern about scientific priority in gene cloning. Given the Stanford biochemists' concern about his reluctance to circulate his plasmid, Cohen might have preferred the clarity of law in proving his contributions to gene cloning to the ambiguities of scientific dependencies. Cohen later acknowledged his view on patenting as a legal method for establishing scientific priority and credit:

> Initially, I was hesitant about going ahead with the patent application . . . but was convinced by Niels Reimers to proceed, and quite happy about that decision . . . As Niels has pointed out to me a number of times, a patent helps to clarify just whose scientific contributions underlie an invention, and the issuing of a patent, especially one that withstands challenges about inventorship, legally establishes the priority of a discovery.[46]

Also, the economic reward of patent royalties was more appealing than complying with what Cohen saw as the exaggerated scientific obligations of Stanford biochemists. Cohen's decision to proceed with the patenting of recombinant DNA technology derived not only from his personal and the institutions' financial interest but also from the complex motives for establishing his scientific credit and seeking moral vindication.

Before Stanford and the University of California officially filed a patent application for recombinant DNA technology, Cohen carefully planned to establish his scientific priority more publicly. In early September 1974, he called the editor of *Scientific American*, Dennis Flanagan, to offer writing an article on the cloning of recombinant DNA molecules and its applications for genetic engineering. The magazine usually commissioned an article by a prominent scientist who launched a new field of research, and in January 1974 Flanagan had already asked Berg to write an article that would discuss the development and current status of genetic engineering with restriction enzymes.[47] Berg had replied in the affirmative, and Flanagan explained to him later in the year: "Cohen, who was not aware that you had been hoping to write an article for us, called me to say that he was currently finishing a technical article for one of the Annual Reviews volumes. He asked if, in view of the fact that he had assembled all the necessary material, we would be interested in a more popular version for *Scientific American*."[48] Berg was delaying the writing of his article because of his involvement in various regulatory and legal issues regarding biosafety of recombinant DNA technology. After Berg received the letter from Flanagan, he met with Cohen and discussed the matter of authorship on the article on genetic engineering work. Berg then replied to Flanagan that he had "discussed [Cohen's] offer to write a *Scientific American* article on the use of restriction enzymes in plasmid research and for cloning DNA molecules." He added: "That's not quite what I would have written about although my article would have certainly discussed that aspect. In any case, since I'm not able to do the article now nor can I realistically promise it to you for the first of the year, perhaps you should accept Cohen's offer."[49]

Soon after Cohen's careful assertion of his inventorship in a more public form, Stanford and University of California filed the recombinant DNA patent application in November 1974. In a move that surprised fellow biomedical researchers who were preoccupied with the public health controversies of recombinant DNA technology, the patent filing put Stanford biochemists, in particular, in an uncomfortable position. Both Berg's leading role in the Asilomar I and II conferences on biohazards and the Stanford biochemists' tightly knit research network magnified the tensions when Cohen and Boyer filed the recombinant DNA patent application. First, Morrow and Robert Helling, Cohen and Boyer's collaborators, refused to abandon their claim to coinventorship, insisting that their coauthorship and scientific credit should be duly acknowledged in the patent filing. Morrow was especially baffled by the patent application's broad claim. In his response to Stanford's request to sign a disclaimer to abandon his inventorship status, Morrow wrote to Cohen and Boyer: "I am not signing this because I have a number of reservations about it . . . Why did you include the use of viral replicons to make functional DNAs when you have not done work on formation of recombinant viruses? Other people worked out viral methods independent of your findings with plasmids."[50] In a meeting at the Stanford Office of Technology Licensing in 1976, Berg raised a similar objection: "Many Stanford scientists contributed to the DNA technology. Why then is Cohen the only Stanford inventor?"[51]

Indeed, the transition from semicommunal ownership to personal ownership in recombinant DNA technology became contentious among those in the early network of researchers. While the Stanford biochemists' collegial customs of reciprocal exchange and sharing made ownership rights ambiguous, both an interest in priority and the appeal of financial gain necessitated the clarity of ownership rights. The legal clarity, however, was hard to achieve in the context of a moral economy of science where research procedures and tools had customarily been exchanged and shared. The crucial exchanges between Stanford biochemists and Cohen and Boyer, for example, caught the attention of several reviewers of the recombinant DNA patent application. Reacting to the application prepared by Rowland, an anonymous patent reviewer wrote to Stanford and University of California patent officials, indicating "other individuals have been involved in the important publications relating to the process, i.e. the earlier work of Mertz, Davis, and Berg."[52] In 1977, a patent examiner at the PTO, Alvin E. Tanenholtz, agreed with this reading of the Stanford patent application, and rejected all of Cohen and Boyer's forty-five inventorship claims. Tanenholtz cited the Stanford biochemists' recombinant DNA work as constituting "prior arts"—notably, the work of Berg's group, including Berg's synthesis of SV40-λdvgal DNA, Mertz and Davis's formation of hybrid SV40-$\lambda dvgal$ DNA, and the Lobban thesis. As Stanford's patent official reported, "In the first [PTO] action, all claims were rejected. The primary substantive rejections were that the claims were either anticipated by or obvious from the teachings of Mertz and Davis (relating to hybridization of SV40 and lambda dv-120 DNA) in view of the letter drafted by the Gordon Conference participants relating to potential biohazards of recombinant DNA published in *Science*."[53]

Rowland countered that Lobban's vector did not contain any foreign genes, and his prophecy should not be seen as a "teaching" of the arts that would be obvious for nonspecialists. With regard to Rowland's argument that Cohen and Boyer's procedure was the only technology-enabling molecular cloning, Tanenholtz replied that Mertz's hybrid SV40-λ*dvgal* plasmid, "duplex fully infectious, covalently closed circular DNA," could be used for cloning SV40 DNA.[54] Rowland further argued that *Eco*RI could have destroyed the replication gene (the O gene) in the plasmid λ*dvgal* necessary for its replication. Rowland's counterargument echoed Cohen's memo circulated around Stanford around the same time. Written as a response to an inquiry by Nicholas Wade, a reporter at *Science*, about the inventorship of recombinant DNA technology, Cohen defended his unique contribution to the cloning of recombinant DNA technology and how it could eventually become a widespread laboratory practice. He acknowledged that Berg's group did plan to clone recombinant DNA. However, as Cohen underlined, it would have been impossible to clone recombinant DNA inserted in *E. coli*:

> As Paul Berg has indicated, concern about possible biohazards related to the SV40 component of the λ*dv*-SV40 molecule that Jackson *et al.* had constructed, led him to decide not to try to clone the molecule in *E. coli*. However, there is no report of success in the cloning of analogous molecules that contain any other fragment of DNA inserted at the λ*dv* site used in the Jackson *et al.* experiments. Apparently the reason for this is that the *Eco*RI cleavage site in λ*dv* is located within the O gene . . . which is essential for replication. Interruption of the continuity of this gene by an inserted DNA fragment prevents λ*dv* from functioning as a replicon.[55]

However, Tanenholtz secured the opinion of the Stanford biochemist Ronald Davis that Mertz's experiment was in fact designed to prevent the destruction of O genes by producing long, polymeric λ*dv* DNA chains that contained the genes. Scientists were also able to show that a λ*dv* plasmid could be used as a cloning vector, suggesting that Mertz and Davis's λ*dv* plasmid might well be able to clone SV40 DNA.[56] Rowland pointed out that Mertz and Davis did not actually put this plasmid into *E. coli* because of the moratorium on tumor-virus recombinant research and argued that they should not "be given credit for work they did not perform."[57] Tanenholtz eventually acquiesced to some of Rowland's counterarguments. One patent lawyer, Albert Halluin at Cetus, another early biotech company, criticized Tanenholtz's review of the recombinant DNA technology. As a lawyer later hired by Stanford to counsel its attempt to patent recombinant DNA technology noted, "Halluin spoke disparagingly about the quality of the PTO prosecution. He specifically noted that Mr. Tanenholtz, the examiner, consistently yielded without much fight to Bert[ram] Rowland's positions. It was his view that neither the Cohen nor the Boyer patents would have survived examination by other examiners in the art, and he backed up his contentions by reference to conversations he has had with various examiners familiar with the case."[58] With no legal action by Lobban, Mertz, or Davis, Tanenholtz ultimately ruled that, although Stanford biochemists' had charted the use of recombinant DNA technology for bacterial transformation and molecular cloning, they neither demonstrated its operability nor specified necessary experimental conditions.

The issue of scientific priority and credit in the patenting of recombinant DNA technology underscored the tension in making recombinant DNA technology proprietary and in carving economically viable intellectual property out of what Stanford biochemists regarded as communal procedures.[59] In a collective context where limited private ownership of research procedures, materials, and tools was assumed, the assertion of proprietorship through patenting raised the question of rights in the network of recombinant DNA research. Some scientists wondered whether recombinant DNA technology, with its broad use in basic molecular biology, should instead be in the public domain. In response to the Stanford OTL's request to review its patent application for recombinant DNA technology, an anonymous patent reviewer wrote:

> A more serious drawback to the patent in my view is that it represents the development of a very basic process in molecular biology; a process that has great implications with respect to basic research. One can, with some justification, argue that this basic process should be left in the public and scientific community domain and not be patented.[60]

Stanford and University of California's attempt to render recombinant DNA technology proprietary challenged the moral economy of science sustained at the former's Biochemistry Department at two interrelated levels. First, by pursuing legal clarity in scientific priority and credit, the patent application challenged the collegial customs of exchange and sharing, which Stanford biochemists regarded as fundamental to the productivity of their work. As scientists in the field of recombinant DNA research were becoming notably sensitive about scientific credit and proprietary interests, improvised and often ambiguous negotiations about ownership rights and uses entailed in scientific exchange became a source of confusion and dispute. Second, the assertion of personal ownership of research procedures and tools undermined the communal practices in the Stanford biochemists' moral economy of science. Even as some Stanford biochemists like Hogness tried to maintain the semipublic ownership of research tools and materials, which they regarded as a key source of creative experimentation, it became increasingly difficult to keep them away from the market economy. As the Stanford geneticist Joshua Lederberg warned, "the possibility of profit—especially when other funding is so tight—would be a distorting influence on open communication and on the pursuit of basic scholarship."[61] With the patenting of recombinant DNA technology, the migration of molecular biologists to studying eukaryotic organisms, the informal circulation of materials and ideas, and the unexpected hybridizations of experimental systems that nurtured recombinant DNA research and technology seemed more insignificant. The flow of scientific ideas, research tools, and materials in the network of recombinant DNA research became more fraught in the mid-1970s with the attraction of financial gain.

Conclusion

In the mid-1970s, the circulation of scientific materials within the community of early recombinant DNA researchers at Stanford came in many guises—as gifts, strategic tools for collaboration, and as thefts. The tale of two plasmids at Stanford—λdv and pSC101—shows how the material "gift" in scientific exchange was often fraught with concerns about priority and reciprocity in scientific relations. When Cohen initially refused to disseminate pSC101 to Hogness due to his concern about scientific credit before publication, the notion of exchanges between Cohen and Stanford biochemists inevitably became contentious. The relationships that Cohen had enjoyed with the biochemists at Stanford eroded as he strategically used his cloning vector to seek alliances, advancement, and scientific priority. As an assistant professor with relatively low scientific status, Cohen faced increasing resentment from Stanford biochemists. By the time the controversy over the dissemination of pSC101 erupted, Cohen's earlier exchanges and strategic associations with the biochemists had become a source of accusations of ingratitude.

The shifting customs of exchange and collaboration at Stanford and the eventual privatization of recombinant DNA technology both proved

crucial to understanding alliances and conflicts in scientific life, power relations among researchers, and large economic changes in biomedical research in the 1970s.[62] As the technological prospects of recombinant DNA
technology expanded rapidly, both for basic research and industrial applications, exchanges of tools, material, and technology in the network of recombinant DNA research became narrow and exclusive; this caused jeal-
ousies, quarrels, and burdens of obligation in the tight-knit community.
On the one hand, the Stanford biochemists relied on a sharing system, in
both their grant monies and materials; the tradition had helped create sol-
idarity with one other. Their collaborative work style and moral economy
of science provided a highly productive platform on which recombinant
DNA technology and its applications emerged. On the other hand, Cohen
was just loosely involved in the Stanford biochemists' scientific life, with
no concrete departmental affiliation either administratively or financially,
and his relationship with them had started as a recipient. When Cohen
made seminal progress in plasmidology and the cloning of recombinant
DNA, the ambiguous dependencies and unceasing sense of obligation ex-
asperated him.

Cohen's decision to proceed with Stanford and the University of California's patent filing for recombinant DNA technology—prompted by
concerns about scientific priority as much as economic calculations—
deepened the Stanford biochemists' concerns about their own scientific
credit and the disruption of their moral economy of science. They recognized that severe competition among scientists often crippled a lively
circulation of ideas and materials, straining their scientific progress. By
arguing for a legal justification for the privatization of semipublic research procedures and tools, the patenting of a basic research technology
like the recombinant DNA work seemed to presage the eventual demise
of a community of scientists working with each other through a collegial
system of exchange. As an early pioneer in recombinant DNA research
who led national discussions on its regulation, Berg felt that the assertion
of personal property rights presented a troublesome image of the scientist
as a thieving merchant, risking public health for profit.

Economic considerations were not the only ones behind Cohen's decision to move ahead with the patenting of recombinant DNA technology.
He had even proclaimed that he would not personally benefit from patent royalties, thus trying to avoid accusations of seeking personal gain (It
should be noted that in 1983 Cohen reneged on his promise to not personally benefit from the recombinant DNA patent).[63] The legal determina-

tion of inventorship would more importantly provide him with an opportunity to seek clarity in scientific priority and thereby free him from what he felt were often ambiguous and exaggerated obligations in the community of recombinant DNA researchers. Stanford patenting officials were also mindful of criticisms that recombinant DNA technology emerged as a research tool as a consequence of scientific exchanges and collaborations between basic biomedical researchers located around Stanford. Stanford patent administrators and Cohen were convinced that patenting was not just a means for private gain; they viewed it as a viable means to demonstrate the relevance of academic work by encouraging commercial development of research results. The mere production of experimental knowledge was no longer enough to prove the usefulness of biomedical research. Commercialization, they argued, would serve the public interest better by promoting medical innovation and economic development.

At another level, there is something historically specific about Stanford's foray into patenting in biomedical research in mid-twentieth-century America. It had become common to observe a fierce race between highly competitive scientists, as the prominent molecular biologist James Watson famously wrote in the late 1960s. However, a concern for money was relatively absent in earlier competition in molecular biology, which was instead driven by the pursuit of scientific credit.[64] Cohen's economic calculations were not initially paramount because he temporarily sacrificed any financial gain; still, this choice reflected the emergence of the privatization of scientific research in American universities. Indeed, institutional arrangements, public policy discussions, and moral justifications surrounding the privatization of scientific research in the academy were rapidly reshaping in the 1970s. This fostered the birth of private, commercial ventures in the biomedical enterprise. Scientists increasingly began to equate the pursuit of money with the pursuit of knowledge, and financial gain became a by-product of medically and socially relevant research. In the next chapter, I show how the private ownership of recombinant DNA technology was advanced as an efficient means to favor public interest by calling for relevance in biomedical research. As we will see, the privatization of recombinant DNA technology was, in fact, a large and important part of the broader reconfiguration of public knowledge in biomedical research, presaging significant changes in the moral positions and scientific life of biomedical researchers

CHAPTER FIVE

Who Owns What? Private Ownership and Public Interest

In November 1974, Stanford University and the University of California (UC) filed a patent application for recombinant DNA technology on behalf of Stanford geneticist Stanley Cohen and UC biochemist Herbert Boyer.[1] From early 1976, as word about the patent filing spread, Cohen and Boyer's claim of the inventorship and ownership of recombinant DNA technology came under relentless scrutiny from fellow scientists, patent examiners, university and government officials, politicians, and the public. At one level, the identification of Cohen and Boyer as the sole inventors of recombinant DNA technology infuriated Stanford scientists who had shared research techniques and exchanged materials like DNAs, enzymes and plasmids with them, including collaborators like the Stanford biochemist John Morrow.[2] At another level, the assertion of private ownership of recombinant DNA technology surprised the biomedical research community, including some university and government officials. Could individual scientists "own" a basic research technology in molecular biology whose development had been funded by taxpayers? Politicians and citizens, already embroiled in public-health controversies during the February 1975 Asilomar II conference on the safety of recombinant DNA research, now became concerned about the prospect of private control of the technology for profit.[3]

In June 1976, in the midst of the controversy over recombinant DNA patenting, Robert Rosenzweig, vice president for public affairs at Stanford University, wrote to Donald Fredrickson, director of the National Institutes of Health (NIH). The letter solicited the agency's position on the "discussions taking place . . . over the wisdom of proceeding with an application for patent protection for discoveries in the area of recombinant

DNA."[4] More specifically, Rosenzweig was concerned whether the NIH, as the primary sponsor of recombinant DNA technology research, would waive its patent title in favor of Stanford and UC. It should be noted that in 1968 the NIH had reinstituted its long-abandoned institutional patent agreement (IPA), by which the agency could waive title to patents arising from support to its grantees should they request it.[5] Such agreement, however, was conditional, and the NIH could revoke the waiver and claim title on several grounds. As scientists and politicians sharply questioned Stanford and UC's claim to private ownership of recombinant DNA technology, national debates about the legal ownership of inventions arising from public support intensified. Indeed, in August 1977, the NIH decided to suspend the transfer of legal ownership of inventions to private investigators and institutions.[6]

This chapter examines how academic administrators and scientists, government officials, and industrialists contested the legal ownership of recombinant DNA technology in the name of the public interest. Despite the central place of recombinant DNA technology and its patenting in the historiography of biotechnology and the commercialization of biomedical research, issues regarding the ownership of recombinant DNA technology and the public interest were hotly contested and has not been adequately acknowledged. In fact, the first Boyer-Cohen patent was only granted in 1980, after six years of heated debate.[7] This historical oversight occurs because the basic tenets of the Bayh-Dole Act, which allowed research universities to claim ownership of inventions arising from federal support, are often implicitly assumed when scholars analyze the commercialization of recombinant DNA technology.[8] The Bayh-Dole Act, however, was enacted only in 1980, and public policy controversies over federal patent policies became increasingly entangled with recombinant DNA patent applications during the 1970s.[9] I contend that an examination of the contentious path to the private ownership of recombinant DNA technology is critical to our understanding of the commercialization of academic research. Indeed, it was in this period that the commercialization of biomedical inventions arising from government support was first proposed, debated, and eventually recognized as one of the key missions of research universities.

At another level, this chapter brings to light the economic and legal transformations surrounding the emergence of biotechnology as a commercial form of biomedical enterprise.[10] As frequently described in standard histories of recombinant DNA technology, this development has been lauded as one of the defining achievements in the history of genetic engineering.[11] The subsequent patenting of Cohen and Boyer's experimental procedures has been regarded as a logical, if not inevitable, step toward the commercialization of biotechnology for the molecular manipulation of life.[12] Boyer's founding of a start-up biotech company, Genentech, with venture capitalist Robert Swanson in 1976 has in turn been taken as an emblematic shift in the scientific life of molecular biologists, whose entrepreneurial ventures into the world of commerce accelerated when Boyer's initial investment of $500 grew to be worth about $37 million with Genentech's initial public offering at the end of 1980.[13] Rather than focusing on the inherent technological power of recombinant DNA technology, I seek to situate its commercialization in the context of the reconfiguration of the public knowledge of the academy and the private knowledge of industry in the 1970s, illustrating how the dazzling emergence of the biotechnology industry depended on the economic and legal reconception of public knowledge.

My analysis of the privatization of recombinant DNA technology builds on historical works that have analyzed the critical role of patenting in the commercialization of biomedical research and the introduction of economic development as a new social mission for research universities from the 1970s.[14] I first examine a broader debate on the shifting role of research universities and the federal government in the nation's economic life amidst the declining economic productivity of American capitalism in the 1970s.[15] The discussion about the relationship between scientific research and economic development in many ways set the stage for the emergence of patenting as a subject of policy debates among scientists, university and government officials, and politicians. Against this background, I examine Stanford's institutionalization of patent management as a site for commercialization of academic research from the late 1960s. In particular, I analyze how Niels Reimers, the founder of Stanford's Office of Technology Licensing (OTL), attempted to claim the proprietary right over recombinant DNA technology on behalf of Cohen and Boyer. Reimers's attempt reflected recent shifts in federal patent policies—especially those of the primary sponsors of recombinant DNA research, the NIH and the National Science Foundation (NSF)—that allowed an agency to waive patent title (ownership) to its grantees or contractors should they request it.[16]

The legal and political debate over the ownership of recombinant DNA technology shows how a small but influential group of government officials and university research administrators—the so-called managers of research—introduced a new legal framework for the commercialization of academic research by linking private ownership and the public interest.[17] Mobilizing a recent ascendancy of economic rationales in (intellectual property) law, they asserted that the "public" ownership of recombinant DNA technology was fundamentally at odds with the promotion of the public interest.[18] On the contrary, they proclaimed that placing inventions in the public domain could only hinder the proper development of research results into viable commercial products. With the redefinition of the public interest in terms of economic development amid the American economic recession of the 1970s, the articulation of a causal link between private ownership and the public interest became one of the fundamental moral assumptions in the commercialization of biomedical research, bringing about a significant shift in the scientific and moral life of academic researchers.[19] The resulting reformulation of federal patent policies, which eventually led to the passage of the Bayh-Dole Act, provided a powerful legal platform for the emergence of commercial biotechnology.[20] A new legal regime for biomedicine was not so much a reactive response to new genetic engineering technologies as a shaping force for biotechnology.[21] This chapter in sum provides a broad historical background in the proliferation of patenting in commercial biotechnology from the 1980s onward. It ends with a brief discussion on "too much private ownership" in biomedical research at Stanford and its implications for commercial biotechnology.

Federal Grant University and the Uses of Knowledge in the 1960s

By the mid-1960s, the rapid expansion of research universities had brought them to greater prominence in the nation's economic and political life. Beginning during World War II, American research universities had come to rely heavily on federal support of scientific research, a relationship based on the public's faith that scientific and technological development would yield broader cultural, social and economic benefits.[22] The large influx of federal money was increasingly connected to the cultural and political mobilization of science, either in the context of the Cold War or the war against dread diseases.[23] Observing the rising influence of the federal government and politics over the affairs of the research university, UC president Clark Kerr once characterized it as the "federal grant university," whose existence and growth increasingly depended on political negotiations over why and how best to support academic research.[24]

For their part, academic scientists and university administrators opportunistically cultivated a loosely held belief among politicians and the public that scientific research promised benefits—that scientific discoveries would provide the foundation for technological innovations and economic growth. In his 1963 *The Uses of the University*, Kerr celebrated the emergence of the "multiversity," through which diverse arrays of social, economic, and political demands had been met. More significantly, Kerr—a former labor economist—underscored that American research universities had evolved into institutions of new economic significance whose contributions had become central to the advance of the "knowledge industry."[25] According to Kerr, research universities could claim that they played a significant role in the transformation of American economy because the production and management of knowledge had become the key to economic prosperity.

Starting in the 1960s, however, this emphasis on the tangible benefits of scientific research for the public at large as a justification of federal support for academic research prompted the emergence of a related set of policy questions about the actual social and economic relevance of academic research.[26] These questions about the public and economic relevance of research coincided with the surfacing of economic and science policy studies that challenged a broadly construed postwar consensus about the causal link between progress in academic research and the growth of industrial innovation—that is, the linear model of scientific research, technological development, and economic growth.[27] For example, the Department of Defense (DOD) Project Hindsight, launched in 1965, concluded that defense support for "basic and undirected science" had not significantly contributed to the development of major weapons systems.[28]

Critiques of governmental patronage of science in the late 1960s began to challenge one of the central justifications for federal support of scientific research. After World War II, a group of scientists and economists had claimed that the market was a poor form of economic organization for maintaining technological innovation in the face of the economic uncertainty and financial risk scientific research entailed.[29] In his widely cited 1958 RAND report, for example, the Nobel laureate economist Kenneth J. Arrow argued that "for optimal allocation [of resources] to invention it would be necessary for the government or some other agency not governed by profit-and-loss criteria to finance research and invention."[30] In other words, market-driven, shortsighted applied research could not make the best use of unanticipated scientific and technological developments. Though the post–World War II system of federal funding fell short of scientists' ambitious demands for public support of basic science, the influx of government funding as an alternative mode of research support was believed to help build a broader and more productive reservoir of scientific and technological resources.[31]

By the late 1960s, however, questions about the economic and social contributions of scientific research increasingly held research universities accountable for the federal support they received for scientific research. In 1969, Stanford president Kenneth S. Pitzer wrote to Robert H. Finch, secretary of the Department of Health, Education, and Welfare (DHEW), articulating the shifting political and economic context for research universities. "The winds of change are evident," he observed. "With other urgent national demands, the growth in federal support characteristic of the past has stopped and in many programs has been reversed. Finally, student unrest has eroded the sense of enthusiasm and support for higher education both in Washington and in the general public."[32] Rebutting critics, Pitzer sought to underscore that "the basic body of advancing knowledge from university research has become the mainspring of our economic growth and of our social development."[33] In an economy that was becoming increasingly dependent on advanced science and technology, he further warned, "[generating] new basic knowledge, for which universities are uniquely equipped, is no longer a luxury."[34]

In a related development, the nation's primary sponsor of biomedical research, the NIH (the research agency of DHEW), was facing increasing political pressure to develop more therapeutic drugs and medical innovations. As the NIH's annual budget grew more than one billion dollars in the mid-1960s, providing about 40 percent of direct support for the nation's health research, politicians wondered "whether the American people are getting their money's worth from the expenditure."[35] President Lyndon Johnson, visiting the NIH in 1966, asked whether "too much energy was being spent on basic research and not enough on translating laboratory findings into tangible benefits for the American people."[36] Patrons of medicine like Mary Lasker and the American Cancer Society also began to call for more goal-oriented and applicable biomedical research, asking the NIH to undertake a national program to develop novel research-based therapies or cancer vaccines.[37] Along with the Project Hindsight report, the series of political attacks on the utility of biomedical research "fell like a bombshell on NIH officials."[38]

In the context of the prospect of a national economic recession and the looming decline in federal support of scientific research in the late 1960s, university administrators and a group of governmental officials recognized that federal support of scientific research was coming under increasing political scrutiny.[39] Policy studies questioning the economic impact of basic research made it increasingly difficult to mobilize the linear model to win public and government support. Though it was hardly surprising that institutions receiving public support would be held accountable for their effectiveness, scientists and university administrators were rather surprised to find that their assertions as to the direct and indirect contributions of their research to the nation's economic prosperity was met with skepticism. Members of the academic community and their patrons, such as government officials, tried to devise pragmatic solutions that would promote the relevance of research in the nation's political and economic life and thus secure its ongoing support.

While some scientists cautioned that the demand for relevance in biomedical research should not dictate research agendas, other scientists and government officials tried to capitalize on this political pressure, claiming that "it is now time to put the same kind and magnitude of effort into the fight against disease as we are committing to the space and atomic energy programs."[40] But academic and government biomedical researchers, whose research had been mainly funded by the NIH, did not have the legal and technological means to develop commercially and medically viable therapeutic drugs. As early as 1962, Kenneth Endicott, director of the National Cancer Institute, one of the largest institutes within the NIH, requested a thorough review of the current DHEW patent policy. He acknowledged that it would be difficult for basic biomedical research organizations to undertake a full-scale drug development because of their lack of capital and technical know-how, and he suggested that the NCI needed access to "industrial know-how to supplement NCI's high level scientific skills and to apply the body of basic research findings we have already built up by the use of the special strengths that industry has, in engineering, mass production, and the management of large-scale plants and polls of industrial manpower."[41]

Private industry, however, was hesitant to collaborate with academic and government researchers whose major work had been supported by public funds. NIH director James Shannon, who had led clinical and pharmacological programs as part of the US antimalarial drug project during World War II, was well aware of the chemical and pharmaceutical industries' concerns about protecting their proprietary knowledge and practices when they participated in federally funded research and development programs.[42] In 1963, Shannon hired Norman Latker, a young attorney from the Air Force, as the NIH's first patent counsel and instructed him to conduct a review of NIH patent policy. In a letter to the DHEW summarizing Latker's findings, Shannon noted that "the drug industry has refused through the Pharmaceutical Manufacturer's Association and in some instances individually to collaborate with our scientists in bringing their drugs to the point of practical application without some guarantee of exclusive patent rights."[43] Shannon suggested that the DHEW needed to revise its patent policy so that the agency and its academic grantees could have the legal means to use their basic findings for developing useful medical and pharmaceutical products.

Within this context of a national debate over the contributions of scientific research to American economic prosperity and public health, federal patent policy emerged as a critical issue for biomedical researchers and government officials. In the late 1960s, two government studies on patent policy, *Problem Areas Affecting Usefulness of Results of Government-Sponsored Research in Medical Chemistry* and *Government Patent Policy Study*, claimed that federal agencies had failed to effectively gain tangible economic returns and health benefits from government-funded research.[44] Both studies depicted an outdated federal patent policy as a critical barrier that hindered the use of a vast pool of inventions arising from publicly funded research. First, these reports pointed out, each government agency dealt with inventions arising from government-sponsored research on an ad hoc basis, leading to a wide disparity in patent practices and regulations.[45] For instance, the DOD had a license policy and usually waived patent titles on DOD-sponsored inventions to industrial contractors and grantees. The DHEW, on the other hand, maintained a title policy, which dictated that the agency should own the patent rights on inventions resulting from its contracts or grants. In the latter case, the title policy was originally intended to build a reservoir of publicly available scientific knowledge.[46]

Deploring such conflicting policies, both reports recommended that the federal government implement a uniform procedure in order to make use of government-owned inventions for the public benefit. NIH's patent counsel Latker, the eventual "father of the Bayh-Dole Act," emerged as a central figure in the crusade for the reformulation of government patent policy.[47] Latker had contributed to the preparation of the government patent policy reports, in which he indicated that the DHEW's requirement to retain patent title posed a major impediment to the commercial development of the NIH's vast pool of inventions. The Federal Council for Science and Technology report singled out the NIH's medical chemistry program as an exemplary failure of DHEW patent policy. This basic research program supported academic researchers who were interested in chemical byproducts with therapeutic potentials. The report criticized the NIH program for failing to attract the participation of the pharmaceutical industry in the commercial development of biochemicals into drugs because it did not waive its patent title in favor of its contractors or grantees. When called before a congressional committee on government patent policy, Latker recalled this instance: "I think in 1963 through 1968 the pharmaceutical industry ran a virtual boycott of all the inventions that were under ownership of the Government. At that time we had no capability of licensing industry on an exclusive basis, and virtually, I think, our entire patent portfolio was dormant."[48]

Latker promoted a new incentive structure for use of inventions arising from government-supported research. He claimed that the DHEW's title policy created inadequate incentives for NIH grantees and contractors, mostly academic universities and hospitals, to develop government-held inventions that might promote the public welfare. From the late 1960s, Latker tried to streamline the DHEW's patent policy in a way that could promote further development work by instituting a new set of IPAs with several universities. Each IPA was intended to encourage the use of DHEW-funded inventions by simplifying the process of requesting that the DHEW waive its patent right in favor of grantees or contractors. In 1968 a dozen universities and medical institutions, including Caltech, Cornell, MIT, Minnesota, Princeton, and Mount Sinai Hospital, arranged the first round of IPAs with the NIH.[49] Latker's efforts to reformulate government patent policy and to introduce market incentives for the development of economically and medically viable products signaled a new relationship between academy, government, and industry in the 1970s.

Patenting at Stanford University

Until the mid-1960s, American research universities had relegated the management of university inventions either to outside organizations, like the nonprofit Research Corporation (RC), or to university-affiliated but independent groups, such as the Wisconsin Alumni Research Foundation (WARF).[50] The RC, for example, became one of the most prominent management organizations for academic patents, as it began in the 1930s to develop an invention administration agreement with a few universities like MIT. The institutional separation between the university and patent management came into being in part as a way to preserve the university's core academic mission.[51] There were ethical concerns regarding conflicts of interest and worries that a commitment to commercial pursuits might interfere with faculty members' educational obligations. Additionally, universities themselves did not have expertise in the management of intellectual property. Thus, they were content to assign their faculties' inventions to these outside organizations in exchange for modest royalties. These patent management organizations in turn devoted their efforts to finding potential industrial companies that might be interested in licensing these patents so as to develop commercially viable products.[52]

With the large influx of governmental support for science and engineering during and after World War II, a set of challenging issues emerged pertaining to the ownership of scientific knowledge and engineering technologies gained through research funded by government grants and contracts. Most universities, prompted by fears of litigation, did not actively pursue the patenting of inventions arising from wartime research. MIT president Karl Compton, for example, warned the RC that it might be difficult to establish inventorship with regard to findings arising from complex and large-scale engineering projects scattered throughout the nation. And as we have already seen, because each government agency had developed its own provisional guidelines in matters of patent ownership, few academic and patent management institutions made systematic attempts to deal with inventions arising from government-sponsored research.[53]

When federal support in science began to be challenged in the mid- and late 1960s, however, academic patenting came to be viewed as a way for research universities to respond to the challenges they faced. University administrators began to show an interest in patenting both as an alternative way to draw financial resources and to demonstrate the economic relevance of academic research. They also realized that their familiarity with their own faculty members and research results enabled them to identify viable inventions more effectively than outside organizations or the government.[54] Therefore, by the late 1960s, some universities tried to acquire technical and legal expertise for managing university inventions from the RC in order to establish their own patent management offices, usually called university technology transfer or licensing offices.[55]

Stanford's establishment of the OTL reflected the early trend of research universities undertaking their own patenting and licensing activities.[56] Reimers, a contract officer in the Stanford Office of Sponsored Projects, initially launched the OTL as a pilot program in 1968. Before being hired at Stanford, Reimers had worked at the Philco-Ford (a subsidiary of Ford Motor Company) as a contract administrator. His contract management work with the DOD, in particular, led him to appreciate the potential economic value of licensing from inventions arising from government funding.[57] Working as a contract manager at Stanford in the mid-1960s, he noticed that faculty members disclosed about thirty inventions per year to Stanford administrators. Unlike private contractors, however, Stanford did not seek patent title and was not involved with licensing activities. Following academic conventions of other research universities, Stanford had relegated these patenting and licensing activities to the RC beginning in 1954.

Stanford's decision to relegate its patent management to the RC, as the historian Rebecca Lowen has shown, evolved with its shifting institutional relationship with private industry.[58] In 1937 Stanford had drawn a patent policy that claimed for the university ownership of inventions resulting from research. This patent policy was established in large part to capitalize on the invention of the klystron in the Department of Physics during the Depression. Stanford secured an exclusive licensing arrangement with the Sperry Gyroscope Company, which by the early 1940s had brought royalty income totaling about thirty thousand dollars (equal to the salaries of five full professors at that time). But this exclusive licensing contract, unusual in an academic context, prompted the chairman of the Department of Physics and director of klystron research, David Locke Webster, to resign from his position in 1939. Frustrated with Sperry's frequent pushing for new inventions, Webster bitterly professed, "science and patents don't mix any more than oil and water."[59] Given this detrimental fallout of its industrial association, Stanford began to reassess its patent policy. Some faculty members argued that a patent should be assigned to an individual faculty member who conceived inventions, whereas others claimed that it should be assigned to the public at large or to the sponsoring corporation or government agency. Stanford's invention administration agreement with the RC in 1954 came into being in this context of controversy over academic patenting. While Stanford relegated patent management to the RC, the university's patent policy still allowed the inventor or the university to retain patent rights.[60]

Drawing on his industrial experience, Reimers noticed that Stanford's arrangement with the nonprofit RC left much to be desired, especially in terms of financial returns. From 1954 through 1967 Stanford received a rather small amount of royalty income from the RC, totaling about $4,500. Reimers claimed that he could make better use of Stanford inventions. His licensing program, Reimers explained, would contribute to the development of academic inventions to further the public good, and the income it generated could provide additional funds to faculty members and the university. With a reserve of $125,000 from the university, Reimers started a pilot program in 1968. In May 1969, Richard W. Lyman, then Stanford's provost, wrote to department chairs about the new program for patent licensing, explaining that any royalty income generated would be divided evenly among the inventor, the department, and the university.[61]

The one-year pilot program earned fifty-five thousand dollars in gross income, and its financial success helped Reimers officially establish Stanford's OTL in January 1970. William Miller, vice president for research at Stanford, circulated the university board of trustees' positive evaluation of Reimer's pilot program in June of that year. The board members generally concurred with Reimer's claim that technology licensing could be a means to commercialize academic research results. They were particularly attracted how patent royalties and licensing income—in contrast to contracts and research grants—might provide unrestricted funds for the university. Mindful of cautious academic attitudes toward patenting, Reimers further insisted that his program had been licensing inventions of high social value, such as a potential cure of viral infection and less ecologically damaging form of insect control. In June 1970, in affirming its support for the newly established OTL, Stanford's board of trustees extended the "privilege, previously enjoyed only by faculty, to all University inventors including staff and students to retain proprietary rights in inventions they may make."[62]

In the early years of its operation, the OTL brought inventions to industries whose competitive edge lay in securing a breakthrough patent, such as the chemical and pharmaceutical industries. The electronics industry, in contrast, was rather inattentive to Reimer's licensing inquiry, in part because the rapid progress of electronics quickly rendered a patent outmoded.[63] One of the earliest profitable Stanford inventions brought to the market by Reimers, for example, was a package of steroid synthesis patents developed by Stanford chemist William Johnson.[64] The OTL arranged the licensing of Johnson's patents to a joint venture for the development of insect pesticides by the Sterling Drug Company and the Chevron Corporation. When news of recombinant DNA technology and its applications in the pharmaceutical industry and agricultural business came to Reimers's attention in 1974, the OTL's expertise in biochemical inventions helped him navigate the shifting landscape of academic patenting.

Making Recombinant DNA Proprietary

By the time Reimers started the technology-licensing program at Stanford in 1968, a serious effort had already emerged to reformulate government patent policy in order to provide a means for universities to exploit government-supported research results. The NIH, for example, had reinstituted its IPA, which allowed the agency's grantees to request the waiver of patent title in favor of the inventor. Under this agreement, any university requesting such a waiver had an obligation to make a serious effort to use the invention for the benefit of taxpayers. Reimers' attempt to pursue the patenting and licensing of recombinant DNA technology reflected this new context of shifting government patent policy, especially that of the NIH. As of 1976, the NIH had a total of fifty-six IPAs, and universities had negotiated twenty-nine nonexclusive licenses and forty-three exclusive licenses with commercial organizations, engendering seventeen joint-funding arrangements with universities and commercial companies.[65] Seen from the perspective of shifts in government patent policy, Reimers's initiative to file a patent application for recombinant DNA procedures might be taken as an acceptable—indeed, welcome—new development for university administrators and government officials who set their sights on the use of government-supported inventions. Reimers's initial inquiry about the possibility of a title waiver was in fact greeted warmly by patent officials like Latker at the NIH. Consequently, some governmental officials became deeply involved with the determination of the legal ownership of recombinant DNA technology.[66]

The OTL's interest in recombinant DNA technology, as Reimers recollects, began when he read an article featuring intriguing Stanford research results.[67] The May 20, 1974, *New York Times* article by Victor McElheny, titled "Animal Gene Shifted to Bacteria: Aid Seen to Medicine and Farm," covered research by Morrow, Cohen and Boyer on the cloning of a gene from a species of African toad, *Xenopus laevis*. In a Stanford University Medical Center news release, Joshua Lederberg—who communicated the *Xenopus* DNA–cloning article to *Proceedings of the National Academy of Sciences, USA*—claimed that "it may completely change the pharmaceutical industry's approach to making biological elements such as insulin and antibiotics."[68] This news caught the eye of Reimers, who was actively soliciting information on new inventions with broader commercial applications from Stanford scientists.

Reimers approached Cohen at Stanford's Department of Medicine with his plan for the licensing of recombinant DNA technology. Cohen initially hesitated to file a patent application, suggesting that recombinant DNA technology, as a basic research procedure in molecular biology, seemed to be rather unsuitable for an invention. As Cohen later acknowledged, his "framework was that one patents devices, not basic scientific methodologies."[69] Moreover, Cohen was hesitant about patenting because he was well aware that his gene cloning work had relied heavily on close interaction with colleagues in the Stanford Biochemistry Department, such as Peter Lobban, Janet Mertz, and John Morrow.[70] Reimers, nonetheless, persuaded Cohen to file a patent application, suggesting that "no invention is made in a vacuum" and the patenting of biochemical processes could serve Stanford inventors and the chemical and pharmaceutical industry well.[71] Reimers also underlined recent shifts in government patent policy, especially Stanford's IPA with the NIH, which encouraged individual scientists to file a patent application on research results arising from government support.

Meanwhile, Reimers hired William Carpenter, a Stanford Business School student, to investigate what would be necessary for the commercial application and licensing of this invention. As Carpenter's brief report acknowledged, the major limiting factor in the application of recombinant DNA–cloning technology to the production of economically and medically viable molecules was the inability to select and isolate genes that could produce useful gene products via molecular cloning. On the one hand, Boyer assumed that the research for gene isolation and selection was inherently "applied" in its nature. Boyer thus concluded that government funding for the development of a gene selection technique would be "far from definite."[72] On the other hand, Cohen believed that there was in fact considerable interest in developing these application techniques. He suggested that scientists and government funding agencies had sufficient incentive to invest their efforts in the development of eukaryotic gene selection and isolation technology. If Cohen's view proved to be right, and if OTL's "only motivation is to see [that] this technology is brought forward to public use and benefit," Carpenter's report concluded, "then licensing a private company would not be necessary at this point."[73]

The conflicting opinions of Cohen and Boyer as to the need for patenting did not keep Reimers from asserting their proprietary claim on recombinant DNA cloning technology. The year's grace period afforded by US patent law was about to end in late 1974, as Cohen and Boyer's first recombinant DNA paper had been published in November 1973.[74] Cohen filed an invention disclosure on a standard Stanford OTL form on June 24, 1974. This invention disclosure, "A Process for the Construction of Biologically Functional Molecular Chimeras," traced the conception of recombinant DNA–cloning technology to a coincidental conversation with Boyer at a Hawaii conference on bacterial plasmids in November 1972, when they decided to collaborate on an experiment that would clone genes in bacteria transmitted through a hybrid plasmid.[75] Stanford University and UC, on the behalf of Cohen and Boyer, filed a patent application with twenty-six claims on November 4, 1974. Bertram Rowland, a patent lawyer hired by Stanford, prepared this patent application. Deciding to narrow the list of the inventors of recombinant DNA technology to Cohen and Boyer, Rowland sent a disclaimer letter to two of their collaborators in the recombinant DNA cloning work, the Stanford biochemist John Morrow and the University of Michigan molecular biologist Robert Helling (who had been visiting Boyer's laboratory when the recombinant DNA work was carried out and was a coauthor on the first and third papers). Rowland asked Morrow and Helling to declare that their roles in the collaboration did not merit granting them the status of coinventors.

Stanford's attempt to establish the private ownership of recombinant DNA technology within a research community that had shared practices and materials magnified tensions between the Stanford biochemists and Cohen and Boyer. Because the former deemed that their contributions to the development of recombinant DNA technology were not properly represented in Stanford's patent application, they began to raise a set of technical issues pertaining to Cohen and Boyer's inventorship claims.[76] On January 23, 1975, Morrow, Paul Berg's graduate student, wrote a letter to Cohen and Boyer to inquire about the scope of the patent application and its implications for the field of recombinant DNA research; in particular, he wondered whether the patent application could interfere with the academic use of recombinant DNA technology. Morrow was baffled by Cohen's assertion of the private ownership of recombinant DNA technology, wondering how it would be "possible for you to patent these methods and plasmids when the work was funded mainly by the NIH?"[77] Morrow and Helling refused to abandon their claims to coinventorship, casting doubt on whether their coauthorship and scientific contributions were duly acknowledged in the patent filing (for more, see chapter 4).

After receiving letters from Morrow and Helling, Cohen was worried that the Stanford patent application might be perceived as an attempt on his part to seek "personal gain from the patent."[78] He asked Rowland to write to Morrow and Helling, clarifying that the patent application was pursued at the initiative of the university, not that of Cohen himself. He further insisted that Stanford should make it clear that its commercializing effort was grounded on the assumption that "any financial benefits derived from this kind of scientific research carried out at a non-profit university with public funds to go to the university, rather than be treated as a windfall profit to be enjoyed by profit-motivated businesses."[79] Other scientists questioned the ethics of allowing individual institutions or scientists to benefit from research funded by taxpayer money and wondered how the hasty patent filing could fulfill "the public service ideals of the University."[80] To appease concerns about the potentially harmful impact of profit-seeking motivations on biomedical research, Cohen agreed to give up his future royalty income; instead, he planned to use any such earnings to establish a "Research Development Fund" to support research grants and fellowships (Cohen eventually decided to keep his royalty income personally, reneging on his early promise).[81]

On the patenting front, in the first Patent and Trademark Office (PTO) action on March 31, 1975, most recombinant DNA process claims were allowed, while other recombinant product claims were rejected. PTO patent examiner Alvin E. Tanenholtz indicated to the Stanford lawyer Rowland that he would grant the patent right for the procedures involved in making recombinant DNA molecules with viral and circular DNA and their cloning in bacterial cells. But he questioned the broad scope of the product claim, in which the patent applicants claimed proprietary rights to biologically active materials composed using recombinant DNA technology, such as recombinant plasmids, new hybrid functional genes, and modified cells containing recombinant plasmids: Cohen and Boyer's cloning work was limited to the propagation of hybrid molecules in a bacterial cell. The examiner requested that its product claim be limited to a bacterial cell, excluding cells of higher organisms. More significantly, he rejected Stanford's proprietary claim on transformed bacterial cells on the ground that they could exist in nature, and so—as a product of nature—they could not be considered a subject of invention. This decision was consistent with the PTO's rejection of Ananda Chakrabarty's modified bacteria according to the "product of nature" doctrine.[82] Tanenholtz notified Stanford that the process claim in bacteria seemed to be granted but that its application to eukaryotes, along with proprietary rights on modified (recombinant-DNA engineered) organisms, seemed rather unlikely to be granted.[83] This partial patent success prompted Boyer and Swanson each to invest $500 of his own money to found a start-up company, Genentech, in April 1976. At the beginning, Genentech existed only on paper: it arranged contractual work with Boyer's laboratory at the University of California, San Francisco.[84]

After filing the first recombinant DNA patent application, Reimers prepared his licensing plan. First drafted on May 14, 1976, it highlighted the potential applications of recombinant DNA technology in various pharmaceutical and agricultural areas. Echoing the enthusiasm in the 1960s about the industrial uses of life, Reimers emphasized that microorganisms engineered via recombinant DNA technology, such as nitrogen-fixation microbes, could mass-produce industrial enzymes and various fertilizers.[85] Given the advantage of using biological over synthetic methods in the production of antibiotics and hormones for medical research and treatment, it seemed likely that the application of recombinant DNA technology would attract large multinational pharmaceutical companies, such as Pfizer, Merck, Upjohn, and Lilly. The guarantee as to the private ownership of recombinant DNA technology would provide a strong incentive for private companies to invest in its medical and industrial development.

Reimers's optimistic projections regarding the imminent industrial applications of recombinant DNA technology faced a major hurdle in 1976, as more detailed and technical questions about the scope of and inventorship claims put forward in Stanford's recombinant DNA technology patent application emerged. At a May 1976 meeting at Stanford, Paul Berg, one of the seminal scientists with a hand in the research, raised concerns about the broad scope of Stanford's patent claims on recombinant DNA technology, criticizing the patent application for "claim[ing] everything."[86] Berg then asked whether scientific credit and inventorship claim was duly acknowledged in the patent application. Cohen defended himself by noting that the narrow definition of inventorship reflected how the patent system worked. Lederberg, unconvinced, expressed doubt as to the "validity of the patent application due to the issue of inventorship." He continued to suggest that "Cohen may not be the only Stanford inventor."[87] The complicated issue of inventorship, especially the crucial exchanges between Stanford biochemists and Cohen and Boyer, prompted PTO examiner Tanenholtz to reject all of Cohen and Boyer's forty-five inventorship claims made in their revised application with regard to recombinant DNA technology (for more, see chapter 4). In the end, the conceptual and technological traffic between Berg's group and Cohen and Boyer's genetic-engineering experiment complicated and prolonged the issuance of the first recombinant DNA patent.[88]

Private Ownership and the Public Interest

Beyond technical issues pertaining to the inventorship of recombinant DNA technology at Stanford, broader questions about its ownership emerged at the national level. In the spring of 1975, Stanford scientists Paul Berg and Charles Yanofsky personally expressed their concerns about Stanford's patenting activities to NIH and NSF officials. At one level, Berg and Yanofsky worried that the university's efforts might seriously undermine Stanford scientists' call for a voluntary moratorium on recombinant DNA research. As the controversy continued, one commentator criticized Stanford's patent attempt, writing bitterly that "a research institution is taking steps to patent strains and procedures derived from work done during the moratorium by people calling for it."[89] Berg and Yanofsky had initially met with Reimers to request that the patent application be withdrawn, given that Stanford scientists already signed the recombinant DNA research moratorium letter in 1974.[90] At another level, Berg and Yanofsky were wondering how an individual scientist or institution could "own" recombinant DNA technology. Berg discussed the ownership issue with NIH director Fredrickson, asking whether it would be possible to privately "own" recombinant DNA technology funded by public funds.[91]

At a meeting in May 1976, Stanford Medical School dean Clayton Rich, Reimers, Cohen, and other Stanford scientists gathered again to discuss Stanford's attempts to patent recombinant DNA technology. Rich first acknowledged that patenting biomedical discoveries should be approached cautiously. He pointed out that, since innovative medical procedures had been generally put in the public domain in accordance with ethical principles held by physicians, the profit motive should not trump obligations toward patients.[92] Some scientists first warned that recombinant DNA technology was such a basic process that it should be left in the "public and scientific community domain and not be patented."[93] Stanford scientists who had worked on recombinant DNA in their own research, such as Berg, Ronald Davis, David Hogness, and Yanofsky, shared this cautious view on academic patenting. To avoid conflicts of interest, Berg, as an organizer of the Asilomar conferences on the regulation of recombinant DNA risk, suggested that, if the patent was granted, Stanford should turn for licensing to a nonprofit organization like the RC rather than to the OTL.

The responses by Stanford scientists to the proposed patenting of recombinant DNA technology signaled not only their unfamiliarity with but also their concerns about the NIH's new patent policy. Cohen and Boyer, who were identified as the inventors of recombinant DNA technology in the patent application, had been funded by the NIH and the NSF for the work that resulted in the three crucial recombinant DNA–cloning papers listed in that application. As Reimers noted, given the recent reinstitution of IPAs at both agencies, "at NIH and NSF, individuals with whom Drs. Berg and Yanofsky talked did not see anything amiss in Stanford seeking such development."[94] Reimers stressed that the OTL's patenting attempt reflected broader changes in attitudes among government officials toward patenting and licensing. Rosenzweig underlined Stanford's willingness to seek patent rights as an owner of recombinant DNA technology, noting that it was the NIH's new patent policy to waive patent title to grantees for further development of new inventions arising from government support.[95]

By early 1976, the biomedical community had begun to raise questions about Stanford's assertion of the private ownership of recombinant DNA technology, as its patenting application became widely known during the Asilomar II conference held in 1975. At the Miles Symposium held at MIT in June 1976, academic scientists sharply questioned the validity of patenting recombinant DNA technology. Cohen was publicly criticized for filing the recombinant DNA patent application for financial gain. Some scientists even questioned whether his "private" ownership of recombinant DNA technology would hinder its application in a wide array of basic biomedical research; others accused him of attempting to control DNA-cloning research through the patent filing.[96] Faced with such hostile public criticisms from his fellow researchers, Cohen conceded to Reimers that "while I am not proposing right now that we call the whole [patenting] thing off, I would have no objection to such a decision by the University."[97]

As Stanford's attempted patenting became a subject of national discussion, Rosenzweig and Reimers realized that they needed to address matters of the ownership of recombinant DNA technology with patent officials and high-level government administrators at the NIH. On May 24, 1976, Reimers had a meeting with Latker and David Eden, a legislative assistant of Betsy Ancker-Johnson at the Department of Commerce.[98] Latker had provided a rationale and a defense for the NIH's new patent policy that would encourage the commercial development of inventions arising from government-funded research. At their meeting, Reimers gave Latker and Eden a consultation copy of a draft of Stanford's May 14 licensing plan for recombinant DNA procedures.

On another front, Rosenzweig wrote to NIH director Fredrickson on June 14, 1976, to solicit the NIH's formal position on Stanford's patent application. Though Rosenzweig's letter has been taken as indicative of Stanford's cautious and prudent approach to academic patenting, it also indicates the real fear that the NIH might revoke its title waiver to Stanford and UC.[99] After his conversation with an NIH official regarding Stanford's recombinant DNA patent application, Rosenzweig circulated a memo to other Stanford administrators and scientists, indicating that "NIH had received 'several' letters inquiring about reports of patent activity in the Recombinant DNA area by Stanford."[100] Rosenzweig emphasized he had "made three points throughout [his] conversation: 1. We [at Stanford] did not want the issue to be handled in such a way as to enable people to believe that we were acting in a secretive, sly, or underhanded way. Subject only to the protection of proprietary information about prospective licensees, we were quite willing to discuss our plans."[101] The letter and telephone exchange indicated that the private ownership of inventions arising from government-supported research remained a hotly contested issue with broad significance for research universities, government officials, private industrialists, and the public. As the NIH's IPA stipulated, the DHEW would repeal its patent title waiver if the grantee would not use the invention in a manner that would promote the public interest.[102] In the end, then, Stanford needed to figure out and negotiate an understanding of "public interest" acceptable to the community of scientists, governmental officials, and industrialists.

By the time Rosenzweig's letter arrived at the NIH in June 1976, the underlying assertion of the NIH IPA that there was a causal link between private ownership and biomedical innovations had become controversial outside the scientific community. As early as 1974, the public interest advocate group Public Citizen, led by Ralph Nader, alleged that IPAs were an unconstitutional transfer of public goods to private parties. They charged that, through IPAs, the federal agency did in fact "grant greater rights than a non-exclusive license ('exclusive rights') to patents and inventions developed under federally-financed research and development contracts, including the authority to grant such exclusive rights at the time of entering into such contracts, without Congressional approval."[103] Public Citizen and a few congressmen brought a civil action suit against Arthur F. Sampson, the administrator of the General Services Administration, which was in charge of the federal executive agency's rules and regulations. As Public Citizen argued in *Public Citizen v. U.S.*:

> all of the plaintiffs and contributors to Public Citizen are harmed as taxpayers and consumers because the patents and inventions have been developed at the taxpayers' expense either by Federal Agencies or with Federal funds and the regulations provide for the issuance of exclusive licenses.... A recipient of an exclusive license will acquire a monopoly with a concomitant effect on prices causing plaintiffs and supporters of Public Citizen as consumers to pay again for an invention which they have already paid for as a taxpayer.[104]

The onset of political controversy over the DHEW's new patent policy prompted some governmental officials and university research administrators, such as Latker at the DHEW, Ancker-Johnson at the Department of Commerce, Howard Bremer at WARF, and Reimers, to discuss how best to respond to recent criticisms of the IPAs. In 1974, with Latker's "intense interest and encouragement," they convened the first national conference for university patent and licensing administrators.[105] Throughout this conference, which led to the establishment of the Society of University Patent Administrators, one significant thread emerged: the importance of private ownership for the subsequent commercial development of biomedical inventions.[106] Mobilizing the ascendancy of market rationales in public policy debates in the 1970s, the attendees insisted that private ownership would serve the public best.[107] In response to Public

Citizen's lawsuit, Reimers claimed that, "notwithstanding Ralph Nader and the Antitrust Division of the Justice Department, you will find that your invention or patent is not a monopoly, with rapacious companies eager to sign licenses with you in order to exploit the public."[108] He argued that if Public Citizen and the "Justice Antitrusters" prevailed, "the US public would be the loser," as it would mean "the loss of research advancements into commerce which could be providing jobs, increasing productivity, enabling better health care, helping our international trade position by competing with new technology to offset our high labor costs, and so on."[109]

Though the court ultimately dismissed the case owing to Public Citizen's lack of standing to sue, recombinant DNA patenting resurrected debates over the NIH IPAs, since high-ranking politicians had begun to ask whether the new government patent policy inappropriately transferred publicly held inventions to private hands in the name of the public interest. Senator Gaylord Nelson of Wisconsin, for example, indicated that he was "preparing to hold hearings on inventions derived from government funds for research that were 'enriching' some universities."[110] Amidst the controversy, Fredrickson decided to solicit a range of opinions on the Stanford and UC recombinant DNA patent application. With Latker's help, he drafted a letter to scientists and nonprofit and commercial organizations in September 1976.[111] Fredrickson recollects that Latker was "zealous to counter the prejudice of many members of the scientific community."[112] Latker claimed that the interest of the public might be best served by Stanford's patenting and licensing efforts and that such licensing would provide a means for the regulation of recombinant DNA research. Fredrickson's letter explained that the NIH IPA program was first introduced in 1953 but that it had rarely been used because of a lack of interest, until Latker revamped it in 1968. He also underlined that the aim of the IPA was not to give away inventions in the public domain. Instead, the purpose was to give universities and researchers modest incentives for transferring underused inventions to the market.

Though some scientists whose views Fredrickson solicited were sympathetic to the notion that NIH IPA could be one means for wide dissemination of the benefits of biomedical research, they questioned whether recombinant DNA technology was an appropriate case for the privatization of basic research development. In his response, the molecular biologist David Baltimore disparaged the recombinant DNA patenting application, arguing that there was "hardly need for more stimulation" in this research area.[113] One scientist even indicated that his grant application could "provide legal evidence that the idea of genetic manipulation was in the common domain for several years prior to the actual experimental success."[114] Berg conceded that, "in spite of my reticence and opposition [to recombinant DNA patenting]," he would be "supportive of the objectives of the policy establishing the IPA agreements, sympathetic to the University's need for financial help."[115] He then stressed that the government should require publicly funded discoveries to be published freely as a way to uphold the norms of open exchange in the scientific community.

Even some start-up biotech companies echoed scientists' reservations about the patenting of such a broad and basic technology as recombinant DNA technology. Ronald Cape, president of Cetus, pointed out in his letter to Fredrickson that "it is inappropriate to attempt to patent something as fundamental as a way of making recombinant DNA molecules."[116] Cape also claimed that private companies were eager to use recombinant DNA techniques. As he noted, "In the past exclusive licenses may have been seen as the only way to motivate industry to make the necessary investment to develop an invention to the point where there would be something to exploit commercially. This is clearly not the case here."[117]

In contrast to academic scientists and small biotech companies, big pharmaceutical companies supported the NIH's effort to provide incentives for using "public" inventions. Jerome Birnbaum, Merck's director in basic biological sciences, agreed that "inventions made at public institutions under Government-sponsored research constituted a valuable national resource," and he praised the DHEW's "enlightened patent policy" for encouraging the fullest use of such inventions for public benefit.[118] C. Joseph Stetler, president of the Pharmaceutical Manufacturers Association, also concurred that an IPA would be far superior to public ownership as a means of disseminating and using inventions arising from government-supported research. He warned that any measures to put recombinant DNA technology in the public domain—thereby making it unpatentable—would be "unacceptable in that patent incentives would not be utilized."[119]

Thanks to the NIH's deliberations over the ownership of recombinant DNA technology, Stanford OTL's attempt to commercialize that technology, especially about its mode of licensing to a commercial entity, was under constant negotiation with governmental officials and private companies. Those who advocated licensing as a crucial economic incentive for commercial development believed that exclusive licensing—licensing to

only one entity and thus maintaining a monopoly in using the patent—
might be the most appropriate form of technology transfer. In a 1976
congressional hearing on government patent policy, Latker emphasized
the need for the private ownership of basic biomedical inventions arising from government funding in order to encourage commercial development. He argued that, "because of the basic nature of the research supported [in case of research grants], any invention that evolved could not
be the specific object of the grant and would always require further development which we would not support."[120] In the case of university patent licensing, Latker advocated that the government should grant universities a patent title waiver with no right to revoke it later. By guaranteeing
private ownership, he emphasized, the government could give universities the "kind of flexibility that they need to have at the negotiating table
with industry to arrive at appropriate licensing arrangements."[121] Without such full ownership, Latker continued, industry would question any
licensing arrangements with universities; private companies would want
to deal with the "real" owner of the invention, the government. Latker
asserted that if there is "anything that you need to get a commitment of
capital, it is certainty."[122]

Reimers had to find a way to facilitate the transfer of new technologies
from the laboratory to the commercial world while overcoming opposition to granting exclusive licenses. In his pursuit of licensing plans with
private companies, Reimers faced the unresolved issue of exclusive licensing. On June 2, 1976, he had a meeting with Robert Swanson, Genentech's
CEO, to discuss a licensing deal for recombinant DNA technology. Swanson indicated that Genentech wanted to secure an exclusive license for recombinant DNA technology so that it could attract investment from venture capitalists. He argued that, without an exclusive license that would
guarantee the monopoly of the technology, private investment could not
be obtained. In return, he would offer an equity stake to Stanford and
UC.[123] Representatives from another pharmaceutical company, Upjohn,
bluntly told Reimers that they would not invest in the development of recombinant DNA–related products without an exclusive license. After a
month's deliberation, Reimers notified Genentech that exclusivity could
not be guaranteed, since the recombinant DNA patent issue had become
a subject of intense public policy discussion.[124]

Indeed, recombinant DNA patenting caused so much controversy
that it threatened to torpedo the revamping of federal patent policy as
a whole. In a phone conversation with Reimers on July 29, 1976, Latker

complained that he was receiving "all the red herrings from the sea of scientists and administrators who are unacquainted with the patent system and its role in commercial development."[125] Latker even worried that the enormous public attention that the recombinant DNA patent applications were drawing would stall the Department of Commerce's attempt to institute a uniform government patent policy. (At that time, Latker was the chairman of the Interagency University Patent Policy Subcommittee, which was responsible for the Federal Procurement Regulations on University Patent Policy.) He told Reimers that "there are fears the proposed government uniform patent policy is in jeopardy."[126]

Amid this heightened public contention over the ownership of recombinant DNA technology, Betsy Ancker-Johnson, as assistant secretary for science and technology at the Department of Commerce, in January 1977 instructed patent office employees to accelerate processing for patent applications involving recombinant DNA research. This news, according to Fredrickson's recollection, appeared "without warning" in the *Federal Register*. The Department of Commerce's notice to the PTO underlined "the exceptional importance of recombinant DNA and the desirability of prompt disclosure of developments in the field."[127] Ancker-Johnson's special instructions would facilitate the process of the stalled recombinant DNA–related patent applications.

The Department of Commerce's intervention with the PTO on behalf of recombinant DNA patent applications arose out of a distinctive view of the relationship between basic scientific discoveries and economically viable inventions. Ancker-Johnson, a patent holder herself, was convinced that private ownership was critical for the further development of inventions arising from government support. As she underlined, the Department of Commerce had an obligation to restructure the incentive system for the use of government-funded research results. Its agency, the PTO, could facilitate the process of recombinant DNA–related patent applications in order to demonstrate how such inventions could significantly contribute to economic growth and medical progress, spurring new commercial products and ventures.

At another level, Ancker-Johnson and other government patent officials like Latker mobilized the theory of the "tragedy of the commons" that was emerging in the 1970s.[128] Articulated in figurative terms by the so-called Chicago school of economics, the tragedy of the commons was used to exemplify how markets could fail in the allocation of public goods like natural resources and intangible goods like intellectual property.[129]

A group of scholars in the new field of law and economics based at the University of Chicago, such as Richard A. Posner and Edmund W. Kitch, claimed that the public interest would be best served by assigning property rights in the areas of public goods where market mechanisms had failed to distribute economic resources efficiently.[130] In 1977, Kitch, one of the leading proponents among Chicago law and economics scholars of the expansion of intellectual property rights, argued that the patent system was designed to increase the economic output of technological innovation by assigning the private ownership of a promising technology to its inventors.[131] While scholars in the previous generation, like Fritz Machlup, had been concerned about economic monopolies engendered by the patent system, Kitch instead argued that the expansion of intellectual property rights was essential to the promotion of economic development. He noted that an invention usually raised a number of possible technical prospects and that only a limited number of these prospects could be developed further into commercial products. The patent system, by giving the control over technological invention in its earliest possible inception, enabled him or her to devote the capital and ingenious effort required to sort out the most promising technical prospect and turn it into a commercial product. In other words, exclusive property rights provided the inventor with a legal platform to bring a promising technology into the marketplace, contributing to economic growth through the "prospecting" effect of the patent system. Kitch also underlined the importance of privatizing government-held inventions by "the granting of exclusive licenses of patents."[132] The proper assignment of property rights from publicly held inventions thus would actually lead to the increase of technological innovation.

In the 1977 report *U.S. Technology Policy*, Ancker-Johnson and David B. Change mobilized Kitch's argument, proclaiming that government-owned patents had been underused and their potential benefits only incompletely realized. The report noted that "much government-funded R&D is not exploited for patentable inventions, so that U.S. taxpayers do not obtain an adequate return on their investment in R&D."[133] Moreover, the current shortcomings of government patent policies had grave implications for the future of US economic strength in a globally competitive market where science and technology were becoming vital national assets. As Ancker-Johnson and Change noted, America was "no longer replacing dying exports with a new wave of innovative exports."[134] The report underlined that, in the context of declining productivity of American

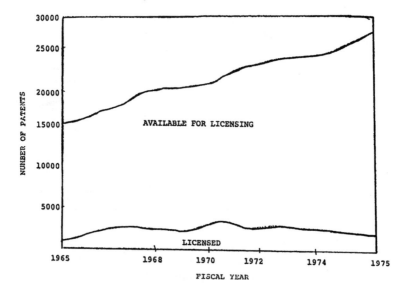

FIGURE 5.1. Underuse of government-owned patents. This graph aimed to demonstrate the underuse of government-owned inventions by private companies. As of 1975, less than 4 percent of twenty-eight thousand government-owned inventions were licensed. Though the number of government-supported inventions had increased, the number of patents that were licensed had been relatively unchanged. Graph from Betsy Ancker-Johnson and David B. Change, *U.S. Technology Policy: A Draft Study*, Office of the Assistant Secretary for Science and Technology, US Department of Commerce, National Technical Information Service, PB-263 806 (March 1977), 73.

capitalism, the promotion of technological innovation through granting private ownership of government-held inventions would be one of the best ways to serve the public interest (see figure 5.1).

This articulation of the causal link between private ownership and the public interest met, however, with considerable political opposition. In February 1977 DHEW secretary Joseph Califano requested Juanita Kreps, secretary of commerce, to withdraw the notice to expedite recombinant DNA–related patent applications so that the DHEW and other federal agencies could "have the opportunity to review Federal policies governing patenting of recombinant DNA research inventions."[135] The Department of Commerce suspended the order through a notice filed in the *Federal Register* in March 1977. At the same time, the Department of Justice circulated a memorandum to federal patent officials that endorsed a new bill prepared by the Senate Health Subcommittee. This new bill intended to reassert the government ownership of recombinant DNA

technology in order to protect the public interest by "(a) put[ting] ownership of all inventions 'useful in recombinant DNA research' that stem
from Government contracts or arrangements, in the United States, subject to specified waiver provisions, and (b) prohibit[ing] the patenting of
all inventions 'useful solely in recombinant DNA research.'"[136] In August 1977, citing political controversies regarding the transfer of "public"
inventions to private inventors, the DHEW Patent Council froze all requests by scientists and their universities to be granted patent rights to
their inventions.[137]

Against the backdrop of emerging opposition to the private ownership of recombinant DNA technology, those who shared the benevolent
vision of private ownership began to build a strong political coalition.[138]
At the Senate hearings on federal patent policy and its monopoly implications, Latker criticized the DHEW's decision to review its IPA, claiming
that more than a hundred biomedical inventions critical to public health
had been frozen.[139] In their defense of IPAs against "the Justice Antitrusters," university patent managers and government patent officials earned
major political support from Senator Bob Dole, who at a press conference in 1978 criticized the DHEW's "stonewalling" of scientists' recent
requests for the ownership of inventions arising from NIH support. Dole
disparaged the DHEW's cautious attitudes, claiming its decision to "effectively suppress these medical breakthroughs is without precedent . . .
Rarely have we witnessed a more hideous example of overmanagement
by the bureaucracy."[140]

In March 1978 NIH director Fredrickson, with support from university
administrators and government patent officials, approved the assignment
of ownership of recombinant DNA technology to Stanford and UC, pend-
ing the DHEW's final policy decision regarding IPAs. As Fredrickson informed the Stanford administrators, "Accordingly, Stanford may proceed
to file recombinant DNA research patent application. You should know
that Federal patent policies are under extensive review by the Executive
Branch and the Congress, and that this may lead to actions affecting the
administration of institutional patent agreements generally and other
conditions for recombinant DNA research inventions specifically."[141]
With powerful political patronage, university and government patent officials' efforts to institute a uniform federal patent policy gained new mo-
mentum. In September 1978 Republican Senator Dole and Democratic
Senator Birch Bayh introduced legislation that allowed federal government agencies to waive patent title to universities, non-profit organiza-

tions, and small businesses. After the 1979 congressional hearing on the "University and Small Business Patent Procedures Act," the so-called Bayh-Dole Act was enacted in 1980, the same year the PTO finally issued the first recombinant DNA patent to Cohen and Boyer and Genentech's public stock offering awed Wall Street.[142] At a time when America embraced private ownership and the market as the solution to an economic downturn, the opening of a legal avenue for academic patenting provided an opportunity for biomedical researchers to forge a new form of biomedical enterprise—biotechnology.

Too Much Private Ownership?

When Latker and Ancker-Johnson's concerted effort to institute a uniform government patent policy was finally legislated, it not only allowed commercialization but also encouraged it as a new social obligation for academic researchers.[143] This shifting legal environment triggered a wave of biotech companies that emerged from an alliance between laboratory scientists and venture capitalists. A rush of academic patenting in biomedical research, however, raised potential problems of building a company based on proprietary claims on basic experimental procedures. In 1982, Stanford administrators and Reimers began to seriously explore how to amend dangerous trends in the pursuit of academic patenting:

> There is concern among the industrial community that technology will be inhibited, slowing the process of commercialization, by the increasing number of institutions with conflicting contractual terms and multiple types of licenses required by a company in order to assemble the rights necessary to bring any given product to market. The burden on industry will be tremendous if it has to deal individually with a large number of universities for rights to commercialize most biotechnology. If current trends continue, it is conceivable that to produce single product a company may need licenses from a half-dozen different universities for as many or more patents. The situation is further exacerbated by the growing desire of universities to receive financial returns from industry for their research results. Universities are also concerned that scientific communication may become inhibited.[144]

More problematic for both academic researchers and industrialists was an increasing trend to file every potential procedure and research technology developed in academic laboratories. Some biotech companies began to question whether this zealous patenting rush would serve the public interest as universities claimed. The report suggested various licensing practices for amending this "too much ownership" situation, such as patent pooling among biotech companies, package licensing between academic and industrial organizations, and preparation of a clearing catalog for exchanging intellectual property assets.

The case of Amgen (abbreviated from Applied Molecular Genetics), the biggest biotech firm in terms of stock market capitalization, offers an important example in which we can examine an industrial perspective on the implications of the patenting of broad research technologies by academic institutions. After the 1980 Supreme Court decision on *Diamond v. Chakrabarty* that permitted the ownership of biologically engineered organisms, some of the biotech companies who successfully developed viable commercial products began to worry about the prospect of granting the recombinant product patent in the early 1980s. These companies, according to Stanford's recombinant DNA–patent licensing arrangement, had to pay 10 percent royalties for basic product sales, and 1 to 0.5 percent royalties for end product (like drugs) sales.[145] In 1984, when the first recombinant DNA product patent (limited to prokaryotic organisms such as bacteria engineered by recombinant DNA procedures) was granted, Amgen faced the prospect of paying a large sum of royalties from the sales of recombinant DNA–based drugs and other products.[146] While the company did not have any approved drug product until 1989, it developed several promising drug candidates using recombinant DNA technology (In the end, Amgen became the single most profitable licensees of recombinant DNA technology, paying about $55 million for royalties, about 21 percent for the total $255 million licensing revenues generated by the recombinant DNA patents for Stanford and UC).[147]

Faced with the prospect of huge financial expenditure, Amgen's CEO, George Rathmann, criticized Stanford's licensing of what he regarded as basic research technologies like recombinant DNA technology. He further claimed that the technology, which was publicly funded by the federal government, should be in the public domain. In a sign of protest, Amgen indicated that it would refuse to pay license royalties in 1984. Amgen signaled that it would not pay any product royalty to Stanford and UC from a profit it earned from the sales of its various drug products manufactured using recombinant DNA technology. Rathmann further indicated that his concerns were not limited to the licensing arrangement but reflected worries about the role of academic institutions in the realm of technology transfer in the biotech industry. He was willing to discuss these matters with Stanford representatives, including President Donald Kennedy. In his conversation with Stanford administrators, Rathmann voiced biotech companies' concerns with the increasing eagerness of academic technology transfer operations. As Stanford's legal counsel Adrian Arima noted to Kennedy in December 1985, "the issues Rathmann is concerned about are real, and have been expressed by Cetus and other biotechnology companies":

> At issue are the scope of Cohen/Boyer vis-à-vis new rDNA techniques, the tendency of universities and companies of patenting every new development (so the field is cluttered by overlapping patents), the aggregate royalty burden from multiple licenses covering a single product, and whether "research" by commercial firms can infringe patents.[148]

In defense of its licensing agreement with Amgen, Stanford OTL contacted the law firm, Finnegan, Henderson, Farabow, Garrett & Dunner in January 1986, to draft a complaint against Amgen's refusal to pay product royalties. Katherine Ku at Stanford OTL also wrote to Kennedy about the OTL's position on Amgen's criticisms. Ku pointed out that Rathmann's assertion that the Cohen-Boyer technology, derived from federally funded research results, should have been in the public domain no longer had legal merit. She noted that the issue of ownership of government-supported inventions had been nationally discussed and resolved in the late 1970s. In fact, Congress had enacted in 1980 the Bayh-Dole Act in order to explicitly encourage commercial development through the university ownership of inventions arising from government-sponsored research.[149]

Rathmann in response argued that the Cohen-Boyer technology transfer did not serve US economic competitiveness. From his perspective, the patenting of recombinant DNA technology did not serve the interest of the public or small business owners as the government had originally intended when it transferred ownership of the technology to Stanford and UC. On the contrary, he noted that the Boyer-Cohen patent was granted only in the United States, and as such was hardly beneficial to American biotech and pharmaceutical companies. He also pointed out that the Cohen-Boyer patent discriminated against small biotech companies. As he put it, though "small biotechnology companies have contributed the most to the industry and yet are required to pay on an equal basis as large companies." This was hardly an outcome that the Bayh-Dole Act envisioned. Its initial legislative rationale was to waive government title *only* to small business and nonprofit organizations in order not to favor big business.[150] But major multinational pharmaceutical companies could subvert the bill's restrictions by building a laboratory and factory for the development and production of recombinant DNA–based products in a foreign country. Even Reimer later acknowledged that the issues Rathmann raised were valid enough. In his 1998 oral history interview, he pronounced that "Whether we licensed it or not, commercialization of recombinant DNA was going forward. As I mentioned, a non-exclusive licensing program, at its heart, is really a tax . . . [b]ut it's always nice to say 'technology transfer.'"[151]

Conclusion

This chapter has situated the debate over the ownership of recombinant DNA technology in the context of two interrelated changes in the research university as a site for knowledge production. At one level, I show how the subtle redefinition of universities' mission toward economic development provides an important context for understanding the commercialization of recombinant DNA technology. In the years leading to the late 1960s, American research universities used the argument that they played a critical role in the nation's economic prosperity to support their appeals for public funding to promote research in science and technology. The post–World War II rhetoric that technological innovation was predicated on the progress of basic science came to be challenged in the late 1960s and early 1970s, however. From the mid-1960s onward, the rising call for relevance and accountability in academic research challenged a loose postwar consensus about the economic and medical contributions of the research university.[152] Accordingly, public enthusiasm for biomedical research began to wane. Lay activists and politicians called for more immediate medical applications to emerge from biomedical research, asking how best to get practical results from the money they spent on research support.[153]

This contentious political context in the early 1970s created a new opportunity for those who were involved in the affairs of biomedical research. As we have seen, a new class of managers of research, such as university research administrators and government officials, emerged as prominent intermediaries between patrons of science and academic researchers. With the prospect of the flattening of federal research budgets, these managers were looking for ways to translate the results of scientific research into tangible economic assets.[154] At a more practical level, the capitalization of new molecular technologies was largely influenced by the industrialization of biomedical sciences and the development of the pharmaceutical industry.[155] These commercialization trends in some areas of biomedicine that had significant ties to industry largely set the framework for using some of the key biomedical technologies mobilizing intellectual property laws.[156]

Within this political and economic context, federal patent policies became a focal point of discussions on regarding the postwar economy of academic research. In an attempt to reassert the economic value of academic research, Stanford's research administrator Niels Reimers, for example, began to refashion himself as a university patent manager. In the context of the heated negotiations over how best to support and mobilize biomedical research for the benefit of the public, patent managers in academic institutions and government agencies opportunistically mediated the relationship between science, government, and industry by introducing a new set of administrative and legal practices, such as IPAs and, eventually, the Bayh-Dole Act in 1980. With the establishment of the private ownership of recombinant DNA technology, the proprietary nature of biomedical knowledge came to characterize the emerging biotechnological complex.

At another level, by examining the privatization of recombinant DNA technology in terms of changes in legal and economic assumptions about public knowledge, this chapter addresses broader implications of the commercialization of academic research for research universities, funding agencies, and the public. As I show, the debate over recombinant DNA technology was an important part of the broader reconfiguration of the ownership of public knowledge in late-twentieth-century American capitalism. At the heart of the debate was a redefinition of what public knowledge meant for the public interest and its relation to the private knowledge industry. With a renewed call for free markets and new patent policies, the articulation of a causal link between private ownership and public benefit proved critical to the privatization of biomedical research.[157] To begin with, the positive conception of property rights as a means to promote technological innovation helped redefine the public interest in terms of economic growth amid the declining productivity of American capitalism in the 1970s. By the time the ownership of recombinant DNA technology was determined and the Bayh-Dole Act was passed, the commercialization of academic research through its private ownership had emerged as a new social obligation for research universities. As a hybrid creation of academic and commercial institutions held together by the moral assumption of a causal relation between private ownership and the public interest, biotechnology has gained a prominent place in the commercialization of biomedical research, making life into a "productive force" in late-twentieth-century American capitalism.[158]

In the end, the limited implementation of the private ownership and control of recombinant DNA technology in the course of its commercialization reflects the historical contingency of the causal link between private ownership and the public interest forged in the 1970s by patent administrators, industrialists, and legal scholars.[159] Stanford administrators, wary of academic and political concerns about its monopolistic use, eventually decided to institute nonexclusive licensing of recombinant DNA technology, making it available to any academic and commercial organizations with a modest yearly licensing fee of ten thousand dollars. As Reimers later notes ironically, this limited enforcement of recombinant DNA patents to avoid complaints about monopoly led to its wide commercial use and huge licensing revenues, with total royalties amounting to $254.3 million during the life of Cohen-Boyer's recombinant DNA patents until 1997.[160] The abstract rendering of the negative relationship between the intellectual commons and technological innovation—the tragedy of the commons—does not fully account for the way scientists, patent administrators, and government officials implemented the privatization of recombinant DNA technology in practice. Looking forward, recognizing the historical genealogy and limitations of the private ownership of academic research and the terms of its arguments and counterarguments can provide a balanced platform for discussing key contemporary issues of the commercialization of biomedical research—for example, the impact of patenting on academic openness, particularly in the exchange of information and materials, and the ethical issues involved in clinical trials and gene patenting by for-profit institutions, to name a just few.[161] In the next chapter, I analyze how Stanford biochemists tried to accommodate the newly legitimized avenue for academic patenting, illustrating the shifting moral economy of science and public obligations of academic researchers in an age of commercial biotechnology.

CHAPTER SIX

Reenvisioning the Biomedical Enterprise in the Age of Commercial Biotechnology

During the hippie era [of the 1960s] people put down the idea of business—they'd say, "Money is bad," and "work is bad," but making money is art and working is art and good business is the best art. — Andy Warhol, 1975[1]

In 1980 a series of catalytic events occurred in the emergence of biotechnology as a key industrial sector. These began on June 16, 1980, when the US Supreme Court, in the landmark case *Diamond v. Chakrabarty*, ruled that genetically modified life forms could be patented. In October of that year, Genentech, a recombinant DNA technology–based company, made an initial public offering of stock on the NASDAQ exchange under the symbol "GENE," making Herbert Boyer a multimillionaire scientist-entrepreneur. The first recombinant DNA patent covering the techniques of the research was then granted by the US Patent and Trademark Office (PTO) on December 2, 1980, six years after Stanford and the University of California filed the patent application on behalf of Stanley Cohen and Boyer. Just ten days after the patent was issued, the Bayh-Dole Act was passed, allowing research universities, small businesses, and nonprofit organizations to claim ownership of inventions arising from federal support. Indeed, by the end of 1980, the institutional and legal framework for the commercialization of academic research was firmly in place.

With their reservations about commercialization, scientists in the Stanford Biochemistry Department seemed to be on the wrong side of history, or on the losing side. They were often portrayed, if not ridiculed, as being "those who were left behind" the biotech boom.[2] Despite their contributions to the development of recombinant DNA research and technology, they did not financially benefit from its commercialization. On the contrary, because of the narrow legal determination of its inventors, their role in its development was not well known, which was often buried under a popularly circulated narrative of Cohen and Boyer's moment of genius at the 1972 Hawaii meeting where they decided to collaborate on their recombinant DNA–cloning experiment. Only some of the scientific community recognized the Stanford biochemists' contribution, especially with Paul Berg's award of the Nobel Prize in 1980 "for his fundamental studies of the biochemistry of nucleic acids, with particular regard to recombinant-DNA." The prize might have been the only consoling event for the Stanford biochemists in 1980 amid the biotech frenzy.[3]

The Stanford biochemists ambivalently witnessed the ascendance of biotechnology. Through their participation in the debate over the patenting of recombinant DNA technology, they observed shifts in the institutional, legal, political, and economic environment of biomedical research in the late 1970s and early 1980s. These changes encompassed, for example, the institutionalization of patent management at research universities, the enactment of the Bayh-Dole Act in 1980, and the flattening of federal support for biomedical research. The distinction between the public knowledge of the university and the private knowledge of industry in biomedicine had been challenged and redefined, as discussed in chapter 5. Moreover, the articulation of private ownership benefitting the public interest provided promoters of biotechnology with a strong economic incentive and moral justification for commercial biotech ventures based on proprietary claims on knowledge. Amid discussions of the privatization of recombinant DNA technology, Berg began to express his softening attitude toward commercial engagements by academic institutions. In his letter to the National Institutes of Health (NIH) director Donald Fredrickson, he conveyed his view of the marketing of biomedical research, agreeing with "the notion that commercial exploitation of this basic research development should be accompanied by a return of some of that wealth (more or less as a tithe) to the institutions from whence it came."[4]

By the late 1970s, more academic institutions and scientists tried to market inventions made in their laboratories. Private venture capitalists, lured by huge financial gains, were ready to invest the necessary capital in biotech firms established by notable scientists. For example, in 1978, Walter Gilbert at Harvard University and Philip Sharp at MIT founded Biogen in Cambridge, Massachusetts, to produce medically useful proteins using recombinant DNA technology. In the same year, Ivor Royston and Howard Birndorf at the University of California, San Diego, founded Hybritech, one of the earliest commercial companies to adopt hybridoma technology; this was a key technology that enabled the large-scale production of medically useful antibodies (monoclonal antibodies).[5] Like many academic researchers, the Stanford biochemists became involved in commercial biotechnology as well, first through the department's Industrial Affiliates Program (IAP) and, more substantially, through the founding of a commercial company, DNAX.[6]

This chapter, by taking account of the Stanford Biochemistry Department's tradition of a communal culture of science, analyzes how those scientists took cautious steps to accommodate the changing legal and economic conditions of biomedical research during the 1980s.[7] First, the biochemists attempted various connections to industry, not just as an attempt to draw financial resources, but, more importantly, as a way to preserve their communal style of scientific research. The departmental IAP was what Stanford biochemists regarded as one of the safest ways to uphold their moral economy of science while developing their connections to industry. DNAX, founded by Stanford scientists Arthur Kornberg, Berg, and Charles Yanofsky, was designed as an alternative form of a biological research company. Kornberg and Berg envisioned it as a freestanding, basic research institute that would commit itself to the advancement of biomedical research, without worries of either a governmental shift in research policy or pressures from private industries striving for profit.[8] My analysis of the diverse motives and rationales behind the Stanford biochemists' participation in the biotech industry illuminates the shifting boundary between the moral and capitalistic economies of science in commercial biotechnology's coming of age.[9]

Therefore, this chapter illustrates a moral dimension of academic scientists' entrepreneurial forays into commercial biotechnology in the 1980s. As scholars in science studies have pointed out, several social, economic, and technological factors influenced the emergence of scientist-entrepreneurs in biomedicine; these included calls for relevance in biomedical research in the 1970s, new financial incentives for venture capital investment, and managerial and technological imperatives (i.e., production and circulation of large-scale biomedical materials) that drove the industrialization of the biomedical enterprise.[10] The emerging link between private ownership and the public interest highlights the importance of a shift that complemented and reinforced those discrete developments; this change involved a positive moral rendering of the privatization of bio-medical research that was believed to bring medical benefits.[11] For ex-ample, in 1977, Boyer passionately argued that his motivation behind the founding of Genentech was to see that recombinant DNA technology "gets transferred to private industry so that public benefits come out as soon as possible."[12] My analysis of the Stanford biochemists' attempts to reenvision their biomedical enterprise through their connection with private industry illustrates the changing moral and political landscape of biomedical research in late-twentieth-century American capitalism.

In the first section of this chapter, I examine how the leveling of federal funding for biomedical research in the mid-1970s introduced fiscal chal-lenges and financial instability in research universities; this was especially true at Stanford where government grants supported a large share of its medical research. I then situate the Stanford biochemists' changing per-spective on industrial support in biomedicine against the backdrop of the shifting federal research economy. The scientists understood that, prop-erly instituted, an industrial affiliation could provide more flexibility for their research than federal support. More importantly, in their attempts to forge a new relationship with industry, the Stanford scientists raised a set of critiques on the various patterns of industrial affiliation in biotechnol-ogy. I examine what they perceived as problematic patterns in the com-mercialization of biomedical research and investigate how they tried to build what they regarded as an "academic" research institute in a corpo-rate setting.

The Stanford biochemists' deliberations on various modes of indus-trial affiliation and commercial enterprise also reflected their effort to preserve their collaborative practices, which they believed crucial to the flexibility and productivity of their research. Reflecting on the proper re-lationship between scientific research and technology transfer in the com-ing of age of commercial biotechnology, they perceptively noticed that industrial support, including IAP, had its own limitations, compared with federal, philanthropic, and other public support. The fate of DNAX, its difficulties in attracting investment and its subsequent acquisition by a major pharmaceutical company, illustrates limitations that academic biol-ogists faced in finding a proper balance between research and commerce. The chapter concludes with an examination of a biotechnology confer-ence in 1980 by the president of Stanford University, Donald Kennedy, to discuss the commercialization of academic research. By analyzing how university administrators and academic scientists tried to deal with the rush to privatize research, I end by probing the changing academic perspectives on the emergence of biotechnology.

The Financial Instability of Stanford's Biomedical Enterprise

The Stanford Medical School was one of the major beneficiaries of federal support in biomedical research after World War II. The school's expansion was heavily underwritten by federal grants; these supported its ambitious relocation from San Francisco to the Stanford University campus in Palo Alto in the late 1950s, as well as its major investment in basic biomedical sciences, such as biochemistry and genetics. When the medical school established the Biochemistry Department in 1959, the provost, Frederick Terman, adopted his postwar expansion strategy, charging faculty salaries and university overhead to the federal government. In terms of its financial arrangement, especially its heavy dependence of federal funding, Stanford's medical school could be characterized as an exemplar of what Clark Kerr called "the federal grant university."[13]

The enormous expansion of federal support of science after World War II left an enduring institutional legacy for research universities, especially for Stanford. Above all, Stanford's heavy dependence on the federal government for financial support made the university vulnerable to major shifts in public and private funding for scientific research. As cultural and political attitudes toward science underwent a profound shift in the late 1960s, biomedical researchers began to be aware of their heightened reliance on the public through government support.[14] Lay activists and medical professionals were growing impatient with the relative lack of tangible clinical and medical progress from basic biomedical research. By the early 1970s, political attitudes regarding biomedical research had shifted as politicians and the public increasingly called for relevance in biomedical research with more tangible medical applications.[15] The Nixon administration's War on Cancer campaign signaled major changes in federal priorities for biomedical research and encouraged some scientists to shift their research agenda toward areas like cancer, heart disease, and human cell biology.

The War on Cancer Act was enacted in 1971 with the intention to increase investment in the one specific disease category of cancer. This raised serious concerns among academic scientists and university administrators about the federal support of biomedical research. For academic researchers who had become increasingly dependent on federal funding, the growing public and political demand for goal-oriented research and immediate medical applications presented an especially challenging environment. To some scientists, the political demand often seemed to be accompanied by a lack of appreciation for the unpredictable nature of scientific research. In order to assuage this pressure on biomedical research, Senator Edward Kennedy helped establish a President's Biomedical Research Panel in early 1974. Its aim was to ensure that the NIH was achieving the proper balance between basic and applied research in its support for biomedicine. The panel's final report warily observed that "current trends in the thinking of the Administration and the public favor disease-targeted research, primarily on cancer and cardiovascular diseases, for which funds are provided in larger amounts than ever before."[16] The report criticized this disease-focused approach because it essentially resulted in the relative neglect of basic biomedical research, which often had "little public visibility and no emotional appeal."[17]

In the early 1970s, the Stanford biochemists worried whether the growing demand for medical applications and accountability might undermine federal funding for some essential areas of biomedical research that could not be directly tied to a tangible medical and economic outcome; these areas included basic research and student training. When Paul Berg, then the chair of the Stanford Biochemistry Department, observed the controversy over the proposed termination of the NIH's training grants for students in biomedical research, he expressed concern about the careers of the department's six incoming graduate students. These included one of his students, Janet Mertz:

> I think they are a very bright group; many of them have had extensive research experience and all of them are eager to conquer the world! The only concern I have is what will happen to these people four to five years hence. If there is in fact a constriction in the opportunities for biochemically trained people at that time, we shall have on our hands the responsibility of trying to figure out what to do with them and how to place them. This doesn't even take into account the problem of what will happen if the [NIH's] Training Grant funds are eliminated.[18]

Indeed, the NIH support for graduate education through its training grant programs to academic institutions had been phasing out for individual awards and much reduced for institutional awards since 1972.[19]

More importantly, the President's Biomedical Research Panel pointed out that the huge influx of federal funding to research universities and medical schools in the previous twenty-five years had created the potential for severe financial instability. By the late 1960s, the federal support of biomedical research totaled approximately half of research grant and contract funds in major research universities. The panel found that, in its support of scientific research, the federal government treated academic institutions "strictly as research outfits" and cared "little about their fate either before or after the research projects it supports are carried through."[20] More problematically, without due consideration of a potential reduction in federal support of science, universities had opportunistically expanded their faculties and facilities using government funds. In consequence, the "potential for crisis" in academic institutions became particularly serious when large research projects or disease-related centers were established, as in the War on Cancer. Any significant reduction or shift in federal funding of biomedical research could thus introduce a distressing financial instability to an academic institution.

What would be the implications for research institutions of a major cutback in federal funding after nearly twenty years of rapid expansion in support? With the imminent prospect of a federal funding reduction amidst the economic recession of the 1970s, the panel's report cautioned the government to be aware of a potential crisis in research universities:

> NIH—and beyond NIH, the federal government—should recognize the dangerous degree of instability introduced in the financial structure and general activities of Universities and Medical Schools by the rapid expansion of federally supported research programs carried out in these institutions.[21]

Stanford administrators were particularly concerned about their medical school's dependence on federal support of research and education. In 1969, 60 percent of the medical school's budget was supported by federal research funds. In fiscal year 1973–74, this had increased to 64 percent (twenty-three million dollars of the medical school's thirty-six million dollar budget was from the federal government).[22] In late 1974, the dean of the medical school, Clayton Rich, circulated a memo that projected a "major deficit" in the next few years due to inflation and the economic recession. Though the Stanford Biochemistry Department had garnered generous federal support (its federal grants amounted to one million dollars in 1975, a two-fold increase from five hundred eight-two thousand dollars in 1966), faculty members increasingly realized that the further expansion of federal support in biomedical research might be unsustainable.

By the late 1970s and early 1980s, the Stanford biochemists were increasingly concerned about the general decline of federal support, which was leading to escalating competition and political demands for relevance in their research. In a 1981 faculty meeting, Kornberg spoke disparagingly about the recent flattening of federal funding, complaining about the growing competition for grants, governmental shifts to smaller grants, and "increased nuisance" in federal funding, including the demand for relevance and accountability in grant applications. For him, these shifts made federal support far less attractive because of the loss of autonomy and flexibility. As Kornberg repeatedly stressed, creative scientific research could thrive only when scientists had freedom to pursue unexpected experimental findings, regardless of other considerations. In addition, the growing emphasis on accountability by the government made it increasingly difficult for scientists to acquire funds for advanced laboratory equipment, and expansion and renovation in research laboratories. This relative lack of flexibility prompted Stanford scientists to seek less-restrictive funding from other sources.[23]

By the late 1970s, Stanford's medical school, which accounted for 42 percent of the university's entire operating budget, seemed to face a financial crisis of its own. The medical school and teaching hospital, as a news article in *Science* put it, were "in the throes of a fiscal and administrative crisis that offers no ready or painless resolution and shows every prospect of worsening."[24] In addition, the medical school's greater emphasis on research (rather than clinical care) underwent a critical alteration due to recent changes in federal healthcare policy. The passage of the Medicare Act in 1965 and the relative leveling federal funds for biomedical research from the early 1970s resulted in the balance of power shifting away from basic biomedical research toward clinical activities.[25] By the late 1970s, when clinicians brought in a major portion of the school's budget through its teaching hospital, their influence loomed large. Consequently, the medical school's identity as a research-oriented institution was challenged.

Kornberg, who had led his research group's move to Stanford from Washington University in St. Louis, with the aim to build a *bio*medical school where biological advances would fundamentally shape the trajectory of medicine, declared in 1977 that "basic science at Stanford is a paper tiger."[26] Though faculty members in the basic biomedical sciences,

such as biochemistry, molecular biology, and genetics, became prominent leaders in their fields of research, the initial plan for the Biochemistry Department and its central role in the development of a research-
oriented medical school never ultimately materialized. By the late 1970s,
the medical school's initial vision to develop an outstanding program of
basic medical science through its integration with biochemistry (Kornberg's group), biology (Charles Yanofsky's group), chemistry, clinical
science (Henry Kaplan's group), and genetics (Joshua Lederberg) had
fallen apart. Kornberg increasingly lamented that the school had drifted
steadily toward emphasizing clinical practice and moneymaking activities.[27] Faced with the unfavorable political climate, the flattening of federal support in biomedical sciences, and a growing clinical orientation in
the medical school, Stanford biochemists began to seriously contemplate
the long-term prospect of their department.

Sharing Patterns in Biotechnology's Coming of Age

The emergence of biotechnology as a commercial enterprise loomed
large as Stanford biochemists discussed how to thrive in the shifting economic and legal environment of biomedical research. At a time when the
nascent biotech industry was drawing huge financial resources from venture capitalists, Stanford biochemists regarded it as a potential source to
support their research and education through patenting and industrial affiliations. In their deliberations, the faculty worried about whether a commercial tie might be compatible with their moral economy of science, in
which the sharing of ideas, materials, and monies contributed to productive collaborations and unexpected experimental findings. Amid the rapidly growing commercial interests of venture capitalists and university researchers in genetic engineering, more academic institutions and scientists
aggressively tried to retain the rights to patents resulting from inventions
that originated in their laboratories. This trend to claim proprietary rights
on discoveries made in an academic setting prompted concerns about the
potential impacts of commercialization on academic institutions; these
possible effects included the increase of secrecy and conflicts of interests.
For example, on December 1980, Stanford president, Donald Kennedy,
called for a meeting among university and biotech industry leaders to
confront "the rush to proprietary control of recombinant DNA research";
he wanted them to consider the importance of the academic culture of
openness and free exchange of ideas.[28]

The proliferation of ownership claims of key research tools and materials in molecular biology posed a particular threat to the work of Stanford biochemists, among whom the semicommunal ownership of ideas, tools, materials, and research monies had been fostered. Indeed, their experience underscored how the pursuit of proprietary interests and patenting impeded the open exchange of research materials among researchers. They saw this worrisome trend to claim proprietary rights of key research techniques and materials accelerating. A few years later, after the first instance of a refusal to disseminate a research material (Cohen's plasmid pSC101) at Stanford, Berg himself encountered what he regarded as another example of the emerging culture of secrecy; in early 1979, he requested some clones from the Stanford biologist Robert Schimke, who was collaborating with Cohen in order to express a functional mammalian protein in bacteria.[29] Schimke, however, was hesitant to distribute his dihydrofolate reductase clones, one of the first plasmid clones that could express mouse genes encoded for an enzyme. Infuriated, Berg considered Schimke's refusal fundamentally detrimental to the scientific enterprise, where collegial and open exchange could ideally foster scientific creativity while regulating competitive feelings. Berg wrote to Schimke:

> Losing my temper when we spoke the other night was a foolish and useless response. I'm sorry it came to that; but let me reiterate that I meant every word I said, it's only the tone I regret ... When I asked you yesterday if I could come over and pick up a culture and possibly some DNA, you said that would be fine, but that you'd have to come over here to get it and would I come over to your lab in the afternoon ... What followed seemed like a run of inconsistent statements: "I have to protect myself"; "When my MUA [Memorandum of Understanding Agreement necessary for recombinant DNA experiments] is approved on Feb. 13 you can have the clone"; "I don't have the clones or the DNA here"; "Stan[ley Cohn] is making the clones available tomorrow, yes he's going to send them to people tomorrow." If Stan was sending them out tomorrow what was all the fuss about our not being able to have them or having to wait until you felt secure ...
>
> By contrast to the experience in this instance I can reflect with satisfaction and appreciation on the openness, cooperation and generosity of Tom Maniatis, Rich[ard] Axel, and many others in the SV40 field. I can't recall even once being refused by another lab when I requested enzyme, a particular mutant, virus or bacterial stocks, etc ... Seeing that kind of behavior in this instance was particularly disappointing. Considering Stan's past behavior perhaps I should not have been so surprised by his stand, but yours seemed out of character.

It's not the proper way to do science; and it's doubly bad because it shows young people around us some of the very worst faces of what they dub the "rat race."[30]

When Stanford biochemists cautiously deliberated on a possible affiliation with private industry, faculty members worried about potential disruptions that a commercial affiliation might create in their particular departmental research culture. In a 1979 faculty meeting, Kornberg reflected on the history of the department's moral economy. Kornberg noted that his teachers, such as Carl Cori, had exhibited resistance to federal support because of the potential danger of government control over science. Kornberg and his colleagues had instituted a departmental fund-pooling practice in order to establish an unrestricted source of money that they could share. By turning the research grants into a semicommunal resource, Stanford biochemists believed they could alleviate governmental intrusion in the affairs of science. The sharing of research materials, instruments, and space had been encouraged not only to save funds but also, more importantly, to foster the productive exchange of scientific ideas. Thus, the ability to control their research directions, while maintaining their communal laboratory culture, was one of the critical considerations when Stanford biochemists looked for another source of support research in the late 1970s.[31] Kornberg emphasized:

> Our traditional pattern of sharing funds and resources has been successful. It accounts in part for the spirit and quality of the department. This pattern can also be defended for its economies in large equipment, expendables, space and technical staff. The scientific interactions generated by this sharing are a priceless dividend ... The pattern has also made possible a greater flexibility in the size and operations of a research group from an ebb, say, during a sabbatical period to a surge when a "breakthrough" comes ...
>
> However, all patterns in our affairs must be continually adapted to the variety of external and internal changes that affect us (evolution is essential for survival). For this reason, I agree that it is appropriate to consider the recent changes imposed on us, our responses to them, and how we should prepare for the future. The principle external changes are the sharp declines in funding resources (federal, foundation, Stanford) and the resultant greater efforts needed to get the money and to account for it ... I would suggest several moves that may improve our financial security *without distorting the traditional pattern*.[32]

As Kornberg saw it, the Stanford biochemists' collective practices had productive effects when they all shared similar research interests on the structure and function of genes and chromosomes in the 1960s. The diversification of their research interests in the 1970s, however, raised the question of whether the sharing pattern would still lead to frequent collaboration and interaction between faculty members and research groups. Even so, they agreed that their sharing practice could provide a greater flexibility in the size and operations of the department, helping to maintain more autonomy at a time when political demands seemed to dictate the direction of research agendas through the distribution of federal grants. The departmental patterns for sharing and interaction had been "crucial to the success of the department" and had fostered the development of recombinant DNA technology. Thus, the Stanford biochemists concluded that they search for new industrial support, but that they should be "strongly committed to preserving and strengthening their research practices in the years ahead."[33]

While deciding to pursue industry as a potential supporter of biomedical research, the resulting resolutions at a faculty meeting underlined that financial considerations should not influence the operation of the department. In terms of their grant pooling practice, they agreed that each faculty member and his group would make a "realistic estimate" of their income and spending. As Stanford biochemists' double-bookkeeping practice had not earlier recorded each group's expenses (see chapter 1), there had been little incentive to reduce inequalities and seek additional funds. However, as calls for accountability and relevance in federal funding intensified, they modified their double-bookkeeping practice in early 1980 so that each group could be accountable for their expenses. The new accounting method was designed to provide "only reasonable estimates" as a guide, rather than a strict financial budget (see figure 6.1).

In terms of patenting and licensing, faculty members took a cautious position. With its need for a clear determination of the inventor, patenting could be potentially detrimental to a collaborative research culture. As the July 1980 faculty meeting minutes indicated, concerns were expressed about "problems connected with patenting discoveries made in the department." While faculty did not oppose patenting as a potential means of financial gain, the pooling of research grant monies at the department made it necessary to have a discussion about the distribution of income from patenting. Stanford's existing policy dictated that the income generated by patenting and licensing should be divided evenly among an inven-

BREAKDOWN BY RESEARCH GROUP

Group Size			Total Direct Costs Awarded*	Monthly Expense Allowance	Actual Expenditures January
6	Baldwin	NIH NSF	24,355 18,843 43,198	3,600	1,862
14	Berg	NIH NCI AC	61,589 34,119 28,080 123,788	10,316	8,504
7	Brutlag	NIH NF DF	29,555 5,296 6,400 41,251	3,438	2,753
12	Davis	NIH NSF USDA AC DF	30,896 26,981 6,716 19,097 6,400 90,090	7,508	5,053
15	Hogness	NIH NSF AC	31,160 29,057 29,147 89,364	7,447	9,415
10	Kaiser	NIH NSF	20,941 13,566 34,507	2,876	2,778
12	Kornberg	NIH NSF	57,830 29,449 87,279	7,273	8,099
8	Lehman	NIH NSF	70,900 23,870 94,770	7,898	2,531
8	Rothman	NIH Bio	14,690 4,030 18,720	1,560	7,418
10	Stark	NCI	54,307	4,526	6,844
			677,274	56,442	55,257

* Salaries have been deducted

FIGURE 6.1. Stanford Biochemistry Department budget 1981. From 1959, Stanford biochemists maintained double-bookkeeping for their research grants, recording grant income from each faculty member. However, they recorded only total expenditures, which made it impossible to know each group's spending amounts (see chapter 1). The budget document changed in 1980 under financial pressure, and the department then recorded each group's individual expenditures so that they could be more financially accountable. Image from "Monthly Financial Statements," 28 April 1981, Paul Berg Papers, Stanford University Archives, SC 358, box 18, folder: Faculty Minutes. Reproduced with permission, Department of Special Collections, Stanford University Libraries, Stanford University.

tor, the department, and the university. At the July meeting, faculty could not reach a decision on "whether or not to have a departmental policy on the distribution of profits from patents and licenses" that differed from Stanford's policy.[34] Nevertheless, when faced with financial instability and lured by the potential for enormous profits from the nascent biotech industry, Stanford biochemists more actively explored alternative modes of support for biomedical research from industry, asking how best to preserve their distinctive research practices amid the coming of age of commercial biotechnology.

The Industrial Affiliation Program

In the Stanford biochemists' search for alternative modes for financial support, the nature of the funding was as important as the total amount. From the late 1960s, increasing demands from funding agencies for relevance and accountability seemed to the Stanford biochemistry faculty to unnecessarily reduce both their flexibility and control. It was in this climate that industry reemerged as an alternative source of financial support. As early as 1974, the editor of *Science*, Philip Abelson, claimed that scientists needed to reconsider their wariness of industrial support:

> [A]t one time, industry was a prime supporter and defender of academic institutions. During the last two decades, however, while universities fell into dependence on government, industry and universities have been estranged. But academia now understands that it cannot count on government for sustained rational behavior ... [I]f anything is clear, it is that we cannot depend solely on the wisdom of politicians in the solution of long-range problems. We must find better ways. A closer cooperation of academic scientists and dynamic elements of industry could lead to effective actions.[35]

In a faculty meeting, Berg suggested that an Industrial Associate Program (IAP) similar to that of the Chemistry Department, might offer additional resources without imposing too many limitations on research or the department's educational missions. The IAP would provide an opportunity for affiliate companies to acquire current information on new trends and potential applications in biochemistry, molecular biology, and related fields. Each affiliate would have a direct liaison with a specific faculty member who would visit the company once a year to introduce recent developments in biomedical research. With an annual fee of about ten thousand dollars, IAP members would have an opportunity to participate in an annual symposium once a year to discuss their research interests with faculty members, students, and postdoctoral fellows. In addition, each affiliate could have access to students and fellows who might consider an opportunity to work for the company.

In late 1979, the Stanford biochemists prepared a draft of guidelines for their IAP in consultation with the Chemistry Department, stressing that the program should not interfere with their core research missions.[36] From the outset, Stanford biochemists indicated that this program should not be considered a consulting agreement and that the affiliates should not expect their liaison faculty members or students to necessarily respond to their research needs. As they underlined, the program was intended to introduce recent advances in molecular biology, as well as potential employees, to each affiliate.

As the IAP guidelines indicated, Stanford biochemists would use its membership fees to "diversify the department's financial support and to minimize the effects of fluctuations in federal research funding, particular for basic research."[37] Above all, the program could provide unrestricted funds that would allow more flexibility than many governmental sources. The Biochemistry Department's IAP was particularly attractive in this respect. In contrast to the restrictions that project-based government grants involved, income from IAPs could provide support for more unusual and risky research projects, graduate student training, or new equipment and laboratory construction. The program in turn would be a new way to provide other career opportunities and funding for its graduates, especially foreign fellows who were not eligible for various forms of support in the United States. It was decided that this program would be a top priority in the department's search for new support.

In early 1980, Stanford biochemists sent invitations to various chemical and pharmaceutical companies that might be interested in learning of recent advances in molecular and cell biology through their IAP. The aim of their program, as the letter indicated, was to "promote the rapid transfer of information, techniques, and ideas" in genetic chemistry and cell biology.[38] The department's IAP began with eleven affiliate companies in September 1980, each contributing an annual membership fee of twelve thousand dollars. The first affiliate member list consisted of major chemical, biotech, and pharmaceutical companies; in 1980, these included Bethesda Research Laboratories, Bristol-Myers Company, Cetus Corpo-

ration, Chevron Research Company, E. I. DuPont, Genentech, General
Electric, Hoffmann-LaRoche, Mallinckrodt, Monsanto Company, P. L.
Biochemicals, Revlon Health Care, Smith Kline & French Laboratories,
Syntex Research, and the Upjohn Company.

The IAP only brought a modest income compared with major federal
grants. This support, as department head A. Dale Kaiser noted in the early
1980s, would not be a major source of income, but it would "act as a cush-
ion" for the department in coming years.[39] From its inception until 1990,
the IAP generated about one hundred sixty thousand dollars per year.[40]
Despite this modest income, the Stanford biochemists contended that the
IAP program, when compared with commercial ties with a company, consulting, or contractual work, did not interfere with their research and educational activities. Instead, the program provided unrestricted funds that
could be used without the unreasonable demands and obligations that
Stanford biochemists sometimes felt were part of the federal or philanthropic grants.

DNAX: An Alternative Form of Biomedical Enterprise

As early as 1976, entrepreneurial biochemists and molecular biologists,
such as Herbert Boyer, William Rutter, and Walter Gilbert, created new
forms of commercial enterprise in biomedicine with the infusion of venture capital. Often capitalizing on the technological potentials of recombinant DNA technology and monoclonal antibody technology, these
companies, epitomized by Genentech and Biogen, would spawn the biotech boom in the early 1980s. When Genentech went public in October
1980, its initial stock offering was valued at about $89 a share, and Boy-
er's stake in Genentech was worth approximately $37 million by the end
of 1980.[41] It was a stunning financial success for a company with no viable product, and this prompted other academic scientists and universities to appreciate commercial potentials in the application of biomedical technologies.[42]

In this biotech boom of the early 1980s, a number of research universities tried to find a prudent way to commercialize inventions from academic laboratories. Stanford and the University of California, Berkeley
(UCB), with their expertise in biomedical research and their proximity to
venture capital firms, such as Kleiner & Perkins, tried to pioneer a novel
corporate structure that would avoid potentially harmful effects that a

university-industry joint commercial venture might have on academic research and education. In their attempt to properly balance academic pursuits and commercial interests, Stanford and UCB researchers devised an organizational separation between a nonprofit research center and a for-profit biotech company. First, they founded a biotech company, Engenics, with start-up capital from six sponsoring companies.[43] In connection with this biotech company, Stanford and UCB created a nonprofit research center, the Center for Biotechnology Research, which was financed by Engenics. The two universities would hold patent rights arising from the center's research activities, and the six sponsoring companies could license those patents with commercial potential.

Engenics, by introducing an equity incentive between the research center and the company, tried to avoid a direct financial link to its research laboratories at the center. Engenics distributed 30 percent of its equity to the center, and the center used the money to support the research projects of faculty members, such as Alan Michaels and Channing Robertson from Stanford's Department of Chemical Engineering and Harvey Blanch, a UCB professor of chemical engineering. The scientists who founded Engenics obtained a 35 percent stake of its equity, and the six sponsoring companies took the remaining 35 percent of its equity in return for their initial investment, which was $2.4 million dollars. With this equity incentive structure, the academic scientists believed they could maintain the center's independence as an academic institution. The company in turn did not control or oversee the center's research projects. As Stanford professor Robertson said, "a great deal of care has been taken to maintain this separation [between the center and Engenics] not only to protect the intellectual sanctity of the situation but to ensure that the Center can operate free from external pressures."[44] Engenics was formally established in September 1981.

Though Stanford biochemists did not participate in the initial wave of commercial biotechnology, one entrepreneur did try to use their expertise in recombinant DNA and genetic chemistry. Alejandro Zaffaroni had maintained a relationship, both as scientist and patron, with the Stanford Biochemistry Department from the late 1960s. He had started his career as a scientist and then founded ALZA, a pharmaceutical company specializing in drug delivery systems. Kornberg had served on the scientific advisory board of ALZA since 1968. As a sign of appreciation, Zaffaroni donated fifteen hundred shares of ALZA stock to Kornberg's department in 1969; this totaled about sixty thousand dollars, which was worth more than three years of a faculty salary at Stanford in the late 1960s.[45] Since 1969, Zaffaroni generously donated gift funds to the biochemistry department.

Zaffaroni became involved with the commercialization of biomedical research in the late 1970s when Harvard scientists asked him to join one of the university's new ventures in biotechnology, the Genetics Institute. At about the same time, the Stanford biochemists became aware of Engenics's plan, especially its commercial arrangement that could maintain scientists' own research interests in a biotech company. In the late 1970s, Engenics invited Kornberg, Berg, and Yanofsky to join its venture into commercial biotechnology. They, however, were not so interested in the bioengineering project that Engenics wanted to develop. Engenics's initial aim was to improve technologies within two or three years for the mass production of bacterial cultures and animal cells in order to isolate and purify useful biologicals produced by genetic engineering. However, impressed by its organizational arrangement that tried to balance research and commercial imperatives, Kornberg did consult with Zaffaroni about whether Engenics could provide him with viable scientific opportunities and financial stability. Intrigued by Kornberg's interest in a biotech venture, Zaffaroni instead suggested to him that it might be better to establish a new biotech company in areas of their own research interests in molecular biology and biochemistry. Led by Kornberg, Berg and Yanofsky also began to seriously consider a new biotech venture.

Intrigued by the commercial, organizational, and scientific prospects in launching a new form of biotechnology, Zaffaroni and Kornberg agreed to form a company by establishing DNAX, an independent research institute, in 1979. Berg and Yanofsky were attracted by Kornberg's portrayal of DNAX as "an enterprise in which topflight science [can be] linked to long-range medical objectives"; the company appeared to be "an attractive entry into the biotech parade."[46] The "Stanford Three"—Berg, Kornberg, and Yanofsky—served on the new research institute's scientific advisory board, soon to be established as the DNAX Research Institute. Berg and Yanofsky, pioneering players in recombinant DNA research in the 1970s, had both been critical of the early biotech ventures engaged in by fellow scientists. To them, the relentless pursuit of patenting often seemed to trump obligations for medical innovations that made the institutional and legal existence of private biotech ventures possible. They shared a belief that a new form of commercial organization for biomedical research and development could be arranged without what they regarded as improper and harmful arrangements between academic scientists and industrialists that were often exacerbated by insatiable profit motivations.

Berg in particular criticized Boyer's establishment of Genentech in an academic setting. He recalled that Genentech's "improper" use of materials and instruments for the company in Boyer's laboratory at the University of California, San Francisco (UCSF), provided a recipe for disastrous personal and institutional conflicts over the ownership of research materials and intellectual property rights:

> [Genentech] created a lot of problems with UCSF, enormous angst: Who did what? And who owned what? And so on and so forth. There were probably suits between UCSF and Genentech about materials. Genentech didn't have a place to work. So Herb Boyer, who was "Genentech," was working at UCSF laboratories, and that, most of us thought, was totally improper.[47]

Genentech decided, in 1978, to establish its own facility in South San Francisco, and hired some UCSF researchers who had been working on the cloning of the human growth hormone gene. Soon, however, allegations emerged that two Genentech researchers, Peter Seeburg and Axel Ullrich, "transferred" some key research materials in gene cloning, including synthetic DNA necessary for producing a human growth hormone, from a laboratory at UCSF to Genentech; they were said to do this without authorization from the lab's chief, Howard Goodman. Both Seeburg and Ullrich had previously worked for Goodman, who was consulting on human growth hormone for Eli Lily, and his laboratory was competing against Genentech and Biogen over the production of the hormone via gene cloning. Genentech's subsequent success in the cloning of the human growth gene after Seeburg and Ullrich's midnight raid of Goodman's lab eventually led a bitter legal battle between Genentech and UCSF, highlighting potential tensions between academic and commercial institutions.[48]

In launching DNAX, the Stanford Three and Zaffaroni aimed to overcome what they identified as three major shortcomings of recently established startup biotech ventures.[49] First, they criticized a commercial strategy of some biotech companies that had focused on the monetization of intellectual property.[50] With no viable product sales, new startup companies tried to generate their revenue by the licensing of processes, which often could be substituted for other processes in a short time. The founders of DNAX believed this licensing strategy to be a "precarious" one that could not guarantee the long-term growth of the company. Second, in regard to the financial strategy of a new biotech venture, they acknowledged that those companies with no immediate commercial products were destined to struggle for additional investments. By selling new equity shares in order to raise additional capital, the owners often diluted the shareholder base, hurting existing shareholders. Third, they cautioned that a significant ownership of a biotech company by one or two major companies could "greatly reduce the freedom of opportunity" in a new venture. They believed the recent trends in the acquisition of a biotech startup by big pharmaceutical and chemical companies would hamper any innovative and independent aspects of new companies.[51]

DNAX's organizational structure was adopted from that of Engenics, which was rather unusual from other commercial ventures in biotechnology. The core of DNAX would have two separate entities: DNAX Ltd. and DNAX Research Institute of Molecular and Cellular Biology as a research subsidiary of DNAX Ltd. The former would function as the entrepreneurial base, identifying business opportunities and building joint ventures with major corporations. A joint venture, a spin-off from DNAX, would focus on product development based on novel processes or technologies arising from the DNAX Research Institute. The corporate joint venture would then commit its financial and technological resources to full-scale production of a commercial product. Though the DNAX Research Institute was wholly owned by DNAX Ltd, the research institute was insulated from commercial joint ventures (see organizational chart in figure 6.2).

The scientific advisory board, which initially consisted of three Stanford scientists, Kornberg, Berg, Yanofsky, and two Harvard scientists, Judah Folkman and Kurt Isselbacher, oversaw DNAX's overall research program and policy guidance. DNAX scientists and technicians, mainly consisting of protein chemists, nucleic acid–cloning and –sequencing chemists, DNA synthetic chemists, animal cell–culture specialists, immune biologists, and associated technicians, would participate in product program teams. They would be responsible for carrying out basic research projects until they could demonstrate the feasibility of a viable product. If a product program team came up with a potentially viable process or product, a management team of DNAX, Ltd, would find a business partner and initiate a joint venture.[52]

This flexible business organization, especially its joint venture structure, was intended to deal with the inherent scientific uncertainty in the

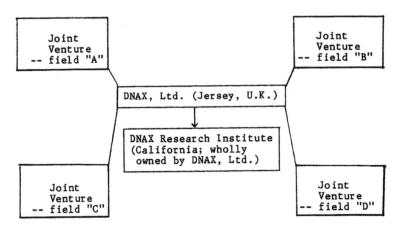

FIGURE 6.2. DNAX organizational chart. DNAX consisted of two separate entities, DNAX Ltd. and the DNAX Research Institute of Molecular and Cellular Biology. The former dealt with business and managerial tasks, and the latter carried out major scientific projects. The development of a potential product was carried out in each joint venture, in scientific consultation with the research conducted. Joint ventures could be formed anywhere in the world. Image from DNAX, "Business Plan," 20 May 1981, p. 10, DNAX Papers, Arthur Kornberg Personal Collections, Stanford University. Reprinted by permission from Arthur Kornberg.

business of biotechnology by enabling DNAX to take advantage of any unanticipated experimental findings. DNAX could form a joint venture with whichever corporate partners it could find based on its technological expertise. As emphasized in the business plan, "the key to DNAX's commercial strategy is *flexibility*. It simply is not known at this time what the course of technological application in the fields of genetic chemistry and cellular biology is going to be. The Joint Venture strategy basically provides a framework for continued growth as long as opportunities can be identified."[53]

If DNAX's flexible organizational structure and commercial strategies were established to avoid three major problems occurring in recent biotech ventures, its separation of managerial and scientific entities was designed to develop an ideal biotech research institution that could maintain healthy interactions with the academic community. In DNAX's mission statement, Kornberg insisted on including the following phrase: DNAX was to provide "fair and appropriate recognition to academic scientists and their institutions for product developments based on their discoveries."[54] Kornberg and Zaffaroni believed recent biotech companies, with their narrow focus on patenting and licensing of academic research, exhibited the failure to develop a long-term relationship with the academic community:

> In addition to the shortcomings of the recently established ventures [outlined above], DNAX believes that most have failed to establish the proper relationship with the academic community which DNAX thinks is so vital for survival in the longer term. DNAX's objective is to develop policies which could become *the model for productive relationships between an industrial enterprise and academic institutions*. DNAX's Scientific Advisory Board will be critical in this regard.[55]

The rapid pace of technological innovation in biotechnology had often made a new technology or patent become quickly obsolete. More problematic was the exchange between academic scientists and industrialists, which was often obstructed by concerns about industrial secrecy. The members of DNAX's scientific advisory board wanted its research institute to be an exemplar in its open interaction with the academic community. Direct communication between academic scientists and industrialists, they underlined, would be critically significant to the business of biotechnology where innovation was based on the unpredictable development of basic biomedical knowledge.

DNAX's insistence on its distinctive organizational features and the academic environment of its research institute were based on both well-intended ethical principles and business considerations. Zaffaroni and Kornberg believed that long-term business survival in biotechnology depended on maintaining the creativity and academic freedom of the scientists involved. If scientists were forced to work on a specific project without the freedom to publish, which was critical to their future career, it would be difficult to attract them to the company. In order to maximize proximity to Stanford's biomedical research community, the DNAX Research Institute of Molecular and Cellular Biology was located in the Stanford Industrial Park in 1980.

In addition to establishing DNAX Ltd. separately from the DNAX Research Institute, the scientific advisors of the latter tried to secure academic appointments for their senior scientists. For example, at an April 1979 faculty meeting in the Stanford Biochemistry Department, Kornberg proposed to appoint Ken-ichi Arai, who was to take a position as a senior scientist at the DNAX Research Institute, as a visiting professor.[56] As Kornberg's former postdoctoral fellow from 1977, Arai had already returned to Japan in 1980 to take a faculty position in the University of Tokyo Medical School. Arai and his wife, Naoko Arai, however, were not satisfied with their situations in Japan. Like other assistant faculty members, Ken-ichi Arai had joined a research group whose agenda was supervised by a senior faculty member. He had then started to work on GTP-binding proteins and their connection with the *ras* oncogene under the supervision of his mentor, Yoshito Kaziro. Naoko Arai, who was also a postdoctoral fellow at Kornberg's laboratory, did not get a formal position at the University of Tokyo, and was working without any salary. Around that time, the ambitious Ken-ichi Arai was trying to organize the Asian Molecular Biology Organization (AMBO) with the help of Kornberg and James Watson, but soon this seemed an impossible task for a young faculty member.[57]

When Kornberg contacted the Arais with a position at DNAX that included Ken-ichi Arai's joint appointment at Stanford's Biochemistry Department, they were tempted by the offer. Kornberg emphasized that DNAX was recognized by the US State Department as an academic center for training and research, which alleviated immigration issues for foreign researchers. Moreover, DNAX would provide generous fellowships for two to three years so that Ken-ichi and Naoko Arai could bring their graduate students or postdoctoral fellows to DNAX, to work on projects of their own choosing. For the Arais, it seemed like a better opportunity than their current situation at the University of Tokyo. With further support by Schering-Plough (twelve million dollars) in the Stanford Industrial Park, DNAX built excellent facilities for protein microsequencing, recombinant DNA technology, and cloning work, both in bacterial and animal cells. DNAX's scientific staff would soon be expanded to one hundred.[58] Attracted by the opportunity to pursue research with generous funds and excellent facilities, the Arais decided to join DNAX in 1981. Kornberg assured them that scientific operations in the DNAX Research Institute would resemble those of the Stanford Biochemistry Department in terms of its open and academic atmosphere.

For its financial arrangements, DNAX took extra care in assembling its incentive and payment structure. In late 1980, Zaffaroni drew up a licensing arrangement with Stanford University for the use of Henry Kaplan's cell lines for the production of human hybridomas. He intended to acquire an exclusive right to Kaplan's cell lines and antibodies so that DNAX could be the sole distributor. Then DNAX could invest in the cell line's further development and sell it to other biotech companies engaged in monoclonal antibody production. In return for this arrangement, Zaffaroni offered Kaplan and Stanford cash royalty payments, or a modest equity stake in DNAX if Stanford or Kaplan preferred that option.[59]

This possible equity arrangement raised concerns among Stanford scientists. Berg wrote a letter to Zaffaroni indicating that he was "concerned by the likelihood that such arrangements could be misinterpreted or generate future embarrassment and conflicts with the university."[60] The apprehension was about the appearance of conflicts of interest. He underlined that part of DNAX's goal was to promote a new model for the interaction between the academy and industry as a way to bring all possible benefits of biomedical research to the public. He also wrote: "in my view the [university's] principle functions are to promote and carry out the scholarly, creative works that generate new knowledge and to promote ways and means of transmitting existing and new knowledge to its students and society at large."[61] Berg reminded Zaffaroni that the DNAX Research Institute was designed to function as an academic research institute inside the biotech venture. Scientists could regard an equity incentive, despite its good intentions, as financial pressure. In the same way, the university's stake in the equity could hamper the open interaction between DNAX researchers and Stanford scientists. Berg let Zaffaroni know that Kornberg and Yanofsky agreed with his opinion that DNAX should avoid profit-motivated incentives, including such an equity arrangement with the university. Berg advocated that the separation between the DNAX Research Institute, commercial enterprise and management, as well as its joint venture programs, would ultimately benefit both the company and the academy, leading DNAX to a scientific and commercial success.

DNAX: Immunogenetic Engineering

When Harvard scientists who were founding the Genetics Institute in the late 1970s approached Zaffaroni in the late 1970s, he asked his employee William P. O'Neill of his pharmaceutical company, ALZA, to evaluate the plan for the new institute. An expert on the new biotech industry, O'Neill had helped prepare a report for the congressional Office of Technology Assessment (OTA) on the current status and future prospect of commercial biotechnology. This OTA report, *Commercial Biotechnology: An International Analysis*, emphasized that the current boom in biotechnology represented a departure from the traditional biotechnology field (such as industrial fermentation or mass-screening drug development) in its foundational dependence on basic biomedical knowledge.[62] As the OTA report summarized, recently launched commercial biotech ventures used their expertise in recombinant DNA technology and hybridoma technology to help manufacture medically useful biomolecules, such as insulin, interferon, and antibodies, on an industrial scale.[63] By the late 1970s, numerous companies had grown out of molecular biology laboratories, including Genentech, Biogen, Genex, Hybritech, and Becton-Dickinson. O'Neill's survey on the emerging biotech industry also pointed out the initial success of Cetus and Genentech, which prompted several major pharmaceutical companies and entrepreneurial ventures to establish research projects based on the premise that "rational bioproduction is now feasible."[64]

After Zaffaroni and Kornberg decided to start a biotech company, they prepared a business plan based on their scientific expertise. In Kornberg's memo of 1980, he indicated that DNAX should be able to demonstrate its unique strength in the new field of genetic engineering. More specifically, he stressed that DNAX had three competitive advantages. First, its advisory scientists included two of the best researchers, Berg and Edgar Haber, in the field of recombinant DNA technology and immunology. Second, because of the expertise in these two areas, the combination of recombinant DNA technology with the hybridoma technology for the production of useful antibodies was likely to be the best potential product for DNAX. Third, AZLA's drug delivery system would provide "dramatic examples of use of hormone, interferon, or vaccines."[65] DNAX's scientific advisory board agreed that the focus of the company would be in the development of genetically engineered antibodies for diagnostic and therapeutic uses. Kornberg also emphasized that DNAX's open atmosphere and excellent resources could attract high quality scientists and venture capitalists. In preparing the business plan, the scientific advisory board characterized its research program as "immunogenetic engineering."

In a belief that scientific discoveries often emerge from unexpected experimental discoveries, DNAX's research was divided into two kinds of programs—basic and project related. DNAX scientific staff members selected the first batch of research programs. In this basic research area, DNAX's scientists would have "complete freedom" to determine their own research programs. The DNAX scientific advisory board selected another batch of research programs for project-related research. These project-related research programs would aim to develop commercially viable procedures or products, which in turn could lead to a joint venture with other commercial entities. In this way, DNAX scientists would have an opportunity to pursue their own research projects, as well as those selected in consultation with the company's scientific management team. DNAX's founders believed that there would be productive interactions between basic and goal-oriented research "in an open academic atmosphere" that could stimulate innovative biomedical discoveries.[66]

DNAX soon established two research divisions, one in molecular biology and the other in immunology. Five research scientists were recruited for the division of molecular biology; they were interested in the genetic regulation of eukaryotic organisms, DNA replication, and immunoglobulin gene cloning. The molecular biology group included Ken-ichi Arai, Naoko Arai, Frank Lee, Kevin Moore, and Gerald Zurawski. All had been postdoctoral fellows in the laboratories of Kornberg and Yanofsky. Two researchers specializing in monoclonal antibody research, Robert Coffman and Timothy Mosmann, were appointed to the division of immunology. To foster the interaction among academic researchers, DNAX scientists could interact with the university-based laboratories of the company's scientific advisory board. In addition, the DNAX researchers were encouraged to publish their work in scientific journals. Though each scientist pursued their research interests, their broadly conceived research projects were expected to contribute to further development in genetic engineering for the production of a wide range of useful proteins.

In 1981, DNAX's scientific advisory board decided to pursue the large-scale production of designer antibodies for its first development-oriented project.[67] This immunogenetic engineering would combine protein production by means of recombinant DNA technology with monoclonal antibody generation, making antibodies designed by genetic engineering.[68] DNAX's scientific research projects in the molecular biology and immunology divisions would significantly contribute to this development-oriented project. For example, Ken-ichi Arai began work on the cloning and expression of mammalian genes. This research might be critical in the development of mammalian cloning vectors (as in the SV40 cloning vectors development in Berg's lab); these might be used for the production of antibodies in mammalian cells. Immunology division scientists worked on the basic mechanisms of monoclonal antibodies against cell surface antigens. The understanding of antibody-antigene configuration dynamics would contribute to the design of antibodies.

DNAX's project to engineer antibodies was indebted to Harvard

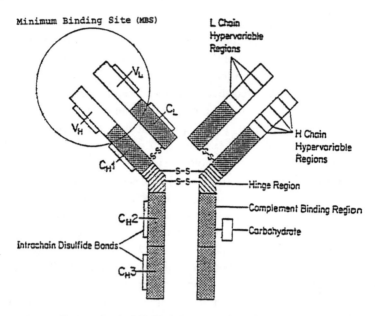

FIGURE 6.3. Antibody molecule. DNAX's immunogenetic engineering project aimed to produce and design an antibody. The specificity of an antibody molecule could be engineered
by the modification of minimum binding sites (MBS) in the antibody molecule, which would
react to antigens. DNAX's immunogenetic engineering project aimed to produce and design
an antibody with the production of MBS of particular immunological functions. Image from
"DNAX's Immunobiology Strategy: Confidential," 18 February 1981, DNAX Papers, Arthur
Kornberg Personal Collections, Stanford University. Reprinted by permission from Arthur
Kornberg.

immunologist Edgar Haber's participation as a member of the company's
scientific advisory board. Haber had recently developed a technique to
suppress human hypertension (high blood pressure) using antibody-
antigen reactions.[69] With Haber's guidance, the development project
began with the identification of a minimum binding site (MBS) peptide,
the smallest peptide or peptide pair that possesses the binding properties
of the corresponding antibody combining site.[70] The DNAX team focused
on the preparation of an MBS peptide that could be chemically modified and mass produced (see figure 6.3). Haber offered exclusive rights of
his promising antibody technology to DNAX for hypertension control. If
the DNAX team could further develop the genetic and immunochemical
technologies for the design and production of antibodies, these could be
used widely in clinical and diagnostic applications.[71] The team predicted
they could develop a wide range of medical products, such as a detoxi-

cant, an imaging enhancer to visualize a blood clot, and therapeutic aids for drug delivery and targeting. In 1982, DNAX made a concrete plan to launch a joint venture for the development of a hypertension control drug.[72] With a definite business plan in hand, Zaffaroni began to present the DNAX plan to potential investors.

DNAX: A Global Enterprise?

In its business plan, DNAX emphasized that its flexible organizational structure would enable the formation of joint ventures anywhere in the world. In addition, the DNAX Research Institute's academic status would make it easier to attract foreign researchers. Their presence at DNAX would in turn help the company maintain a global research network that could facilitate future joint ventures. Its business plan anticipated this global approach: "DNAX intends to be a truly international enterprise, locating its joint ventures and new research operations in optimal locations globally to accomplish research, development, production, and marketing."[73]

In order to secure initial investment capital, the DNAX management team visited pharmaceutical companies all over the world. As DNAX was incorporated on the Isle of Jersey (in the English Channel) to minimize taxes, its initial investment was restricted to foreign investors, though American investors could later have limited opportunities. In late July 1981, Zaffaroni, Kornberg, and Haber first visited a Japanese company. In that meeting with business leaders and scientists, Kornberg explained DNAX's recent advances in molecular biology and commercial ventures. He presented the company's scientific expertise based on his research in enzymology, Berg's research on recombinant DNA technology, and Yanofsky's gene-expression studies. Zaffaroni discussed DNAX's business plan, as well as its focus on recombinant DNA technology and cell-fusion techniques for the production of therapeutic molecules.

By the time of this visit, Japan's major pharmaceutical companies had established research and development programs in biotechnology with the country's own scientists and industrialists. According to one survey, about two hundred university and government research laboratories had started to establish genetic engineering research teams by 1981. In their attempt to use advanced genetic engineering methods, such as recombinant DNA and hybridoma technologies, these research laboratories in ac-

ademic institutions and pharmaceutical companies usually hired Japanese
scientists who had worked in advanced laboratories at overseas universities and companies. However, the Japanese laboratories rarely formed a
corporate arrangement with foreign biotech companies. Hiroyuki Mat-
sumiya, president of the Biosystems International of Tokyo, ascribed this
lack of international cooperation to concerns about industrial secrecy. As
he put it in 1981, "secrecy is the name of the game in Japanese Industry."[74]
DNAX's attempt to form a joint venture with a Japanese company revealed a more complicated challenge than they had anticipated.

The DNAX team had a specific plan and purpose for their Japan trip.
With Haber's research on hypertension control, DNAX wanted to form
a joint venture with Suntory Ltd. in Osaka. Suntory, Japan's major liquor
company, began to diversify its business into genetic engineering by establishing the Suntory Institute for Biomedical Research in 1979. The Suntory Institute recruited prominent Japanese molecular geneticists, such
as Shoji Matsubara, from the City of Hope National Medical Center in
the United States; in 1977, scientists at the Suntory Institute had participated in the first bacterial synthesis of human growth hormone (soma-
tostatin), along with Genentech scientists. Ken-ichi Arai, in particular, was
interested in biochemist Teruhisa Noguchi's research at the Suntory Institute. Noguchi's team had isolated a sequence of peptides that could di-
minish hypertension. Arai had contacted Noguchi with regard to DNAX's
interest in designer antibodies that could bind enzymes and resulted in
lowering blood pressure. He also explained Haber's recent elucidation of
the mechanism of hypertension control through enzyme-antibody modification. Arai emphasized that with DNAX's development in the general
technology for immunogenetic engineering, Noguchi's peptides could be
modified in a way that would be clinically effective. According to Arai,
some scientists at the Suntory Institute were sympathetic to DNAX's antibody project.

When the DNAX team visited the Suntory Institute, Haber, Kornberg,
and Zaffaroni presented their business plan on hypertension and proposed a joint venture. Some scientists at the Suntory Institute, however,
were not so impressed by DNAX's plan. Moreover, Noguchi, who was
"proud of assembling many of key scientists of Japan as advisors for Suntory," such as Yuichi Yamamura, Den-ichi Mizuno, Hayaishi, and Yoshiaki
Miura, was disappointed by Kornberg and Haber's scientific lectures. Noguchi mentioned that Kornberg's talk on DNAX's research projects was
full of "old-fashioned" technologies. With respect to Haber's antibody re-

search proposal, Noguchi indicated that he needed further clinical trial results to see whether it could move beyond "philosophy." Noguchi was further taken aback by Zaffaroni's hasty joint venture proposal that Suntory should work under ALZA's guidance in order to gain access to advanced drug delivery systems. Noguchi indicated that his scientific team was collaborating with the Honjo Chemical Company and had already made significant progress. His team was also considering a collaborative project with Takeda, the largest pharmaceutical company in Japan with much better scientific and financial resources than DNAX. Even Aria wrote to Kornberg: "please tell Alex [Zaffaroni], Japan is rather advanced than he thinks."[75]

Noguchi also offered broadly construed critiques of America's biotechnology boom. He thought that American biotech companies were busy selling themselves to the public equity market before they even developed a viable product. Citing Genentech as an example, Noguchi criticized its president, Robert Swanson, as a man "who sold the rainbow," as opposed to one "who got the rainbow." Noguchi noticed that Wall Street, after a brief period of publicity and hype, had developed a "generally cold attitude for biotechnology venture"; this made it difficult for other biotech companies like Biogen to raise money.[76] He wondered how DNAX, with its unusual organizational structure and a small number of scientific staff, could attract a substantial private investment without any proprietary biomedical technologies. By contrast, Suntory, which had received an initial research contract from the Health and Welfare Department of the Japanese government, was not yet concerned about its own funding.

As the trip to Japan illustrated, the DNAX team had difficulties convincing potential investors of the value of its business plan. It did not yet own key technologies that could guarantee a certain profit and draw capital investment. DNAX's organizational structure, which separated the traditional connection between research and development, seemed odd to industrialists, who were not convinced that this separation would be beneficial. In the end, DNAX's travel to Japan, England, and France ended without much success. "Never before had I been turned down for research support," Kornberg recollected about this worldwide business trip.[77] Instead, Zaffaroni managed to acquire an initial investment of four million dollars from Swiss bankers. With this initial capital, DNAX appointed its first scientific staff, Ken-ichi Arai, Naoko Arai, and Gerald Zurawski (a postdoctoral fellow at Yanofsky's lab), and began its research in 1981.

DNAX's plan to establish its research institute as an autonomous center for basic research faced considerable financial problems. As DNAX failed to secure a joint venture with Suntory, Ken-ichi Arai instead decided to bring Japanese graduate students or postdoctoral fellows with him who might be interested in his research on the construction of a mammalian cloning vector. With limited funding, he was initially able to bring only three graduate students from the University of Tokyo. Soon he gathered five more postdoctoral fellows in his group at DNAX. Though the company had an academic status, its plan to grant formal degrees was not approved. Ken-ichi Arai also did not have any formal faculty position at a university in the United States. He therefore improvised an arrangement with Kornberg and Kaziro, his teacher at the University of Tokyo, to supervise graduate students. He urged Kornberg to help institute a graduate program within the DNAX Research Institute. Ken-ichi Arai indicated that the Roche Institute, which operated within a major pharmaceutical company (Hoffmann–La Roche) would be a good model. Without a graduate program, he wrote to Kornberg that DNAX might become "just one of [the] gene-splicing companies with typical industrial atmosphere."[78]

By late 1981, DNAX's initial vision of creating a research enterprise in an industrial setting with an academic atmosphere faded as it rapidly burned through its initial four million dollar capital without acquiring any additional investments. In January 1982, Zaffaroni approached Robert P. Luciano, president of a major US pharmaceutical company, Schering-Plough. Other major pharmaceutical firms, such as Eli Lilly and Merck, shifted their work into molecular biology in order to break new ground in rational drug design, moving away from traditional chemical-screening projects. Schering-Plough grew interested in building its own research division in molecular biology and genetic engineering. As new participants in biotechnology, major pharmaceutical companies were not sure about how to accommodate the field, especially with its peculiar mix of research and development, into their business organization. However, they decided that acquiring a start-up biotech company, which had pioneered an institutional rearrangement among academic scientists, industrialists, and venture capitalists, could become a convenient way for a big pharmaceutical firm to gain expertise in the field of genetic engineering.[79]

On January 13, 1982, Zaffaroni had a meeting with Berg, Kornberg, O'Neill, and Yanofsky in order to prepare for their meeting with Luciano. They discussed the financial distress of DNAX and decided to sell their company to Schering-Plough. Once DNAX had been purchased, it existed as a freestanding subsidiary of Schering-Plough, and Luciano claimed that his company was now squarely at "the forefront of recombinant DNA technology and immunology research, the two principal pathways to the development of future therapeutic agents."[80] DNAX's management team and the scientific advisory board, on which Berg, Kornberg, and Yanofsky still served, sought to maintain the research institute's previous organization and academic culture. DNAX's scientific advisors and Schering-Plough, however, decided to move away from ambitious immunogenetic engineering. Instead, in consultation with Ken-ichi Arai, DNAX's management team decided to shift its research focus to T cells and cytokines. By expressing clones of growth factors isolated by immunologists, DNAX researchers contributed to the elucidation of T-cell signal pathways and the function of cytokines.

Ken-ichi Arai, however, was disappointed by the changing fate of the DNAX Research Institute. For example, he complained about Schering-Plough's approach to research management, including the introduction of performance appraisals. J. Allan Waitz, the new president of DNAX, requested these as a way to evaluate the current status and progress of scientists' research projects. Arai questioned whether this kind of evaluation would be effective, and acknowledged that he felt the "present form of 'performance appraisal' is too 'project oriented' and 'too technical' for the purpose of reviewing principal/research scientists."[81] Waitz slowly learned that some of the distinctive aspects of DNAX, especially its commitment to an open and academic environment, were intended to not only to foster an ideal environment for research but also to offer useful strategies to the business of biotechnology. After eight years of a career as a research scientist in DNAX, Ken-ichi Arai returned to the University of Tokyo in 1989 to be his mentor's successor. On his return to Japan, he promoted biotechnology in Japan, and his connection with Schering-Plough helped the company's entry into the Japanese market, contributing to the emergence of biotechnology as a global enterprise.

Exclusive Industrial Contract?

From the late 1970s, Stanford biochemists experimented with new forms of the biomedical enterprise. In addition to the IAP and DNAX, another form of support was proposed in December 1981. In a faculty meeting, the Stanford biochemist James Rothman notified other Stanford biochemistry faculty members that Smith Kline and French Labs (SKF) wanted to discuss the possibility of developing an industrial liaison with the department. George Poste, vice president and director of research at SKF, wanted to create a research exchange program as a first step, so that the company could bring its scientists "on short notice to pick up freshly developed techniques to incorporate into SKF."[82] Poste also expressed his opinion that "minimal commitment programs like IAP are a transitory form of relations, and will be short-lived generally."[83] Thus, he underlined that it would be "absolutely essential" that SKF have an exclusive connection to the department. Poste indicated that with a five-year exclusive contract, SKF would provide substantial resources that could fund the expansion of the Biochemistry Department, including facility renovations and new instrumentation, in addition to research support.

The proposed industrial support to the Biochemistry Department, with SKF's insistence on exclusivity, raised serious concerns among Stanford biochemists, which prompted them to discuss the nature and implications of this sort of financial backing. The insistence on exclusivity in SKF's proposal alarmed a young faculty member, Douglas Brutlag, who wrote a confidential letter to A. Dale Kaiser, the current department chair, declaring that he would be against the proposed arrangement. He further suggested that this exclusive arrangement would "jeopardize any relationships that our department as a whole have with our other Industrial Affiliates and [that] it would also have with other companies."[84] Kaiser subsequently drafted a confidential memorandum titled, "Creation of a Substantial Financial Relation with a Company," circulating it to biochemistry faculty members. In this memo, he expressed his concerns on the growing commercial ties of the department with private industry, especially the department's recent involvement with DNAX. His first concern was that their commercial connections, if they became substantial, would reduce the department's ability to obtain federal funding. Grant review committees or study sections, whose aims were to evaluate basic biomedical research according to its scientific merit, would not view the department's commercial emphasis favorably. The ties with industry, especially through a contract with a single company, could threaten the department's core mission of research and education.[85]

Kaiser further emphasized that commercial support, compared to federal funding, had been too much goal-oriented, short-term, and rarely stable. Though the department had largely depended on federal agencies for its financial support, it was able to insulate itself from "unreasonable demands by the multiplicity of our grants and by the peer review system" through its moral economy of research. If a single company provided a large portion of support, the department would become vulnerable to demands from that company. This potential loss of autonomy in research would not only tarnish the "reputation of independence which I think our Department presently enjoys" but also make it difficult to attract the best students and postdoctoral fellows. This exclusive form of commercial support would exacerbate potential problems of the department's continuing large financial connection with a company. Kaiser instead suggested that the department's current IAP, which had "few strings attached and leaves our initiative and reputation intact," would be a more prudent approach in its search for alternative support for biomedical research.[86] Stanford biochemists decided in the end that they would not arrange an exclusive SKF contract.

Pajaro Dunes "Biotechnology" Conference

As a number of biotech ventures emerged from academic laboratories with the infusion of capital, university administrators became increasingly concerned in the late 1970s and early 1980s about potential conflicts with academic commitments. Stanford University and UCSF, among the first academic institutions that triggered the emergence of commercial biotechnology, included faculty who uneasily observed the frenzy of marketing gene modification and cloning technologies, as well as the potentially detrimental impact on academic culture. The controversy over the ownership of the human growth hormone gene between Genentech and UCSF that erupted in 1978 was one of the most illustrious of such examples, leading to an erosion of personal and public trust in academic researchers who developed zealously proprietary attitudes toward knowledge and tools of scientific significance. Stanford president, Donald Kennedy, called for a meeting in December 1980 for university and biotech industry leaders to deal with "the rush to proprietary control of recombinant DNA research."[87] Kennedy soon sent letters to five major research universities, inviting Derek Bok of Harvard, Marvin Goldberger of Caltech, Paul Gray of MIT, and David Saxon of the University of California. He extended his invitation to major biotech and pharmaceutical companies, such as Genentech, Syntex, Gillette, DuPont, Eli Lilly, and Cetus. In 1982 the heads of five major research universities and eleven corporations convened at the Pajaro Dunes Conference Center in California in order to "consider the opportunities and problems offered by developing university-industry relationships in the field of biotechnology."[88]

The "Biotechnology Conference" marked the beginning of the discussion on the increasing commercialization of academic research for institutions dedicated to higher learning. The conference, despite repeated requests for admission from the press, students, and representatives from public interests group, was closed. Kennedy noted that the small size of this exclusive meeting facilitated university and industry leaders having "full and frank" discussion at an early stage of the commercialization of academic research.[89] In the age of dwindling government support, Kennedy felt that the future of higher education increasingly depended on the willingness and ability of universities to capitalize on their research while preserving core academic values. After all, some Stanford scientists and administrators had played a catalytic role in the commercialization of biomedical research through the patenting of recombinant DNA technology as a way to demonstrate medical relevance of its research and to draw additional financial resources.

The conference proposal prepared by the office of the Stanford president began by noting, "there has been a dramatic acceleration of commercial interest in the kinds of research that have been thought of as 'basic' research which is restricted almost entirely to university settings."[90] Kennedy added that this new commercial interest was fueled by "the possibility that there may be significant losses in government support for basic research." The chance of losing government funding would result in scientists and universities seeking new sponsors, and genetic engineering was a major focus of this rising commercial interest. Participants of this biotech conference were asked to consider the following key questions about the proper mode of collaboration between universities and industry. If the trend in the reduction of federal funding for academic research increased, should the university regard a commercial tie with industry as a major alternative source of support? Should the university enter a proprietary activity itself or delegate this to other nonprofit institutions? Would increased interaction with private industry be beneficial to both the practitioners of science and other members of the universities? More significantly, what effects would the increasing commercial practices in universities have on the central missions of higher education, such as the creation and dissemination of knowledge and the education of its students?

First, the Stanford president's conference paper acknowledged that recent technological developments, the prospect of decline in federal funding, and changes in federal policies on intellectual property all enabled academic researchers to engage with private industry through a wide array of arrangements. Scientists emerged as major players in a new pattern of affiliations between university researchers and private industry. University administrators noted that the result was a variety of alliances in biotechnology between the academy and industry. These included:

> large grants from single firms to university laboratories, with commitments to exclusive licensing; an array of equity consulting agreements between individual faculty members and firms; still stronger linkages between faculty members and firms that involve the migration of entire programs of research into the proprietary sector, often with a level of supervision by the faculty member that approaches line management; active programs of technology licensing on the part of universities; and others.[91]

In reflecting on these diverse forms of affiliation, the participants of the conference noted that there was a distinctive pattern in the biotech industry that stemmed from its heavy dependence on securing intellectual property rights of broad research procedures, tools, and materials. Most academic scientists and commercial biotech researchers tended to make proprietary claims as early and broadly as possible on promising procedures or technologies in molecular biology, even before any concrete and obvious usefulness had been determined. Biotechnological inventions usually derived from basic research, biotech leaders pointed out, and these novel procedures, materials, and technologies often required continual investment for full commercial use and development. The focus on intellectual property was thus developed as a prominent business strategy for securing the infusion of venture capital at early stages of research and development. As a substantial amount of funds were required for commercial development, a startup biotech company often needed a series of infusions of capital; with each investment, the number of the company's initial share increased through share split, while the value of the company reflected the additional capital investment. This share split made the value of the company's initial stake balloon with subsequent capital investment. Eventually, this culminated in the company offering its public stock at Wall Street.

For example, Boyer first acquired 25,000 shares of Genentech stock when the company received the capital investment of $100,000 from

Kleiner & Perkins, a venture investment firm, which then itself acquired 20,000 shares of Genentech stock. These initial stocks were spilt with each successive capital investment, and by the time Genentech offered 1.1 million shares on the NASDAQ exchange, Boyer owned 925,000 shares. When Genentech's stock went on the market, its price skyrocketed from $35 to $80, closing at $71 a share on October 14, 1980. Boyer's stock holding valued about $65 million in the first day of Genentech's public stock offering. Kleiner & Perkins, which by then possessed 938,000 shares purchased at an average of $1.85 per share, demonstrated the huge financial reward a venture capital investment could bring in a high-tech industry like biotechnology.[92]

The financial operation of the biotech industry thus contributed to its propensity to focus on basic research in molecular biology, as the Stanford president's conference paper emphasized:

> New economic incentives [in the biotech industry], especially a pattern of capitalization in which large changes in value are associated with successive generations of investment . . . [have] placed a premium on the early possession of valuable intellectual property, and [have] pushed the zone of corporate interest into increasingly "basic" research areas.[93]

The huge economic incentive for commercializing "basic" findings in biotechnology presented both an opportunity and challenge for academic institutions and researchers. At one level, it enabled scientists in basic biomedical research who otherwise would be indifferent to commercial development to reap enormous financial gains, while the infusion of capital could contribute to the acceleration of medical innovations. At another level, academic institutions and scientists wondered whether this commercializing of basic research related to genetic engineering could be accomplished without damaging "principles [that] accord with basic science—openness, interaction, conflict-of-interest rules."[94] From the perspective of academic researchers, the tendency in biotechnology to patent every novel procedure and technology at an early stage posed a serious threat to academic work. Privatization of key tools and materials would not just hinder their dissemination for wider use in biomedical research and development; the research community as a whole would suffer from secrecy stemming from concerns about securing priority and proprietary rights of key procedures and materials.

Participants at the biotech conference tried to discuss how to institute a proper mode of commercialization in a way that could preserve some valuable aspects of academic research. They were broadly concerned whether this drive toward privatization might

> promote secrecy that will harm the progress of science, impair the educational experience of students and postdoc fellows, diminish the role of the university as a credible and impartial source, interfere with the choice by faculty members of the scientific questions they pursue, or divert the energies of faculty members and the resources of the university from primary obligations to teaching or research.[95]

The participants debated intensely, for example, whether academic institutions would need special conflict-of-interest rules when faculty members were engaged with proprietary ventures, and they wondered whether universities could grant a company an exclusive license for profit. The proliferation of industry contracts, increasing interest in patents and licenses, and faculty members' founding of a biotech firm challenged academic institutions and scientists to confront such questions.

In the end, the discussions at this conference provided a framework for navigating the uncharted territory that lay between the world of research and commerce. However, academic and industry leaders could not generate uniform principles or ethical codes that could guide the biotech enterprise in the academy. While they agreed that it would be beneficial for both to engage in the commercialization of biomedical research for medical innovations and public benefit, they conceded that it might be too early to set any concrete guidelines. Instead, university administrators and biomedical researchers were left to determine their own arrangements with private industry on an ad hoc basis. This then led to confusion about the proper mode of the academy's affiliation with industry. Some scientists, faced with their institution's cautious and often conflicting policies toward commercialization, even decided that it would be too difficult to engage in a biotech venture while trying to maintain their academic obligations. For example, the prominent biochemist and Nobel laureate Walter Gilbert announced that he would resign from Harvard in 1982 to focus on Biogen, the major biotech firm he had cofounded.[96] In their attempts to take advantage of the commercial prospects of biotechnology, university administrators and biomedical researcher had to find a way to fulfill their obligations not only to the academic community for creative research but also to the public for medical innovations.

Conclusion

The Stanford biochemists' attempts at various forms of industrial affiliations and ventures in the late 1970s and early 1980s reveal how their views of commercialization shifted as they reassessed its potential benefits and risks against the backdrop of their particular research culture. In their attempt to seek additional funding from private industry, they were primarily concerned about how to maintain their practices of sharing and semicommunal ownership of ideas and materials that contributed to a collegial and productive experimental life. The particular design of the IAP and DNAX institutional arrangements, which could keep corporate interests at bay, reflected the Stanford biochemists' cautious approach to commercialization. They were convinced that the long-term business success of a biotech venture lay in its ability to arrange a productive relationship between academicians and industrialists, and they asserted their steadfast focus on openness and flexibility in scientific research. The biochemists critically assessed the pitfalls of the growing biotech industry, including its hasty monetization of intellectual property in basic research and its escalating effect on secrecy amid fierce competition. In his appraisal of the DNAX venture, Kornberg underscored how the fate of the biotech industry depended on its maintenance of academic culture and openness inside a biotech company:

> [Secrecy] makes even less sense in industry than in academia. In a competitive academic field, the disclosure of a reagent or a procedure, or the hint of success in some direction, may lead another scientist to reproduce and publish a result quickly to gain priority for an important discovery. By contrast, discoveries in the industrial world far exceed the resources needed to pursue them. What matters most is making a shrewd choice of which discovery to develop, because each of the costly and time-consuming hurdles of clinical testing, regulatory approval, quality control, and marketing is crucial in the success of a product.[97]

At the same time, the prospect of the general decline in federal support, along with the demands during the 1970s for relevance and accountability in biomedical research, made commercial affiliations seem relatively less intrusive to some other scientists. Increasingly inflexible government grant arrangements prompted Stanford biochemists to seek unrestricted funding from other sources. Their establishment of the IAP, for example, was introduced in this context of the shifting nature of federal support for biomedical research. More importantly, a new legal regime for academic patenting, starting with government's institutional patent agreements and later formalized by the Bayh-Dole Act in 1980, encouraged academic researchers to commercialize their new discovery or invention in order to accelerate medical innovations and gain profit for scientists and universities. The shifting moral landscape of private ownership in academic research in turn prompted academic researchers to reconceive their commercial engagements, reinforcing changing economic, legal, and political developments toward the commercialization. Stanford biochemists' changing perspective on the nature and pattern of federal funding, along with federal policies encouraging commercialization for the public interest, helped them see industrial support and commercial engagement in a more morally positive light.

In seeking to explain the diverse motives and rationales of academic scientists' participation in the biotech industry, historians and sociologists have tended to stress advances in biomedical research, legal shifts allowing academic patenting, and the relative flattening of federal funding during this period.[98] These scientific, legal, and economic changes in the 1970s gave academic institutions and researchers a systematic incentive for making proprietary claims on scientific knowledge, which prompted the rise of the so-called entrepreneurial university whose key mission was to promote economic development.[99] Some academic leaders and scientists, however, were hesitant to accommodate commercialization. As discussed at the first biotech conference proposed by the Stanford president, concerns focused on the erosion of academic values, such as the free and open exchange of ideas, conflicts of interest between professors and students as they become involved in commercial ventures, and the waning spirit of free inquiry in the academy.[100]

As my analysis of the Stanford biochemists' attempt to reenvision their biomedical enterprise in the coming of age of biotechnology shows, a sharp distinction between "open" academic and "secret" industrial research, however, had become increasingly difficult to maintain because of the shifting political and moral economy of biomedical research in late-twentieth-century American capitalism. The traditional sharing practices in the Stanford Biochemistry Department could initially be maintained in part because of a post–World War II public and political consensus for government support in basic research.[101] Once that consensus disintegrated in the 1970s, scientists had to accommodate diverse political concerns and economic interests through their engagements with government, industry, and the public. At a time when capitalistic conceptions of knowledge took a strong hold among policy leaders and politicians, university administrators and some scientists were drawn to commercial engagements with industry through the privatization of biomedical research. With a new legal regime for academic patenting firmly in place, private industry seemed to provide a better alternative mode for supporting science, with less regulation and restriction compared to government grants.

Seen from this perspective, commercial biotechnology was increasingly perceived by academic researchers and university administrators as an alternative way of doing science, with more financial rewards and supposed social benefits. Scientists' pursuit of money, which was thought to spur medical innovations for public benefit, became a moral calling for a scientific vocation in the age of biotechnology. Stanford biochemists' struggle to maintain DNAX as an alternative type of biotechnology, and its subsequent acquisition by a major pharmaceutical company, proved that their particular vision of their basic and applied medical science was far from guaranteed. As they realized, the fate and shape of their research enterprise was inescapably linked to political, economic, and moral conditions. During the beginning of the pursuit of biotechnology, the Stanford biochemists' relationships with the public and their sponsors were increasingly mediated through the circulation of capital.

Conclusion

Biotechnology in the second half of the twentieth century transformed the research university, bringing it to markets under a new intellectual property regime and a transforming moral landscape in the academy. In the early 1970s, a creative community of biomedical researchers in the San Francisco Bay Area developed gene manipulation and cloning technologies, including recombinant DNA technology, in order to investigate more complex and medically relevant objects at the molecular level. These molecular technologies emerged from university laboratories, whose institutional arrangements and disciplinary trajectories were shaped by large public funds and expectations of medical benefits. In this context, a call for relevance in biomedical research, genetic engineering developed as one of the key technologies applied to agriculture, medicine, pharmacy, and industry.

Some entrepreneurial scientists, research administrators, and industrialists subsequently mobilized the potential of recombinant DNA technology by launching a commercial venture, claiming private ownership of the technology and its benefit to the public at large. This transition of genetic engineering into commercial biotechnology was further mediated by market incentives introduced by shifting institutional arrangements, such as university patent management and federal patenting policies, which granted legal privileges to academic institutions and investigators for proprietary claims for publicly supported research. When start-up biotech companies like Genentech awed Wall Street in the early 1980s with its dazzling initial stock offering, biotechnology reached its zenith, creating a new high-tech industry with millionaire scientist-entrepreneurs. With the advent of a knowledge-based economy, generating new profits and medical innovations became another public obligation of academic researchers and institutions.

Biotechnology grew into a field where scientists, research administrators, and industrialists challenged, broke, and eventually redrew the boundary between academic research and industry by linking the biological and the medical, the scientific and the commercial, and the public and the private. These connections among the scientific, economic, legal, and moral aspects of biotechnology were relentlessly forged by its early participants and promoters. Through their laboratory experiments, they promised immediate medical breakthroughs; through their claim of proprietary knowledge, they sought clarity of scientific credit and economic reward; and through their commercial biotech ventures, they preached the broad public benefit that would result in economic growth. Yet, as critics of commercial biotechnology argued, these promoters fell short of their claims of the benevolence of private ownership. The critics thought those biotechnologists undermined a creative community of scientists through their assertion of proprietary attitudes toward knowledge, subverted due scientific credit and reward through their pursuit of legal clarity for inventorship, and confused private gain with the public interest amid their enormous financial gain. Biotechnology, in other words, also became an area where attempts and efforts to preserve academic research flourished.

By analyzing the shifting institutional, scientific, legal, and moral transformations that attended the commercialization of biomedical research in the latter half of the twentieth century, I have admittedly adopted the perspective of Stanford biochemists who protested key aspects of commercial biotechnology. These dissenters, contrary to the opinion of many contemporary commentators and later historians, were not conservative scientists refusing to venture into medically useful research while clinging to pure science; nor were they reactionaries who were personally bitter about lost scientific credit and economic opportunities.[1] Instead, they had a sophisticated and nuanced understanding of what commercial biotechnology brought to their laboratory and institutional life as creative academic researchers. First, the growing proprietary interests disrupted their sense of a moral economy of science that was based on semicommunal conceptions of knowledge and tools, which thus threatened the productivity of their work. Second, legal assertions of private ownership of academic research reconceived the research university's obligations toward economic development, challenging views of public knowledge and its relation to the public interest. Third, moral justifications for privatization that equated private gain to the public interest seemed to undercut public trust and investment in academia, while shifting vocational aspirations of scientists toward entrepreneurial biotech ventures.

In order to fully appreciate these Stanford biochemists' view of commercial biotechnology, I have shown how their laboratory life and their moral economy of science critically depended on the cultural, institutional, and political status of biomedicine, whether it was mobilized to fight the "war against disease" or reconfigured to better enable the commercialization of academic research. Within the context of the post–World War II rise of biomedicine as a major research enterprise, Stanford University established its Biochemistry Department as a major biomedical research department focused on the genetics and biochemistry of DNA, the material focus of biomedical research at the time. The "DNA Department" at Stanford developed its own style of research management and a collaborative culture. These emerging patterns of sharing tools, techniques, and ideas, along with scientists' wariness of goal-directed federal funding, fostered pooling practices (both monies and materials) among research groups as a way to sustain their research productivity and flexibility during a time of enormous expansion of federal support for biomedical research. Initially, the Stanford scientists' conception of open science had relied on a particular postwar arrangement between biomedical research and the government. By mobilizing the medical aspirations of lay activists and politicians, biomedical researchers were able to garner a bounty of federal funding, which in turn sustained the "public" nature of their scientific knowledge in the academy.

The increasing demand for practical applications and medical relevance in biomedical research by voluntary health activists and politicians from the late 1960s in turn changed the financial and material conditions of the experimental life of Stanford biomedical researchers. By the early 1970s, both the economic utility of public knowledge for industrial and medical development and the political support for biomedical research were being challenged. Indeed, a different set of economic and political ideas surfaced about the relationships among the intellectual commons, intellectual property, and use of scientific knowledge. During the stagflation of the 1970s, as federal research policies increasingly demanded tangible and immediate medical applications and technological innovation through privatization, proprietary concerns began to permeate the scientific life of biomedical researchers; the world of academic research became increasingly competitive. While Stanford biochemists' culture of sharing regulated these competitive attitudes and concerns over scientific priority and credit, the emergence of the legal avenue for privatization of scientific knowledge and its moral justification challenged the existing moral economy of science at Stanford. Instead, as some Stanford administrators and scientists took advantage of an economic opportunity created by the National Institutes of Health's (NIH) institutional patent agreement for commercializing recombinant DNA technology, potential tensions and competitive feelings were amplified.

On a broader level, the commercialization of recombinant DNA technology was at the center of the reconfiguration of public knowledge and the academy, and private knowledge and industry in the 1970s. Institutional and legal rearrangements in the academic institutions (Stanford and the University of California) and the federal government (NIH) brought crucial changes in the economic, legal, and moral landscape for the research university in the 1970s and 1980s. I have contended that the seemingly "inevitable" link between recombinant DNA technology and commercial biotechnology, often put forward by proponents of its commercialization that assumed commercial potentials of genetic engineering, has obscured this significant shift in the relation between public and private knowledge, and its relation to the public interest in scientific enterprise. This, in turn, changed the post-World War II institutional arrangement between academy, government, and industry that had sustained a reservoir of public knowledge. A group of government patent officials and university patent managers, echoing industry's broad criticism of the federal government's ineffective patent policy and management, began to argue that a large amount of biomedical knowledge in the public domain had not been used for the public good, because no legal ownership stake had been permitted by the NIH. According to this argument, the public knowledge of the academy whose work had been nourished by taxpayers' money had ironically harmed the public interest because the research was in the public domain and thus not patentable. That lack of private ownership and control, according to proponents of biotechnology, had hampered industry's willingness to invest in medical innovations.

Market-based ideas about knowledge and the articulation of private knowledge benefiting the public through economic and medical innovation in the 1970s increasingly mediated political and public demands for medical outcomes. As private ownership of knowledge (through patenting) came to be viewed as a new way to liberate biomedical discoveries for public benefit, Stanford administrators and scientists began to experiment with various forms of industrial associations, such as patenting, licensing, and consulting, and even founded a venture biotech company. This changing view of public knowledge as an impediment to medical innovations was reflected in the changing patent laws that concerned inventions arising from government support. Stanford and the University of California's patent filing on recombinant DNA technology took advantage of the NIH's new institutional patent agreement that allowed private ownership of inventions supported by the taxpayer's money. Consequently, key academic and government officials who were involved in the commercialization of recombinant DNA technology, such as Niels Reimers, Norman Latker, and Betsy Ancker-Johnson, refined their political economic argument on private knowledge and built a political coalition in order to gain broad support for the privatization of academic research. Seen from this historical perspective, the "sell-out" narrative prevalent in the current historiography of the commercialization of academic research narrowly relegates university administrators and academic scientists' commercializing endeavors to inherent technological potentials of biomedical technologies, such as genetic engineering and genomics, and to their pervasive profit motives.[2] Instead, the privatization of biomedical research was one of the significant outcomes as academic scientists, university administrators, and government officials tried to adjust the research enterprise, on the basis of the positive effect of private ownership, to the shifting scientific, economic, and legal conditions surrounding research universities of the 1970s.[3]

As the boundary between the public knowledge of universities and private knowledge of industry shifted during the age of commercial biotechnology, Stanford biochemists in turn wondered whether their traditional patterns of sharing and exchange could be maintained. For example, grant pooling had been devised to maintain research autonomy in the age of federal support in biomedicine. As calls for accountability and relevance in federal funding intensified, the biochemists modified their double-bookkeeping practice so that each group could be accountable for their expenses. More important, the realignment of the moral economy of science within the financial regime of the increasingly capitalistic biomedical enterprise affected the Stanford biochemists; they changed their attitudes toward commercialization, seeking a "proper" format for the interaction between academy and industry in the 1980s. The biochemists believed that industrial support, if properly instituted, could provide research funds in a way that could sustain their moral economy of science. Industrial support seemed to possess fewer restrictions than the demands for medical relevance accompanying federal support. It was in this context that industry reappeared as an alternative source of financial support for scientific research. For example, instead of seeking an exclusive

company contract, the Stanford biochemists set up an industrial affiliates program to generate unrestricted funding. In their affiliation with
the biotech industry, they helped establish an independent research institute for commercial biotechnology. This more positive moral rendering of private ownership in scientific research, articulated by a group of
university patent managers, entrepreneurial-scientists, and governmental
officials, indeed justified a set of expectations about the commercialization of research results arising from government or public support, making commercialization a new public obligation for academic researchers.

The institutional and moral reconfiguration of the research university
gave biotechnology a particular legal and moral form. The United States
Congress eventually passed the Bayh-Dole Act in 1980, allowing universities to claim a proprietary right of inventions that resulted from governmental support. The broader shift in the conception of intellectual
property and its relationship to the public interest meant that the nature
of federal funding also underwent a profound change in the early 1980s.
Biomedical researchers faced more public calls for medical relevance because of demands for practical solutions. Federal funding agencies in turn
tried to implement more accountability and to request tangible medical
applications in return for their research support. Scientists' obligations
to their supporters (including the public at large) meant that they were
expected to engage in more tangible research results, such as patents or
drugs. Biotechnology, with its particular market-based assumption of a
causal relation between private ownership and the public interest, has introduced commercialization as a new mandate for academic institutions
and researchers, bringing financial rewards and supposed social benefits.
Scientist-entrepreneurs' pursuit of profit, a by-product of their zeal to
spur medical innovations for public benefit, became a moral calling for a
scientific vocation in the age of commercial biotechnology.

In conclusion, this book has examined the coproduction of biotechnology and the American entrepreneurial university.[4] Promoters of
biotechnology—ambitious scientists, a group of academic and government research administrators, and private investors—transformed the research university through their relentless pursuit of privatization. The research university in turn provided a valuable intellectual asset to both
academic scientists and private entrepreneurs through its legal arrangements with the federal government that allowed ownership of academic
research. The resulting entrepreneurial university became an arena in
which scientist-entrepreneurs competed for optimum medical contribu-

tion and economic profit. They linked private profit and public benefit in the growing field of biotechnology, shaping a new moral landscape for the privatization of academic research. Indeed, the dazzling emergence of the biotechnology industry was dependent on the legal reconceptualization of public knowledge, as well as on the economic and moral reshaping of commercial ventures by academic institutions and researchers in an age of the ascendency of market values and rationales. In this respect, the commercialization of recombinant DNA technology through private ownership reflected particular scientific, economic, legal, and institutional contexts for biomedical research unique to capitalism in the United States. As for future research, the global and comparative perspective on the history of biotechnology can help us analyze alternative views on the scientific, economic, legal, and moral dimensions of commercial biotechnology.[5]

Abbreviations of Archival Sources

ADK   A. Dale Kaiser Papers, Stanford University Archives, SC 356

AK   Arthur Kornberg Papers, Stanford University Archives, SC 359

CGB   Carl G. Baker Papers, Office of NIH History Collections, National Institutes of Health, Bethesda, MD

DNAX   DNAX Papers, Arthur Kornberg Personal Collections, Stanford University

DSH   David S. Hogness Papers, Personal Collections, Stanford University

FET   Frederick E. Terman Papers, Stanford University Archives, SC 160

JEM   Janet E. Mertz Papers, Personal Collections, University of Wisconsin, Madison

NJL   Norman J. Latker Papers, Personal Collection, Bethesda, MD

PB   Paul Berg Papers, Stanford University Archives, SC 358

PEL   Peter E. Lobban Papers, Personal Collection, Los Altos, California

SNC   Stanley N. Cohen Papers, Personal Collections, Stanford University

SUOTL   Stanford University Office of Technology Licensing Archives, Stanford University.

WS   John E. Wallace Sterling Papers, Stanford University Archives, SC 216

Notes

Introduction

1. Paul Berg, in Tom Abate, "Scientists' 'Publish or Perish' Credo Now 'Patent and Profit,'" *San Francisco Chronicle*, August 13, 2001.

2. David A. Jackson, Robert H. Symons, and Paul Berg, "Biochemical Method for Inserting New Genetic Information into DNA of Simian Virus 40: Circular SV40 DNA Molecules Containing Lambda Phage Genes and Galactose Operon of *Escherichia coli*," *Proceedings of the National Academy of Sciences, USA* 69 (1972): 2904–9.

3. For Cohen and Boyer's first gene-cloning experiment, see Stanley N. Cohen, Annie Chang, Herbert W. Boyer, and Robert B. Helling, "Construction of Biologically Functional Bacterial Plasmids *In Vitro*," *Proceedings of the National Academy of Sciences, USA* 70 (1973): 3240–44. For their intraspecies gene cloning, see Annie C. Chang and Stanley N. Cohen, "Genome Construction between Bacterial Species *In Vitro*: Replication and Expression of *Staphylococcus* Plasmid Genes in *Escherichia coli*," *Proceedings of the National Academy of Sciences, USA* 71 (1974): 1030–34. For their interspecies gene cloning, see John F. Morrow, Stanley N. Cohen, Annie Chang, Herbert W. Boyer, Howard M. Goodman, and Robert B. Helling, "Replication and Transcription of Eukaryotic DNA in *Escherichia coli*," *Proceedings of the National Academy of Sciences, USA* 71 (1974): 1743–47.

4. Victor K. McElheny, "Animal Gene Shifted to Bacteria: Aid Seen to Medicine and Farm," *New York Times*, May 20, 1974, 1.

5. "Gene Transplanters," *Newsweek*, June 17, 1974, 54.

6. Joshua Lederberg, *Stanford University News Service*, May 20, 1974, quoted in Sally S. Hughes, "Making Dollars out of DNA: The First Major Patent in Biotechnology and the Commercialization of Molecular Biology, 1974–1980," *Isis* 92 (2001): 545.

7. For a detailed analysis of the role of the Boyer-Cohen recombinant DNA–cloning patents in the commercialization of biology, see Hughes, "Making Dollars out of DNA," 541–75.

8. Susan Wright, *Molecular Politics: Developing American and British Regulatory Policy for Genetic Engineering, 1972–1982* (Chicago: University of Chicago Press, 1994).

9. Charles Petit, "The Bold Entrepreneurs of Gene Engineering," *San Francisco Chronicle*, December 2, 1977, 2, quoted in Hughes, "Making Dollars out of DNA," 414.

10. Genentech is abbreviated from **Gen**etic **En**gineering **Tech**nologies. For a historical account of the rise of biotech companies in the 1970s and 1980s, see Stephen S. Hall, *Invisible Frontiers: The Race to Synthesize a Human Gene* (New York: Atlantic Monthly Press, 1987); Cynthia Robbins-Roth, *From Alchemy to IPO: The Business of Biotechnology* (Cambridge: Perseus Publishing, 2000); and Sally S. Hughes, *Genentech: The Beginnings of Biotech* (Chicago: University of Chicago Press, 2011).

11. Hughes, *Genentech*, 158.

12. Ibid.

13. Keith Yamamoto, "Faculty Members as Corporate Officers: Does Cost Outweigh Benefit?," in William Whelan and Sandra Black, eds., *From Genetic Experimentation to Biotechnology: The Critical Transition* (New York: John Wiley & Sons, 1982), 195–201.

14. Martin Kenney, *Biotechnology: The University-Industrial Complex* (New Haven: Yale University Press, 1986); Sheldon Krimsky, *Biotechnics and Society: The Rise of Industrial Genetics* (New York: Praeger, 1991); Wright, *Molecular Politics*; and Henry Etzkowitz, *MIT and the Rise of Entrepreneurial Science* (London: Routledge, 2002).

15. Robbins-Roth, *From Alchemy to IPO*; and Hughes, *Genentech*.

16. Robert Teitelman, *Gene Dreams: Wall Street, Academia, and the Rise of Biotechnology* (New York: Basic Books, 1989); Daniel S. Greenberg, *Science, Money, and Politics: Political Triumph and Ethical Erosion* (Chicago: University of Chicago Press, 2001); Sheldon Krimsky, *Science in the Private Interest: Has the Lure of Profits Corrupted Biomedical Research?* (Lanham: Rowman & Littlefield, 2003); and Philip Mirowski, *Science-Mart: Privatizing America Science* (Cambridge: Harvard University Press, 2011).

17. I am indebted to recent work suggesting that the boundary between public and private knowledge has been shifting constantly and that the difference between the cultures of academic science and the biotechnology industry has been overemphasized. See Paul Rabinow, *Making PCR: A Story of Biotechnology* (Chicago: University of Chicago Press, 1996); Angela N. H. Creager, "'What Blood Told Dr Cohn': World War II, Plasma Fractionation, and the Growth of Human Blood Research," *Studies in History and Philosophy of Biological and Biomedical Sciences* 30 (1999): 377–405; Nicholas Rasmussen, "The Forgotten Promise of Thiamin: Merck, Caltech Biologists, and Plant Hormones in a 1930s Biotechnology Project," *Journal of the History of Biology* 32 (1999): 245–26; and Steven Shapin, "Who Is the Industrial Scientist?: Commentary from Academic Sociology

and from the Shop-Floor in the United States, ca. 1900–ca. 1970," in Karl Grandin, Nina Wormbs, Anders Lundgren, and Sven Widmalm, eds., *The Science-Industry Nexus: History, Policy, Implications* (New York: Science History Publications, 2004), 337–63.

18. A wave of economic analyses pointed out the slower growth of productivity in the 1970s, esp. a substantial drop in the growth of labor productivity between 1973 and 1976. See Edward F. Denison, *Accounting for Slower Economic Growth: The United States in the 1970's* (Washington, DC: Brookings Institution, 1979); and William J. Baumol, Sue Anne Batey Blackman, and Edward N. Wolff, *Productivity and American Leadership: The Long View* (Cambridge: MIT Press, 1989). Though Sally Hughes has situated the patenting of recombinant DNA technology in the context of university technology transfer, she does not explain how a particular form of technology transfer, patenting, was instituted for recombinant DNA technology; see Hughes, "Making Dollars out of DNA." As Daniel Kevles has demonstrated, this particular arrangement for the private ownership of biomedical research has been strengthened as the scope of intellectual property rights in the life sciences was extended to include a broader range of recombinant DNA molecules and life forms in the 1980s. See Daniel J. Kevles, "Ananda Chakrabarty Wins a Patent: Biotechnology, Law and Society, 1972–1980," *Historical Studies in the Physical and Biological Sciences* 25 (1994): 111–35.

19. Daniel J. Kevles, "Renato Dulbecco and the New Animal Virology: Medicine, Methods, and Molecules," *Journal of the History of Biology* 26 (1993): 409–42; Jean-Paul Gaudillière, "The Molecularization of Cancer Etiology in the Postwar United States: Instruments, Politics and Management," in Soraya de Chadarevian and Harmke Kamminga, eds., *Molecularizing Biology and Medicine: New Practices and Alliances, 1910s–1970s* (Amsterdam: Harwood Academic Publishers, 1998), 139–70; and Angela N. H. Creager, "Mobilizing Biomedicine: Virus Research between Lay Health Organizations and the U.S. Federal Government, 1935–1955," in Caroline Hannaway, ed., *Biomedicine in the Twentieth Century: Practices, Policies, and Politics* (Amsterdam: IOS Press, 2008), 171–201. For the development of penicillin, see Robert Bud, *Penicillin: Triumph and Tragedy* (Oxford: Oxford University Press, 2007). For a historical analysis on the postwar debates about government involvement in science, see Daniel J. Kevles, "The National Science Foundation and the Debate over Postwar Research Policy, 1942–1945: A Political Interpretation of Science—the Endless Frontier," *Isis* 68 (1977): 5–26; Nathan Reingold, "Science and Government in the United-States since 1945," *History of Science* 32 (1994): 361–86; and Daniel L. Kleinman, "Layers of Interests, Layers of Influence: Business and the Genesis of the National-Science-Foundation," *Science Technology & Human Values* 19 (1994): 259–82. On the history of the federal support for biomedical research, see Stephen P. Strickland, *Politics, Science, and Dread Disease: A Short History of United States Medical Research Policy* (Cambridge: Harvard University Press, 1972).

20. Until recently, historians have focused on scientists' increasing control over

life and its representation in reductionistic terms, such as molecules or information. Some of these studies are overwhelmingly tainted by historians' concern over the potential eugenic applications of molecular biology, casting a narrow cultural perspective on the control of life. The power over life, however, has multiple dimensions. See Creager, "Mobilizing Biomedicine."

21. As Angela Creager shows, voluntary health groups' anticipation of medical miracles played a critical role in the expansion of biomedical research. The biomedical side of the molecular approach to life has underscored an alternative cultural and political genealogy of research in the post–World War II years. See Creager, *Life of a Virus*, esp. chapter 5, "The War against Polio," 141–84.

22. Eric Vettel situates the origins of biotechnology in this broad political climate, esp. noting countercultural critics of science and technology. However, by characterizing its technical development as motivated primarily by medical and economic applications, his analysis obscures the epistemological dynamics of biological experimentation that were fundamental to the scientific development of biotechnology. See Vettel, *Biotech*.

23. For a pioneering work in the history of molecular biology after the 1960s, see Michel Morange, "The Transformation of Molecular Biology on Contact with Higher Organisms, 1960–1980: From a Molecular Description to a Molecular Explanation," *History and Philosophy of the Life Sciences* 19 (1997): 369–93.

24. For a critical analysis of this triumphant narrative in the Human Genome Project, see Michael Fortun, "The Human Genome Project and the Acceleration of Biotechnology," in Arnold Thackray, ed., *Private Science: Biotechnology and the Rise of Molecular Sciences* (Philadelphia: University of Pennsylvania Press, 1998), 182–201.

25. Morange, "Transformation of Molecular Biology."

26. The migration of molecular biologists into the biology of higher organisms in the late 1960s and early 1970s has been mentioned in the recollections of prominent scientists. See Sherwood Casjens and Colin Manoil, "1991 Thomas Hunt Morgan Medal: Dale Kaiser," in "*Genetics Society of America 1991 Records, Proceedings and Reports*," supplement, *Genetics* 128 (1991): s12–s13; Charles Yanofsky, "What Will be the Fate of Research on Prokaryotes?," *Cell* 65 (1991): 199–200; and Masayasu Nomura, "Switching from Prokaryotic Molecular Biology to Eukaryotic Molecular Biology," *Journal of Biological Chemistry* 284 (2009): 9625–35.

27. Hans-Jörg Rheinberger, *Toward a History of Epistemic Things: Synthesizing Proteins in the Test Tube* (Stanford: Stanford University Press, 1997); Hannah L. Landecker, *Culturing Life: How Cells Became Technologies* (Cambridge: Harvard University Press, 2007); Angela N. H. Creager, "Adaptation or Selection? Old Issues and New Stakes in the Postwar Debates over Bacterial Drug Resistance," *Studies in History and Philosophy of Biological and Biomedical Sciences* 38 (2007): 159–90; Doogab Yi, "Cancer, Viruses, and Mass Migration: Paul Berg's Venture into Eukaryotic Biology and the Advent of Recombinant DNA Research and

Technology, 1967–1974," *Journal of the History of Biology* 41 (2008): 589–636; and Mathias Grote, "Hybridizing Bacteria, Crossing Methods, Cross-checking Arguments: The Transition from Episomes to Plasmids (1961–1969)," *History and Philosophy of the Life Sciences* 30 (2008): 407–30.

28. The origins of this standard history can be traced to one of the inventors of recombinant DNA technology at Stanford, Stanley Cohen. See Stanley N. Cohen, "The Manipulation of Genes," *Scientific American* 233 (1975): 24–33; and Susan Wright, "Recombinant DNA Technology and Its Social Transformation, 1972–1982," *Osiris* 2 (1986): 303–60.

29. Jackson, Symons, and Berg, "Biochemical Method for Inserting New Genetic Information into DNA of Simian Virus 40"; and Peter E. Lobban and A. D. Kaiser, "Enzymatic End-to-End Joining of DNA Molecules," *Journal of Molecular Biology* 78 (1973): 453–71. The biomedical implications of Berg's recombinant DNA research had been so original and far-reaching for biomedical research that he was awarded the Nobel Prize in 1980 "for his fundamental studies of the biochemistry of nucleic acids, with particular regard to recombinant-DNA." See "The Nobel Prize in Chemistry 1980," Nobelprize.org, at http://nobelprize.org /nobel_prizes/chemistry/laureates/1980/. For Stanford biochemists' involvement in recombinant DNA research, see Harrison Echols and Carol Gross, *Operators and Promoters: The Story of Molecular Biology and Its Creators* (Berkeley: University of California Press, 2001); Hughes, "Making Dollars out of DNA"; Timothy Lenoir, "Biochemistry at Stanford: A Case Study in the Formation of an Entrepreneurial Culture" (unpublished manuscript, April 21, 2002), Microsoft Word file; and Michel Morange, *A History of Molecular Biology* (Cambridge: Harvard University Press, 2000). For Stanley Cohen's acknowledgement of his indebtedness to Stanford biochemists, see Stanley N. Cohen, "The Transplantation and Manipulation of Genes in Microorganisms," in *The Harvey Lectures* (New York: Academic Press 1980), 173–204.

30. Peter E. Lobban, "The Generation of Transducing Phage *In Vitro*," essay for third PhD examination, Stanford University, 6 November 1969, Peter E. Lobban Papers, Personal Collection, Los Altos, CA.

31. Morrow et al., "Replication and Transcription of Eukaryotic DNA in *Escherichia coli*."

32. David C. Mowery, Richard R. Nelson, Bhaven N. Sampt, and Arvids A. Ziedonis, eds., *Ivory Tower and Industrial Innovation: University-Industry Technology Transfer Before and After the Bayh-Dole Act in the United States* (Stanford: Stanford University Press, 2004); and Roger L. Geiger, *Knowledge and Money: Research Universities and the Paradox of the Marketplace* (Stanford: Stanford University Press, 2004).

33. Stuart W. Leslie, *The Cold War and American Science: The Military-Industrial-Academic Complex at MIT and Stanford* (New York: Columbia University Press, 1993); AnnaLee Saxenian, *Regional Advantage: Culture and Competition*

in Silicon Valley and Route 128 (Cambridge: Harvard University Press, 1994); Rebecca S. Lowen, *Creating the Cold War University: The Transformation of Stanford* (Berkeley: University of California Press, 1997); and C. Stewart Gillmor, *Fred Terman at Stanford: Building a Discipline, a University, and Silicon Valley* (Stanford: Stanford University Press, 2004).

34. Stanford's ties to its two powerful patrons—the federal government and the private industry—had become so strong by the mid-1960s that the cultural critics of science and technology often assailed the university for being one of the most active participants in the so-called academic-industrial-military complex. Leslie, *Cold War and American Science*; Paul Forman, "Behind Quantum Electronics: National Security as Basis for Physical Research in the United States, 1940–1960," *Historical Studies in the Physical and Biological Sciences* 18 (1987): 149–229; and Peter Galison, Bruce Hevly, and Rebecca Lowen, "Controlling the Monster: Stanford and the Growth of Physics Research, 1935–1962," in Peter Galison and Bruce Hevly, eds., *Big Science: The Growth of Large-Scale Research* (Stanford: Stanford University Press, 1992), 46–77.

35. Gillmor, *Fred Terman at Stanford*; and Eric J. Vettel, "The Protean Nature of Stanford University's Biological Sciences, 1946–1972," *Historical Studies in the Physical and Biological Sciences* 35 (2004): 95–113.

36. Roger L. Geiger, *Research and Relevant Knowledge: American Research Universities since World War II* (New York: Oxford University Press, 1993), esp. 173–90. By 1964, the NIH's support for university research was $401 million, which was twice that of the Department of Defense's spending for university research, $198 million.

37. David C. Mowery and Bhaven N. Sampat, "University Patents and Patent Policy Debates in the USA, 1925–1980," *Industrial and Corporate Change* 10 (2001): 781–814.

38. I would like to thank an anonymous referee for the University of Chicago Press for pointing out this issue. See Jean-Paul Gaudillière and Ilana Löwy, eds., *The Invisible Industrialist: Manufactures and the Production of Scientific Knowledge* (London: Routledge, 1998); and for a historical analysis of the industrialization of the pharmaceutical industry, see essays in the special issue, "How Pharmaceuticals became Patentable in the Twentieth Century," in *History and Technology* 24:2 (2008), esp. Jean-Paul Gaudillière, "How Pharmaceuticals became Patentable: The Production and Appropriation of Drugs in the Twentieth Century," *History and Technology* 24 (2008): 99–106.

39. Scholars at the University of Chicago, such as Richard A. Posner and Edmund W. Kitch, claimed that the public ownership of scientific research actually hindered the ability to produce useful economic innovations. See Richard A. Posner, *Economic Analysis of the Law* (Boston: Little Brown, 1973). For an overview of the law and economics perspective on intellectual property rights, see Edmund W. Kitch, "The Nature and Function of the Patent System," *Journal of Law*

and Economics 20 (1977): 265–90. For a historical analysis of the rise of the law and economics movement, see Steven M. Teles, *The Rise of the Conservative Legal Movement: The Battle for Control of the Law* (Princeton: Princeton University Press, 2008). For a critical analysis of the theory of the tragedy of the commons in intellectual property rights, see James Boyle, *Shamans, Software, and Spleens: Law and the Construction of the Information Society* (Cambridge: Harvard University Press, 1996); and Michael A. Heller, "The Tragedy of the Anticommons: Property in the Transition from Marx to Markets," *Harvard Law Review* 111 (1998): 621–88.

40. For an excellent overview of the commercialization of science with an emphasis on shifts in funding and the organization of science in the latter half of the twentieth century, see Philip Mirowski and Esther-Mirjam Sent, "The Commercialization of Science, and the Response of STS," in Edward J. Hackett, Olga Amsterdamska, Michael Lynch, and Judy Wajcman, eds., *Handbook of Science, Technology and Society Studies* (Cambridge: MIT Press, 2007), 635–89; Mirowski, *Science-Mart*; and Elizabeth P. Berman, *Creating the Market University: How Academic Science Became an Economic Engine* (Princeton: Princeton University Press, 2011).

41. See essays in the special issue, "How Pharmaceuticals became Patentable in the Twentieth Century," in *History and Technology* 24:2 (2008).

42. For example, Hughes, "Making Dollars out of DNA"; and Charles Weiner, "Patenting and Academic Research: Historical Case Studies," in Vivian Weil and John W. Snapper, eds., *Owning Scientific and Technical Information: Value and Ethical Issues* (New Brunswick: Rutgers University Press, 1989), 87–109. Historian Daniel J. Kevles duly acknowledges that US patenting policy aims not only to promote the commercialization of private invention but also to preserve public interest by encouraging the dissemination of trade secrets. See Daniel J. Kevles, "Principles, Property Rights, and Profits: Historical Reflections on University/Industry Tensions," *Accountability in Research* 8 (2001): 12–26. For an analysis on the political coalition between university patent managers and federal government officials and their role in the passage of the Bayh-Dole Act, see Elizabeth P. Berman, "Why Did Universities Start Patenting? Institution-Building and the Road to the Bayh-Dole Act," *Social Studies of Science* 38 (2008): 835–71.

43. Mowery and Sampat, "University Patents and Patent Policy Debates in the USA." At the center of this practice of limited academic patenting was a widely shared attitude among scientists and industrialists about the proper role of academic research in the production and use of scientific knowledge. As Adrian Johns acknowledges, the traditional norms of science, such as openness and free exchange of ideas, were forcefully articulated in the mid-twentieth century when leading scientists and economists debated the relationship between open science and intellectual property. In the 1940s, critics of patenting like Michael Polanyi argued that, just as a monopoly of economic resources would stifle a nation's economy, so would a monopoly of scientific ideas stifle the scientific community. See Adrian

Johns, "Intellectual Property and the Nature of Science," *Cultural Studies* 20:2/3 (2006): 145–64. See also Robert K. Merton, *The Sociology of Science: Theoretical and Empirical Investigations*, ed. Norman W. Storer (Chicago: University of Chicago Press, 1973), esp. "The Normative Structure of Science," 267–78.

44. Gaudillière and Löwy, *Invisible Industrialist*; and P. Roy Vagelos and Louis Galambos, *Medicine, Science, and Merck* (Cambridge: Cambridge University Press, 2004).

45. On the rise of American conservatism and the ascendency of market ideology, see George H. Nash, *Conservative Intellectual Movement in America since 1945* (New York: Basic Books, 1976). For the reconfiguration of public policy in terms of economic rationales in the 1970s, see Mark A. Smith, *The Right Talk: How Conservatives Transformed the Great Society into the Economic Society* (Princeton: Princeton University Press, 2007). For the ideological shift in the 1970s in federal support in biomedical research, see Doogab Yi, "The Scientific Commons in the Marketplace: The Industrialization of Biomedical Materials at the New England Enzyme Center, 1963–1980," *History and Technology* 25 (2009): 69–87; and Soraya de Chadarevian, "The Making of an Entrepreneurial Science: Biotechnology in Britain, 1975–1995," *Isis* 102 (2011): 601–33.

46. Hughes, "Making Dollars out of DNA"; Kenney, *Biotechnology*; and Mowery et al., *Ivory Tower and Industrial Innovation*.

47. Robert E. Kohler, "Moral Economy, Material Culture and Community in *Drosophila* Genetics," in Mario Biagioli, ed., *The Science Studies Reader* (London: Routledge, 1999), 243; and Robert E. Kohler, *Lords of the Fly:* Drosophila *Genetics and the Experimental Life* (Chicago: University of Chicago Press, 1994). The concept of moral economy was introduced by Edward P. Thompson as a way to explain how peasants in eighteenth-century England held an alternative view of the economics of exchange and work; see Edward P. Thompson, "The Moral Economy of the English Crowd in the Eighteenth Century," *Past and Present* 50 (1971): 76–136. For a somewhat different use of the concept of moral economy that emphasizes how certain cultural values and sensibilities, such as trust, civility, and curiosity inform a certain form of scientific enterprise like seventeenth-century empiricism, see Lorraine Daston, "The Moral Economy of Science," *Osiris* 10 (1995): 3–24; and Steven Shapin, "The House of Experiment in Seventeenth-Century England," *Isis* 79 (1988): 373–404. On the discussion of the exchange of biological materials as gift exchange, see Warwick Anderson, "The Possession of Kuru: Medical Science and Biocolonial Exchange," *Comparative Studies in Society and History* 42 (2000): 713–44; and Hannah Landecker, "Sending Cells Around: How to Exchange Biological Matter," paper presented in *The Moment of Conversion: Exchange Networks in Modern Biomedical Science*, Department of Anthropology, History, and Social Medicine, UC San Francisco, April 4–5, 2003.

48. For recent work in this line of analysis, see Angela N. H. Creager and Gregory J. Morgan, "After the Double Helix: Rosalind Franklin's Research on Tobacco

Mosaic Virus," *Isis* 99 (2008): 239–72; and Bruno J. Strasser, "The Experimenter's Museum: GenBank, Natural History, and the Moral Economies of Biomedicine," *Isis* 102 (2011): 60–96.

49. See Kenney, *Biotechnology*; Krimsky, *Biotechnics and Society*; and Wright, *Molecular Politics*. For a criticism of this corruption narrative, see Rabinow, *Making PCR*; and Shapin, "Who is the Industrial Scientist?" For studies that point to the continuity between academic and commercial biomedical enterprise before and after the recombinant DNA revolution in molecular biology, see Angela N. H. Creager, "Biotechnology and Blood: Edwin Cohn's Plasma Fractionation Project, 1940–1953," in Arnold Thackray, ed., *Private Science: Biotechnology and the Rise of Molecular Sciences* (Philadelphia: University of Pennsylvania Press, 1998), 39–62; Rasmussen, "Forgotten Promise of Thiamin"; and Robert Bud, *The Uses of Life: A History of Biotechnology* (Cambridge: Cambridge University Press, 1993).

50. Merton, *Sociology of Science*, esp. chapter 3, "The Normative Structure of Science," 267–78. For example, see Jennifer Washburn, *University, Inc.: The Corporate Corruption of American Higher Education* (New York: Basic Books, 2005).

51. Kohler, *Lords of the Fly*; Soraya de Chadarevian, "Of Worms and Programmes: *Caenorhabditis elegans* and the Study of Development," *Studies in History and Philosophy of Biological and Biomedical Sciences* 29 (1998): 81–105; Creager and Morgan, "After the Double Helix": and Strasser, "Experimenter's Museum."

52. For an insightful analysis of the scientific life in the age of commercial biotechnology, see Steven Shapin, *The Scientific Life: A Moral History of a Late Modern Vocation* (Chicago: University of Chicago Press, 2008)

53. Social scientists have debated whether the enactment of the Bayh-Dole Act in 1980, which allowed research universities to file a patent for research funded by the federal government, signified a crucial shift in the commercialization of academic research. See Mowery et al., *Ivory Tower and Industrial Innovation*.

54. Arthur Kornberg, *The Golden Helix: Inside Biotech Ventures* (Sausalito: University Science Books, 1995).

Chapter One

1. Arthur Kornberg to Robert Alway, 22 October 1958, Arthur Kornberg Papers, Stanford University Archives, SC 359 (hereafter AK), box 31, folder: 1958.

2. Arthur Kornberg, *For the Love of Enzymes: The Odyssey of a Biochemist* (Cambridge: Harvard University Press, 1989), 177.

3. Arthur Kornberg to Warren Weaver, Rockefeller Foundation, 19 March 1958, AK, box 31, folder: 1958.

4. Nathan Reingold, "Science and Government in the United States since 1945," *History of Science* 32 (1994): 361–81; Roger L. Geiger, *Research and Rele-*

vant Knowledge: American Research Universities since World War II (Oxford: Oxford University Press, 1993); and Angela N. H. Creager, "Mobilizing Biomedicine: Virus Research between Lay Health Organizations and the U.S. Federal Government, 1935–1955," in Caroline Hannaway, ed., *Biomedicine in the Twentieth Century: Practices, Policies, and Politics* (Amsterdam: IOS Press, 2008), 171–201.

5. John E. Wallace Sterling, "The Importance of Research in the Stanford University School of Medicine," n.d. (ca. 1956–58), John E. Wallace Sterling Papers, Stanford University Archives, SC 216 (hereafter WS), box 63.

6. Rebecca S. Lowen, *Creating the Cold War University: The Transformation of Stanford* (Berkeley: University of California Press, 1997); Eric J. Vettel, "The Protean Nature of Stanford University's Biological Sciences, 1946–1972," *Historical Studies in the Physical and Biological Sciences* 35 (2004): 95–113; and Timothy Lenoir, "Biochemistry at Stanford: A Case Study in the Formation of an Entrepreneurial Culture" (unpublished manuscript, April 21, 2002), Microsoft Word file.

7. Geiger, *Research and Relevant Knowledge*, esp. 173–90. For a historical analysis of the immediate post–World War II debate on the federal support for biomedical research, see Daniel M. Fox, "The Politics of NIH Extramural Program, 1937–1950," *Journal of the History of Medicine and Allied Sciences* 42 (1987): 447–66.

8. Peter Keating and Alberto Cambrosio, *Biomedical Platforms: Realigning the Normal and the Pathological in Late-Twentieth-Century Medicine* (Cambridge: MIT Press, 2003), 72.

9. Robert E. Kohler, *From Medical Chemistry to Biochemistry: The Making of a Biomedical Discipline* (Cambridge: Cambridge University Press, 1982); and Angela N. H. Creager, "Wendell Stanley's Dream of a Free-Standing Biochemistry Department at the University of California, Berkeley," *Journal of the History of Biology* 29 (1996): 331–60.

10. Angela N. H. Creager, *The Life of a Virus: Tobacco Mosaic Virus as an Experimental Model, 1930–1965* (Chicago: University of Chicago Press, 2002); and Soraya de Chadarevian, *Designs for Life: Molecular Biology after World War II* (Cambridge: Cambridge University Press, 2002).

11. For an analysis of an early modern form of experimental life and its particular social and political associations, see Steven Shapin and Simon Schaffer, *Leviathan and the Air-Pump: Hobbes, Boyle, and the Experimental Life* (Princeton: Princeton University Press, 1989). For a comparative analysis of the organization of the biochemistry departments in the San Francisco Bay Area, see Simcha Jong, "How Organizational Structures in Science Shape Spin-off Firms: The Biochemistry Departments of Berkeley, Stanford, and UCSF and the Birth of the Biotech Industry," *Industrial and Corporate Change* 15 (2006): 251–83.

12. Robert E. Kohler, *Lords of the Fly: Drosophila Genetics and the Experimental Life* (Chicago: University of Chicago Press, 1994); and Lorraine Daston, "The Moral Economy of Science," *Osiris* 10 (1995): 3–24.

13. J. E. Wallace Sterling, "A Statement to the Board of Trustees," June 1953, p. 2, WS, box 63.

14. "USPHS Grants," WS, box 62, folder: 9.

15. "Stanford Medical School Becomes a True University School," 1956, WS, box 63.

16. Kohler, *From Medical Chemistry to Biochemistry*.

17. George Beadle's postwar Caltech project with Linus Pauling aimed to bring biology and chemistry together. See Lily E. Kay, *The Molecular Vision of Life: Caltech, the Rockefeller Foundation, and the Rise of the New Biology* (Chicago: University of Chicago Press, 1993).

18. Ibid.

19. For Beadle's role in the articulation of biochemical genetics, see Robert E. Kohler, "Systems of Production: *Drosophila, Neurospora*, and Biochemical Genetics," *Historical Studies in the Physical and Biological Sciences* 22 (1991): 87–130; and Paul Berg and Maxine Singer, *George Beadle, an Uncommon Farmer: the Emergence of Genetics in the 20th Century* (Cold Spring Harbor: Cold Spring Harbor Press, 2003).

20. George Beadle's research at Stanford, see Lily E. Kay, "Selling Pure Science in Wartime: The Biochemical Genetics of G. W. Beadle," *Journal of the History of Biology* 22 (1989): 73–101.

21. Frederick E. Terman, memorandum "Biochemistry," Frederick E. Terman Papers, Stanford University Archives, SC 160 (hereafter FET), box 4, folder: Biochemistry, 1956–59.

22. C. Stewart Gillmor, *Fred Terman at Stanford: Building a Discipline, a University, and Silicon Valley* (Stanford: Stanford University Press, 2004); and Lenoir, "Biochemistry at Stanford." This chapter focuses on the Biochemistry Department, as the Genetics Department took off rather slowly. The Genetics Department initially shared its lab space and instruments with the Biochemistry Department. For a historical analysis of the establishment of the Genetics Department at Stanford, see Joseph A. November, *Biomedical Computing: Digitizing Life in the United States* (Baltimore: Johns Hopkins University Press, 2012), esp. chapter 5, "Martians, Experts, and Universitas: Biomedical Computing at Stanford University, 1960–1966," 220–68.

23. Stuart W. Leslie explores the role of the federal government, as well as Terman's role, in the shaping of the engineering school at Stanford in his *The Cold War and American Science: The Military-Industrial-Academic Complex at MIT and Stanford* (New York: Columbia University Press, 1993). By examining Stanford trustees' and administrators' fear of governmental intervention in the academy and their promotion of commercial ties with private industry, Rebecca Lowen presents an ideological rendering of technology transfer at Stanford University. See Lowen, *Creating the Cold War University*. On attempts to "import" the so-called Terman model of industrial and regional development, see Stuart W. Les-

lie and R. H. Kargon, "Selling Silicon Valley: Frederick Terman's Model for Regional Advantage," *Business History Review* 70:4 (1996): 435–72; and Margaret Pugh O'Mara, *Cities of Knowledge: Cold War Science and the Search for the Next Silicon Valley* (Princeton: Princeton University Press, 2005). On the role of Stanford in the rise of Silicon Valley, see AnnaLee Saxenian, *Regional Advantage: Culture and Competition in Silicon Valley and Route 128* (Cambridge: Harvard University Press, 1994).

24. Frederick E. Terman, "Why Do We Do Research?," an address at the seventeenth annual Stanford business conference, 22 July 1958, p. 22, FET, box 1, folder: 56.

25. Geiger, *Research and Relevant Knowledge*.

26. Creager, "Mobilizing Biomedicine." Because of the enormous expansion of federal support for biomedicine, philanthropic foundations such as the Rockefeller Foundation found alternative niches for their distinctive patronage, shifting their support more toward traditional welfare programs in agriculture and public health. See Robert E. Kohler, *Partners in Science: Foundations and Natural Scientists, 1900–1945* (Chicago: University of Chicago Press, 1991).

27. George B. Darling, "Can We Pay for Our Medical Schools?," *Atlantic Monthly* (June 1950): 38–42.

28. For example, the Johns Hopkins University Medical School introduced a five-year education program to reinvigorate basic biomedical sciences in its curriculum. See Peter Lee, *Medical Schools and the Changing Times: Nine Case Reports on Experimentation in Medical Education, 1950–60* (Evanston: Association of American Medical Colleges, 1962)

29. *Stanford Daily*, 27 November 1957, 1.

30. For a glimpse of Harvard's attempt to institute molecular biology, see recollections by scientists in John Inglis, Joseph Sambrook, and Jan A. Witkowski, eds., *Inspiring Science: Jim Watson and the Age of DNA* (Cold Spring Harbor: Cold Spring Harbor Laboratory Press, 2003).

31. On Salvador Luria's years at MIT, see Rena Selya, *Salvador Luria's Unfinished Experiment: The Public Life of a Biologist in a Cold War Democracy* (PhD dissertation, Harvard University, 2002).

32. For a history of molecular biology from the perspective of the so-called Phage Group at Caltech, see John Cairns, Gunther S. Stent, and James D. Watson, eds., *Phage and the Origins of Molecular Biology* (Cold Spring Harbor: Cold Spring Harbor Laboratory Press, 1966)

33. Renato Dulbecco to Caltech faculty members, 19 December 1961, James Frederick Bonner Papers, Caltech Archives, unprocessed.

34. Vettel, "Protean Nature of Stanford University's Biological Sciences," 96.

35. Confidential: Minutes of the Committee on the Search for a Head of Biochemistry, 22 January 1957, WS, box 62. Others who participated in this meeting were Loren Chandler, John Luetscher, Jr., Carl Noller, and William Steere.

36. Creager, "Wendell Stanley's Dream of a Free-Standing Biochemistry Department at the University of California, Berkeley."

37. Confidential: Minutes of the Committee on the Search for a Head of Biochemistry, 22 January 1957, WS, box 62.

38. On the development of enzymology at the NIH and Kornberg's NIH experience as a research biochemist, see Buhm Soon Park, "The Development of the Intramural Research Program at the National Institutes of Health after World War II," *Perspectives in Biology and Medicine* 46 (2003): 383–402.

39. Kornberg, *For the Love of Enzymes*; and Robert E. Kohler, "Walter Fletcher, F. G. Hopkins, and the Dunn Institute of Biochemistry: A Case Study in the Patronage of Science," *Isis* 69 (1978): 331–55.

40. Frederick E. Terman to Arthur Kornberg, 7 May 1957, FET, box 4, folder: Biochemistry, 1956–59.

41. Ibid.

42. Arthur Kornberg, MD, "Biochemistry at Stanford, Biotechnology at DNAX," an oral history conducted in 1997 by Sally Smith Hughes, PhD, Regional Oral History Office, Bancroft Library, University of California, Berkeley, 1998, 19.

43. Renato Dulbecco to Caltech faculty members, 19 December 1961, James Frederick Bonner Papers, Caltech Archives, unprocessed.

44. For a nuanced analysis of the discovery of DNA and its place in the historiography of molecular biology, see de Chadarevian, *Designs for Life*, esp. chapter 6, "Locating the Double Helix," 164–98. For a personal recollection, see James D. Watson, *The Double Helix: A Personal Account of the Discovery of the Structure of DNA* (New York: Atheneum, 1968).

45. Arthur Kornberg, "The Biologic Synthesis of Deoxyribonucleic Acid: Nobel Lecture, December 1959," *Nobel Lectures, Physiology or Medicine 1942–1962* (Amsterdam: Elsevier Publishing Company, 1964), 668.

46. Frederic L. Holmes, *Meselson, Stahl, and the Replication of DNA: A History of "The Most Beautiful Experiment in Biology"* (New Haven: Yale University Press, 2001).

47. Kornberg, "Biologic Synthesis of Deoxyribonucleic Acid," 668.

48. Arthur Kornberg, "Summary of Research Plans for Department of Biochemistry," AK, box 31, folder: 1958.

49. For a historical analysis of the consolidation of biochemistry and genetics during Jacques Monod and François Jacobs's collaboration at the Pasteur Institute during the 1960s, see Angela N. H. Creager and Jean-Paul Gaudillière, "Meanings in Search of Experiments and Vice-Versa: The Invention of Allosteric Regulation in Paris and Berkeley, 1959–1968," *Historical Studies in the Physical and Biological Sciences* 27 (1996): 1–89. At Cambridge, the language of information mediated the institutional marriage between biochemists and molecular biologists. See Soraya de Chadarevian, "Sequence, Conformation, Information: Biochemists and Molecular Biologists in the 1950s," *Journal of the History of Biology* 29 (1996): 361–86.

50. Kohler, *From Medical Chemistry to Biochemistry*.

51. Frederic L. Holmes, *Between Biology and Medicine: The Formation of Intermediary Metabolism* (Berkeley: University of California Press, 1992).

52. For Severo Ochoa's building of molecular biology in Spain, see María J. Santesmases, "Severe Ochoa and the Biomedical Sciences in Spain under Franco, 1959–1975," *Isis* 91:4 (2000): 706–34.

53. Kornberg, *For the Love of Enzymes*, 29–58.

54. Though Kornberg could not elucidate the intermediary pathway of ATP synthesis, he unexpectedly discovered coenzyme NAD (a major respiratory coenzyme). He in turn showed how this transient metabolic intermediate NAD enhanced cellular respiration through its phosphorylation and cleavage.

55. Lily E. Kay, *Who Wrote the Book of Life? A History of the Genetic Code* (Stanford: Stanford University Press, 2000), 235–93.

56. Maxine Singer, "Leon Heppel and the Early Days of RNA Biochemistry," *Journal of Biological Chemistry* 278 (2003): 47351.

57. Kornberg, *For the Love of Enzymes*, 59–87. Two new techniques were particularly indispensable in this endeavor: a radioactive isotope of phosphorus or carbon enabled Kornberg to follow molecular pathways of nucleotide synthesis; and an ion-exchange chromatographic technique, initially developed for the separation of fissionable materials for the atomic bomb, provided a powerful means to sort out a mixture of nucleotides.

58. Ibid., 134–69. For key papers on the enzymatic synthesis of DNA, see Robert I. Lehman, Maurice J. Bessman, Ernest Simms, and Arthur Kornberg. "Enzymatic Synthesis of Deoxyribonucleic Acid. I. Preparation of Substrates and Partial Purification of an Enzyme from *Escherichia coil*," *Journal of Biological Chemistry* 233 (1958): 163–70; and Arthur Kornberg, "Biologic Synthesis of Deoxyribonucleic Acid," *Science* 131 (1960): 1503–8.

59. Paul Berg and E. James Ofengand, "An Enzymatic Mechanism for Linking Amino Acids to RNA," *Proceedings of the National Academy of Sciences, USA* 44 (1958): 78–86; and Paul Berg, PhD, "A Stanford Professor's Career in Biochemistry, Science Politics, and the Biotechnology Industry," an oral history conducted in 1997 by Sally Smith Hughes, PhD, Regional Oral History Office, Bancroft Library, University of California, Berkeley, 2000, 42–43.

60. Hans-Jörg Rheinberger, *Toward a History of Epistemic Things: Synthesizing Proteins in the Test Tube* (Stanford: Stanford University Press, 1997).

61. Michael Chamberlin and Paul Berg, "Studies on DNA-directed RNA Polymerase: Formation of DNA-RNA Complexes with Single-Stranded φX 174 DNA as Template," *Cold Spring Harbor Symposium on Quantitative Biology* 23 (1963): 67–75.

62. William Wood and Paul Berg, "The Effect of Enzymatically Synthesized Ribonucleic Acid on Amino Acid Incorporation by a Soluble Protein-Ribosome System from *Escherichia coli*," *Proceedings of the National Academy of Sciences of the USA* 48 (1962): 94–104.

63. Melvin Cohn, "Contributions of Studies on the β-galactosidase of *Escherichia coli* to Our Understanding of Enzyme Synthesis," *Bacteriological Review* 21 (1957): 140–68, and "The Wisdom of Hindsight," *Annual Review of Immunology* 12 (1994): 1–62. On the development of protein synthesis research at the Pasteur Institute from the conjugation of Monod's enzymatic adaptation work and Jacob's bacterial genetic research, see Creager and Gaudillière, "Meanings in Search of Experiments and Vice-Versa."

64. David S. Hogness, Melvin Cohn, and Jacques Monod, "Studies on the Induced Synthesis of β-galactosidase in *Escherichia coli*," *Biochimica et Biophysica Acta* 16 (1955): 99–116; and David S. Hogness, "Induced Enzyme Synthesis," *Reviews of Modern Physics* 31 (1959): 256–68.

65. A. Dale Kaiser and David S. Hogness, "Transformation of *Escherichia coli* with Deoxyribonucleic Acid Isolated from Bacteriophage-λ*dg*," *Journal of Molecular Biology* 2 (1960): 392–415.

66. A. Dale Kaiser, "The Production of Phage Chromosome Fragments and Their Capacity for Genetic Transfer," *Journal of Molecular Biology* 4 (1962): 275–87; and Marcel Weber, "Representing Genes: Classical Mapping Techniques and the Growth of Genetic Knowledge," *Studies in History and Philosophy of Biological and Biomedical Sciences* 29 (1998): 295–315. As Marcel Weber points out, Kaiser's mapping provided a biochemical evidence to confirm Seymour Benzer's recombination maps.

67. Robert L. Baldwin, "Boundary Spreading in Sedimentation-Velocity Experiments," *Biochemical Journal* 65 (1957): 490–512.

68. A. Dale Kaiser and Robert L. Baldwin, "A Relation between Dinucleotide and Base Frequencies in Bacterial DNAs," *Journal of Molecular Biology* 4 (1962): 418–19.

69. Holmes, *Meselson, Stahl, and the Replication of DNA*.

70. This analysis of the moral economy of science at the level of material culture is indebted to Kohler's astute examination of the *Drosophila* genetics community. I thank Angela Creager who pushed this analysis further, helping me appreciate the role of material culture in laboratory life and beyond. See Kohler, *Lords of the Fly*; and Angela N. H. Creager, "Tracing the Politics of Changing Postwar Research Practices: The Export of 'American' Radioisotopes to European Biologists," *Studies in History and Philosophy of Biological and Biomedical Sciences* 33 (2002): 367–88.

71. Arthur Kornberg to A. Dale Kaiser, "Department Finances," 10 September 1979, AK, box 5, folder: 1979.

72. Berg, "A Stanford Professor's Career in Biochemistry, Science Politics, and the Biotechnology Industry," 90.

73. Kornberg, "Biochemistry at Stanford, Biotechnology at DNAX." Section on "More on the Department's Distinctive Operational System."

74. Kornberg, *For the Love of Enzymes*, 180

75. Arthur Kornberg to A. Dale Kaiser, 9 February 1956, A. Dale Kaiser Papers, Stanford University Archives, SC 356, box 1, folder: Correspondence.

76. Kornberg, "Biochemistry at Stanford, Biotechnology at DNAX," section on "Communal Structure."

77. Gillmor, *Fred Terman at Stanford*, 376.

78. Paul Berg, interview with the author, October 2005.

79. Kornberg, *For the Love of Enzymes*, 185.

80. Arthur Kornberg, "Editorial: Science is Great, but Scientists are Still People," *Science* 257 (1992): 859.

81. Kornberg, "Biochemistry at Stanford, Biotechnology at DNAX." On the emergence of postwar federal research economy, see Geiger, *Research and Relevant Knowledge*.

82. Edwin J. Cohn, "History of the Development of the Scientific Policies of the University Laboratory of Physical Chemistry Related to Medicine and Public Health, Harvard University: A Memorandum on the Unwisdom of Projects and Reports (January 1952)," in Edwin J. Cohn, *A Collection of Pamphlets: Published to Record an Experiment in Organization for Research, Training, and Development in Science Basic to Medicine and Public Health* (Cambridge: Harvard University Printing Office, 1952), 8.

83. For Stanford trustees' concerns on the government control over the academy, see Lowen, *Creating the Cold War University*. One of the most vocal opponents of governmental funding of research and education among Stanford trustees was Herbert Hoover.

84. Kornberg, "Biochemistry at Stanford, Biotechnology at DNAX."

85. Arthur Kornberg to A. Dale Kaiser, "Department Finances," 10 September 1979, AK, box 5, folder: 1979.

86. For an excellent historical overview on the development of the biology of higher organisms, see Michel Morange, "The Transformation of Molecular Biology on Contact with Higher Organisms, 1960–1980: From a Molecular Description to a Molecular Explanation," *History and Philosophy of the Life Sciences* 19 (1997): 369–93.

87. Dean Robert R. Sears (Humanities and Sciences) to President J. E. Wallace Sterling, "Needs for the Development of Biology Under a New Department Head," 25 April 1966, Richard W. Lyman Papers, Stanford University Archives, SC 215, box 38, folder 29–30: Biochemistry and Biological Sciences.

88. Berg, "A Stanford Professor's Career in Biochemistry, Science Politics, and the Biotechnology Industry," 56.

89. David Hogness to faculty members, Department of Biochemistry, "Agenda and Notes for Meeting about Department: Space," 15 December 1969, Paul Berg Papers, Stanford University Archives, SC 215, box 17, folder: Faculty Meeting Minutes, 1968–69.

90. Ibid.

91. Ibid.

92. Ibid.

93. See chapter 2 on the development of recombinant DNA research and technology, where I analyze the transformation of Berg's animal virus experimental systems as an object of scientific research (tumorigenesis) into a pioneering tool that led to the emergence of genetic engineering.

94. My examination of Stanford biochemists' culture of sharing will later help explain how the patenting of recombinant DNA technology challenged their communal practices. Historians have noted that Stanford biochemists, such as Paul Berg and Arthur Kornberg, were opposed to the patenting of recombinant DNA technology. For example, historian Sally Hughes shows Berg's leading role in the heated biohazard controversy regarding recombinant DNA research and the negligence of Berg's contribution to the development of recombinant DNA technology led to his "personal animosity" and "personal grievance." However, more was at stake than the issue of scientific credit. Sally S. Hughes, "Making Dollars out of DNA: The First Major Patent in Biotechnology and the Commercialization of Molecular Biology, 1974–1980," *Isis* 92 (2001): 560.

Chapter Two

1. As Angela Creager shows, voluntary health groups' anticipations of medical miracles played a critical role in the expansion of biomedical research. The biomedical side of the molecular approach to life has underscored an alternative cultural and political genealogy of biomedical research in the post–World War II years. See Angela N. H. Creager, *The Life of a Virus: Tobacco Mosaic Virus as an Experimental Model, 1930–1965* (Chicago: University of Chicago Press, 2002), esp. chapter 5, "The War against Polio."

2. For the history of molecular biology in the post–World War II era, see Lily E. Kay, *Who Wrote the Book of Life?: A History of the Genetic Code* (Stanford: Stanford University Press, 2000); Michel Morange, *A History of Molecular Biology* (Cambridge: Harvard University Press, 2000); Soraya de Chadarevian, *Designs for Life: Molecular Biology after World War II* (Cambridge: Cambridge University Press, 2002); and Creager, *Life of a Virus*.

3. For the broader role of model systems in natural and social scientific disciplines, see Angela N. H. Creager, Elizabeth Lunbeck, and M. Norton Wise, eds., *Science without Laws: Model Systems, Cases, Exemplary Narratives* (Durham: Duke University Press, 2008).

4. Jacques Monod and François Jacob, "General Conclusions: Teleonomic Mechanisms in Cellular Metabolism, Growth, and Differentiation," *Cold Spring Harbor Symposia on Quantitative Biology* 21 (1961): 363.

5. For recollections by prominent scientists, see Sherwood Casjens and Colin Manoil, "1991 Thomas Hunt Morgan Medal: Dale Kaiser," in "*Genetics Society of America 1991 Records, Proceedings and Reports*," supplement, *Genetics* 128 (1991):

s12–s13; Charles Yanofsky, "What Will Be the Fate of Research on Prokaryotes?," *Cell* 65 (1991): 199–200; and Masayasu Nomura, "Switching from Prokaryotic Molecular Biology to Eukaryotic Molecular Biology," *Journal of Biological Chemistry* 284 (2009): 9625–35.

6. Eric Vettel situates the origins of biotechnology in this broad political climate that called for medically useful research. However, by characterizing its technical development as motivated primarily by medical and economic applications, Vettel's analysis obscures the epistemological dynamics of biological experimentation fundamental to the scientific development of biotechnology. See Vettel, *Biotech.*

7. Francis Crick, "Molecular Biology and Medical Research," *Journal of the Mount Sinai Hospital* 36 (1969): 187, quoted in de Chadarevian, *Designs for Life*, 346.

8. Crick, "Molecular Biology and Medical Research."

9. Paul Berg to Arthur Kornberg, 3 April 1970, Paul Berg Papers, Stanford University Archives, SC 358 (hereafter PB), box 2, folder: General Correspondence, 1970, I–O.

10. Paul Berg's venture into eukaryotic biology was intertwined with the molecularization of cancer research in the 1970s. See Jean-Paul Gaudillière, "The Molecularization of Cancer Etiology in the Postwar United States: Instruments, Politics and Management," in Soraya de Chadarevian and Harmke Kamminga, eds., *Molecularizing Biology and Medicine: New Practices and Alliances, 1910s–1970s* (Amsterdam: Harwood Academic Publishers, 1998), 139–70. On the development of genetic approach in cancer research in the era of recombinant DNA technology, see Michel Morange, "From the Regulatory Vision of Cancer to the Oncogene Paradigm, 1975–1985," *Journal of the History of Biology* 30 (1997): 1–29; and Joan H. Fujimura, *Crafting Science: A Sociohistory of the Quest for the Genetics of Cancer* (Cambridge: Harvard University Press, 1996).

11. Hans-Jörg Rheinberger, *Toward a History of Epistemic Things: Synthesizing Proteins in the Test Tube* (Stanford: Stanford University Press, 1997). For an historical analysis of the role of plasmid research in the development of recombinant DNA–cloning technology, see Angela N. H. Creager, "Adaptation or Selection? Old Issues and New Stakes in the Postwar Debates over Bacterial Drug Resistance," *Studies in History and Philosophy of Biological and Biomedical Sciences* 38 (2007): 159–90.

12. On the generation of hybrid experimental systems, see Robert E. Kohler, *Lords of the Fly: Drosophila Genetics and the Experimental Life* (Chicago: University of Chicago Press, 1994).

13. For an infrastructural approach in the history of biotechnology, see Hannah Landecker, *Culturing Life: How Cells Became Technologies* (Cambridge: Harvard University Press, 2007). For a standard history of recombinant DNA technology, see Susan Wright, "Recombinant DNA Technology and Its Social Transformation, 1972–1982," *Osiris* 2 (1986): 303–60; Sally S. Hughes, "Making Dollars out of

DNA: The First Major Patent in Biotechnology and the Commercialization of Molecular Biology, 1974–1980," *Isis* 92 (2001): 541–75; Sheldon Krimsky, *Biotechnics and Society: The Rise of Industrial Genetics* (New York: Praeger, 1991); and Martin Kenney, *Biotechnology: The University-Industrial Complex* (New Haven: Yale University Press, 1986). For a broader historical examination of the industrial uses of life through various chemical and biological techniques (or "zymotechnology"), see Robert Bud, *The Uses of Life: A History of Biotechnology* (Cambridge: Cambridge University Press, 1993). For the history of agricultural biotechnology, see Jack Ralph Kloppenburg, *First the Seed: The Political Economy of Plant Biotechnology, 1492–2000* (Cambridge: Cambridge University Press, 1988).

14. For Stanford biochemists' involvement in recombinant DNA research, see Harrison Echols and Carol Gross, *Operators and Promoters: The Story of Molecular Biology and Its Creators* (Berkeley: University of California Press, 2001); Hughes, "Making Dollars out of DNA"; Timothy Lenoir, "Biochemistry at Stanford: A Case Study in the Formation of an Entrepreneurial Culture" (unpublished manuscript, April 21, 2002), Microsoft Word file; and Michel Morange, *A History of Molecular Biology* (Cambridge: Harvard University Press, 2000).

15. Peter E. Lobban, "The Generation of Transducing Phage *In Vitro*," essay for third PhD examination, Stanford University, 6 November 1969, Peter E. Lobban Papers, Personal Collection, Los Altos, CA (hereafter PEL).

16. Daniel J. Kevles, "The National Science Foundation and the Debate over Postwar Research Policy, 1942–1945: A Political Interpretation of Science—the Endless Frontier," *Isis* 68 (1977): 5–26.

17. Angela N. H. Creager, "Mobilizing Biomedicine: Virus Research between Lay Health Organizations and the U.S. Federal Government, 1935–1955," in Caroline Hannaway, ed., *Biomedicine in the Twentieth Century: Practices, Policies, and Politics* (Amsterdam: IOS Press, 2008), 171–201.

18. Creager, *Life of a Virus*. On the postwar rise of *bio*medicine, see Peter Keating and Alberto Cambrosio, *Biomedical Platforms: Realigning the Normal and the Pathological in Late-Twentieth-Century Medicine* (Cambridge: MIT Press, 2003).

19. Daniel J. Kevles, "Renato Dulbecco and the New Animal Virology: Medicine, Methods, and Molecules," *Journal of the History of Biology* 26 (1993): 409–42.

20. Angela N. H. Creager and Jean-Paul Gaudillière, "Experimental Arrangements and Technologies of Visualization: Cancer as a Viral Epidemic, 1930–1960," in Jean-Paul Gaudillière and Ilana Löwy, eds., *Heredity and Infection: The History of Disease Transmission* (London: Routledge, 2001), 203–41.

21. Renato Dulbecco, "Production of Plaques in Monolayer Tissue Cultures by Single Particles of an Animal Virus," *Proceedings of the National Academy of Sciences, USA* 38 (1952): 747–52.

22. Renato Dulbecco, "Basic Mechanisms in the Biology of Animal Viruses: Concluding Address," *Cold Spring Harbor Symposia on Quantitative Biology* 27 (1962): 519.

23. Nadine Peyrieras and Michel Morange, "The Study of Lysogeny at the Pasteur Institute (1950–1960): An Epistemologically Open System," *Studies in History and Philosophy of Biological and Biomedical Sciences* 33 (2002): 419–30.

24. André Lwoff, "Lysogeny," *Bacteriological Reviews* 17 (1953): 270.

25. Jean J. Weigle and Max Delbrück, "Mutual Exclusion Between an Infecting Phage and a Carried Phage," *Journal of Bacteriology* 62 (1951): 301–18; Esther M. Lederberg, "Lysogenicity in *E. coli* K-12," *Genetics* 36 (1951): 560–69; and Elie L. Wollman, "Bacterial Conjugation," in John Cairns, Gunther S. Stent, and James D. Watson, eds., *Phage and the Origins of Molecular Biology* (Cold Spring Harbor: Cold Spring Harbor Laboratory Press, 1966), 216–25.

26. Echols and Gross, *Operators and Promoters*, 49–51.

27. A. Dale Kaiser and François Jacob, "Recombination between Related Temperate Bacteriophages and the Genetic Control of Immunity and Prophage Localization," *Virology* 4 (1957): 509–21.

28. Ibid., 509.

29. Peyrieras and Morange, "The Study of Lysogeny at the Pasteur Institute (1950–1960)." Harold Varmus points out the significance of lysogeny in the development of the genetics of cancer; see Harold Varmus, "The Pastorian: A Legacy of Louis Pasteur," in Gorge F. Woude and George Klein, eds., *Advances in Cancer Research* (New York: Academic Press, 1996), 1–16.

30. François Jacob to Arthur Kornberg, 13 September 1960, Arthur Kornberg Papers, Stanford University Archives, SC 359 (hereafter AK), box 1.

31. Marguerite Vogt and Renato Dulbecco, "Virus-Cell Interaction with a Tumor-Producing Virus," *Proceedings of the National Academy of Sciences, USA* 46 (1960): 365.

32. Renato Dulbecco, "Transformation of Cells in vitro by Viruses," *Science* 142 (1963): 932–36.

33. Allan M. Campbell, "Episomes," *Advances in Genetics Incorporating Molecular Genetic Medicine* 11 (1962): 101–45.

34. Dulbecco, "Transformation of Cells in vitro by Viruses," 369.

35. Gaudillière, "Molecularization of Cancer Etiology in the Postwar United States."

36. Dulbecco, Temin, and Baltimore shared the 1975 Nobel Prize for their achievement in animal virus and cancer research. Baltimore also had worked as a research associate at Salk with Dulbecco from 1965 to 1968, working on poliovirus. For David Baltimore's cancer research, see Shane Crotty, *Ahead of the Curve: David Baltimore's Life in Science* (Berkeley: University of California Press, 2001). For the politics of biomedical research during the Nixon administration, see Rena Selya, *Salvador Luria's Unfinished Experiment: The Public Life of a Biologist in a Cold War Democracy* (PhD dissertation, Harvard University, 2002).

37. Vogt and Dulbecco, "Virus-Cell Interaction with a Tumor-Producing Virus," 369.

38. Gaudillière, "Molecularization of Cancer Etiology in the Postwar United States"; and Creager, "Mobilizing Biomedicine."

39. On the sense of a crisis in molecular biology in the 1960s, see Michel Morange, "The Transformation of Molecular Biology on Contact with Higher Organisms, 1960–1980: From a Molecular Description to a Molecular Explanation," *History and Philosophy of the Life Sciences* 19 (1997): 369–93.

40. Gunther S. Stent, *The Coming of the Golden Age: A View of the End of Progress* (Garden City: Natural History Press, 1969), 57.

41. Pnina G. Abir-Am, "The Politics of Macromolecules: Molecular Biologists, Biochemists, and Rhetoric," *Osiris* 7 (1992): 164–91.

42. Edward O. Wilson, *Naturalist* (Washington, DC: Island Press/Shearwater Books, 1994), 218–25.

43. H. Chantreene to John Kendrew, 15 August 1967, quoted in de Chadarevian, *Designs for Life*, 48n25.

44. P. G. Abir-Am, "The First American and French Commemorations in Molecular Biology—from Collective Memory to Comparative History," *Osiris* 14 (1999): 324–70; Cairns, Stent, and Watson, eds., *Phage and the Origins of Molecular Biology*; and Jacques Monod, Ernest Borek, and André Lwoff, eds., *Of Microbes and Life* (New York: Columbia University Press, 1971).

45. James D. Watson, *Molecular Biology of the Gene* (New York: W. A. Benjamin, 1965).

46. Crick, "Molecular Biology and Medical Research," 187, quoted in de Chadarevian, *Designs for Life*, 346.

47. Solomon Garb, *Cure for Cancer, a National Goal* (New York: Springer Publishing, 1968), 120.

48. Daniel J. Kevles, "Pursuing the Unpopular: A History of Courage, Viruses, and Cancer," in Robert B. Silvers, ed., *Hidden Histories of Science* (New York: New York Review of Books, 1995), 69–114.

49. Richard A. Rettig, *Cancer Crusade: The Story of the National Cancer Act of 1971* (Princeton: Princeton University Press, 1977).

50. Daniel J. Kevles, *The Physicists: The History of a Scientific Community in Modern America* (New York: Knopf, 1977), 413–16; and Creager, "Mobilizing Biomedicine."

51. Vettel, *Biotech*.

52. Susan B. Spath, *C. B. Van Niel and the Culture of Microbiology, 1920–1965* (PhD dissertation, University of California, Berkeley, 1999). On the articulation of the prokaryote / eukaryote distinction, see Roger Y. Stanier and C. B. van Niel, "The Concept of a Bacterium," *Archiv Fur Mikrobiologie* 42 (1962): 17–35.

53. Angela N. H. Creager, "Mapping Genes in Microorganisms," in Jean-Paul Gaudillière and Hans-Jörg Rheinberger, eds., *From Molecular Genetics to Genomics: Mapping Cultures of Twentieth Century Genetics* (London: Routledge, 2004), 9–41.

54. François Jacob and Elie L. Wollman, *Sexuality and the Genetics of Bacteria* (New York: Academic Press, 1961).

55. Paul Berg, PhD, "A Stanford Professor's Career in Biochemistry, Science Politics, and the Biotechnology Industry," an oral history conducted in 1997 by Sally Smith Hughes, PhD, Regional Oral History Office, Bancroft Library, University of California, Berkeley, 2000, 62.

56. Paul Berg, "Viral Oncogenesis and Other Problems of Regulation," American Cancer Society grant proposal, 26 November 1968, p. 4 (emphasis added), PB, box 16, folder: ACS Grant.

57. For Sydney Brenner's work, see Soraya de Chadarevian, "Of Worms and Programmes: *Caenorhabditis elegans* and the Study of Development," *Studies in History and Philosophy of Biological and Biomedical Sciences* 29 (1998): 81–105. For Jacob's mouse developmental genetics research, see Michel Morange, "François Jacob's Lab in the Seventies: The T-Complex and the Mouse Developmental Genetic Program," *History and Philosophy of the Life Sciences* 22 (2000): 397–411. On Benzer's behavioral genetics research, see Jonathan Weiner, *Time, Love, Memory: A Great Biologist and His Quest for the Origins of Behavior* (New York: Knopf, 1999).

58. Toby A. Appel, *Shaping Biology: The National Science Foundation and American Biological Research, 1945–1975* (Baltimore: Johns Hopkins University Press, 2000), 252–55.

59. Draft, "A Proposal for a National Program on the Molecular Biology of Human Cells," n.d. (ca. 1971–72), p. 3, PB, box 4, folder: NSF Human Cell Program.

60. A. Dale Kaiser, "Cell Transformation by Polyoma Virus and SV40," 6 June 1967, A. Dale Kaiser Papers, Stanford University Archives, SC 356 (hereafter ADK), box 10, Lecture note: Biochemistry.

61. Berg, "Viral Oncogenesis and Other Problems of Regulation."

62. John F. Watkins and Renato Dulbecco, "Production of SV40 Virus in Heterokaryons of Transformed and Susceptible Cells," *Proceedings of the National Academy of Sciences, USA* 58 (1967): 1396–403.

63. Heiner Westphal and Renato Dulbecco, "Viral DNA in Polyoma- and SV40-Transformed Cell Lines," *Proceedings of the National Academy of Sciences, USA* 59 (1968): 1158–65.

64. Joseph Sambrook, Heiner Westphal, P. R. Srinivas, and Renato Dulbecco, "Integrated State of Viral DNA in SV40-Transformed Cells," *Proceedings of the National Academy of Sciences, USA* 60 (1968): 1288–95.

65. Berg, "Viral Oncogenesis and Other Problems of Regulation."

66. Paul Berg, "Viral Genome in Transformed Cells," *Proceedings of the Royal Society of London Series B: Biological Sciences* 177 (1971): 65.

67. Ibid., 65.

68. François Cuzin, Marguerite Vogt, Marianne Dieckmann, and Paul Berg, "Induction of Virus Multiplication in 3T3 Cells Transformed by a Thermosensitive

Mutant of Polyoma Virus. II. Formation of Oligometric Polyoma DNA Molecules," *Journal of Molecular Biology* 47 (1970): 317–33.

69. Ibid., 331.

70. Ibid., 332.

71. Berg, "Viral Oncogenesis and Other Problems of Regulation."

72. Paul Berg, "Viral Oncogenesis and Other Problems of Regulation," 1970 (continuation application), PB, box 16, folder: ACS Grant.

73. Berg, "A Stanford Professor's Career in Biochemistry, Science Politics, and the Biotechnology Industry," 62.

74. Charles Yanofsky, B. C. Carlton, D. R. Helinski, J. R. Guest, and U. Henning, "On Colinearity of Gene Structure and Protein Structure," *Proceedings of the National Academy of Sciences, USA* 51 (1964): 266–72; and Marcel Weber, "Representing Genes: Classical Mapping Techniques and the Growth of Genetical Knowledge," *Studies in History and Philosophy of Biological and Biomedical Sciences* 29 (1998): 295–315. Yanofksy's gene-protein system was indebted not only to Seymour Benzer's fine-structure gene mapping but also to Vernon Ingram's recent biochemical work on sickle-cell hemoglobin amino acid composition. For Benzer's mapping research, see Frederic L. Holmes, "Seymour Benzer and the Definition of the Gene," in Peter J. Beurton, Raphael Falk, and Hans-Jörg Rheinberger, eds., *The Concept of the Gene in Development and Evolution: Historical and Epistemological Perspectives* (Cambridge: Cambridge University Press, 2000), 115–55. For the development of protein sequencing as a tool for molecular genetics, see Soraya de Chadarevian, "Sequence, Conformation, Information: Biochemists and Molecular Biologists in the 1950s," *Journal of the History of Biology* 29 (1996): 361–86.

75. John Carbon, Paul Berg, and Charles Yanofsky, "Studies of Missense Suppression of Tryptophan Synthetase A-Protein Mutant A36," *Proceedings of the National Academy of Sciences, USA* 56 (1966): 764–71.

76. Paul Berg, "Suppression: A Subversion of Genetic Decoding," in *The Harvey Lectures, Series 67* (New York: Academic Press, 1974), 247–72.

77. Creager, "Mapping Genes in Microorganisms," 29.

78. Watkins and Dulbecco, "Production of SV40 Virus in Heterokaryons of Transformed and Susceptible Cells."

79. Paul Berg, "Nobel Lecture 1980: Dissections and Reconstructions of Genes and Chromosomes," in Tore Frängsmyr, ed., *Nobel Lectures: Chemistry 1971–1980* (Singapore: World Scientific Publishing, 1993), 391.

80. Berg, "Viral Oncogenesis and Other Problems of Regulation," 1970 (continuation application), p. 5, PB, box 16, folder: ACS Grant.

81. The dynamics of biological experimentation that unexpectedly brings out a research technology from an object of research is discussed in Rheinberger, *Toward a History of Epistemic Things*; and Creager, "Adaptation or Selection?"

82. For the prevalent use of the use of "vector" in bacterial systems, see Haruo

Ozeki and Hideo Ikeda, "Transduction Mechanisms," *Annual Review of Genetics* 2 (1968): 245–78.

83. Paul Berg, "Viral Oncogenesis and Other Problems of Regulation, 1971 (continuation application), p. 7 (emphasis added), PB, box 16, folder: ACS Grant.

84. Ibid.

85. Kenichi Matsubar and A. Dale Kaiser, "Lambda Dv: An Autonomously Replicating DNA Fragment," *Cold Spring Harbor Symposia on Quantitative Biology* 33 (1968): 769–75.

86. Berg, "Viral Oncogenesis and Other Problems of Regulation," 1970 (continuation application), p. 6, PB, box 16, folder: ACS Grant.

87. Ibid.

88. Alfred D. Hershey, ed., *The Bacteriophage Lambda* (Cold Spring Harbor: Cold Spring Harbor Laboratory, 1971); and Allan M. Campbell, "Bacteriophage λ as a Model System," *Bioessays* 5 (1986): 277–280.

89. Alfred D. Hershey, L. Ingraham, and E. Burgi, "Cohesion of DNA Molecules Isolated from Phage λ," *Proceedings of the National Academy of Sciences, USA* 49 (1963): 748–55.

90. Hans B. Strack and A. Dale Kaiser, "On Structure of Ends of Lambda DNA," *Journal of Molecular Biology* 12 (1965): 36–49.

91. A. Dale Kaiser and Ray Wu, "Structure and Function of DNA Cohesive Ends," *Cold Spring Harbor Symposia on Quantitative Biology* 33 (1968): 729–34: and Lisa Onaga, "Ray Wu as Fifth Business: Deconstructing Collective Memory in the History of DNA Sequencing," *Studies in History and Philosophy of Biological and Biomedical Sciences* 46 (2014): 1–14.

92. Arthur Kornberg to A. Dale Kaiser, "Department Finances," 10 September 1979, AK, box 5, folder 1979 (emphasis added).

93. Mehran Goulian, Arthur Kornberg, and Robert L. Sinsheimer, "Enzymatic Synthesis of DNA: 24. Synthesis of Infectious Phage φX174 DNA," *Proceedings of the National Academy of Sciences, USA* 58 (1967): 2321–28.

94. It was not yet known that *Eco*RI enzyme produced cohesive ends when the Berg's group tried to synthesize SV40-λ*dvgal* recombinant DNA molecules in 1971. Lobban's terminal transferase method thus enabled Berg's group to combine *Eco*RI-cut SV40 DNA and λ*dvgal* DNAs.

95. David A. Jackson, Robert H. Symons, and Paul Berg, "Biochemical Method for Inserting New Genetic Information into DNA of Simian Virus 40: Circular SV40 DNA Molecules Containing Lambda Phage Genes and Galactose Operon of *Escherichia coli*," *Proceedings of the National Academy of Sciences, USA* 69 (1972): 2904–9.

96. Creager, "Mapping Genes in Microorganisms," 29; and Thomas D. Brock, *The Emergence of Bacterial Genetics* (Cold Spring Harbor: Cold Spring Harbor Laboratory Press, 1990).

97. Lobban, "Generation of Transducing Phage *In Vitro*"; and Echols and

Gross, *Operators and Promoters*. For Lobban's role in the advent of recombinant DNA technology, see John Lear, *Recombinant DNA: The Untold Story* (New York: Crown Publishers, 1978), 44–46.

98. Peter E. Lobban, *An Enzymatic Method for End-to-end Joining of DNA Molecules* (PhD dissertation, Stanford University, 1972).

99. Ibid., 6.

100. Peter Lobban, interview with the author. Lobban recalls that he even speculated that he might be able to create DNA copies of messenger RNA for DNA splicing if he could find an enzyme that could transcribe reversely in his original PhD proposal. Since the existence of reverse transcriptase was not yet proven, one of his PhD proposal examiners, Robert Baldwin, insisted that Lobban's proposal on the use of a hypothetical enzyme for genetic engineering should be removed. A. Dale Kaiser discussed Lobban's interest in problems in antibody synthesis (interview with the author, October 10, 2005).

101. Lobban, *An Enzymatic Method for End-to-End Joining of DNA Molecules*, 128–29.

102. Hans-Jörg Rheinberger, "Beyond Nature and Culture: A Note on Medicine in the Age of Molecular Biology," *Science in Context* 8 (1995): 252–53.

103. P. Roy Vagelos and Louis Galambos, *Medicine, Science, and Merck* (Cambridge: Cambridge University Press, 2004).

104. Paul Rabinow, *Essays on the Anthropology of Reason* (Princeton: Princeton University Press, 1996), 99.

105. Kevles, *Physicists*; and Creager, "Mobilizing Biomedicine."

106. Creager, *Life of a Virus*.

107. On the technical implementation of life, see Rheinberger, *Toward a History of Epistemic Things*; Creager, "Adaptation or Selection?"; and Hannah L. Landecker, *Culturing Life: How Cells Became Technologies* (Cambridge: Harvard University Press, 2007).

Chapter Three

1. David A. Jackson, Robert H. Symons, and Paul Berg, "Biochemical Method for Inserting New Genetic Information into DNA of Simian Virus 40: Circular SV40 DNA Molecules Containing Lambda Phage Genes and Galactose Operon of *Escherichia coli*," *Proceedings of the National Academy of Sciences, USA* 69 (1972): 2909.

2. Hannah Landecker, *Culturing Life: How Cells Became Technologies* (Cambridge: Harvard University Press, 2007). Landecker discusses what she calls an "infrastructural approach" in the history of biomedical research, showing how certain materials and technologies reconfigure the condition of life forms, shifting our conceptions of biological life and cultural existence as human beings.

3. Peter E. Lobban, "The Generation of Transducing Phage *In Vitro*," essay for third PhD examination, Stanford University, 6 November 1969, Peter E. Lobban Papers, Personal Collection, Los Altos, CA.

4. Stanford University, Invention Disclosure, "A Process for Construction of Biologically Functional Molecular Chimeras," Inventors: Stanley N. Cohen and Herbert W. Boyer, 24 June 1974, Stanford University Office of Technology Licensing Archives, Stanford University.

5. For Cohen and Boyer's first gene-cloning experiment, see Stanley N. Cohen, Annie Chang, Herbert W. Boyer, and Robert B. Helling, "Construction of Biologically Functional Bacterial Plasmids *In Vitro*," *Proceedings of the National Academy of Sciences, USA* 70 (1973): 3240–44. For their intraspecies gene cloning, see Annie C. Chang and Stanley N. Cohen, "Genome Construction between Bacterial Species *in vitro*: Replication and Expression of Staphylococcus Plasmid Genes in *Escherichia coli*," *Proceedings of the National Academy of Sciences, USA* 71 (1974): 1030–34. For their interspecies gene cloning, see John F. Morrow, Stanley N. Cohen, Annie Chang, Herbert W. Boyer, Howard M. Goodman, and Robert B. Helling, "Replication and Transcription of Eukaryotic DNA in *Escherichia coli*," *Proceedings of the National Academy of Sciences, USA* 71 (1974): 1743–47.

6. For a philosophical analysis of the dynamics of experimental systems in biological experimentation see Hans-Jörg Rheinberger, *Toward a History of Epistemic Things: Synthesizing Proteins in the Test Tube* (Stanford: Stanford University Press, 1997).

7. There is ample literature on this, largely following a teleological narrative in Cohen and Boyer's patent application for recombinant DNA technology. See Susan Wright, "Recombinant DNA Technology and Its Social Transformation, 1972–1982," *Osiris* 2 (1986): 303–60; and Sally S. Hughes, "Making Dollars out of DNA: The First Major Patent in Biotechnology and the Commercialization of Molecular Biology, 1974–1980," *Isis* 92 (2001): 541–75.

8. Finnegan, Henderson, Farabow, Garrett & Dunner, LLP, *Opinion Regarding Validity, Enforceability and Infringement Issues Presented by the Cohen and Boyer Patents*, prepared for Leland Stanford Junior University (confidential opinion of counsel, 9 August 1985), 116. The quotation continues as follow: "Attempts may be made to prove that dimerization of the lambda-dv vector inevitably occurs and that those skilled in the art would have known to use the dimer for transformation even though not described by Mertz and Davis. Such attempts, in our opinion, will likely be viewed as improper hindsight reconstructions."

9. For an illuminating approach in looking at the system of exchange for the analysis of issues in scientific collaboration, competition, and priority, see Angela N. H. Creager and Gregory J. Morgan, "After the Double Helix: Rosalind Franklin's Research on Tobacco Mosaic Virus," *Isis* 99 (2008): 239–72.

10. See note 5.

11. Morton Mandel and Akiko Higa, "Calcium-Dependent Bacteriophage DNA Infection," *Journal of Molecular Biology* 53 (1970): 159–62.

12. Lobban handed Mertz this modified method to test with λ*dv* plasmids. Janet Mertz, "Calcium Chloride Transformation," Laboratory Notebook, December 1970, Janet E. Mertz Papers, Personal Collections, University of Wisconsin, Madison (hereafter JEM).

13. Kenichi Matsubara and A. Dale Kaiser, "Lambda dv: An Autonomously Replicating DNA Fragment," *Cold Spring Harbor Symposia on Quantitative Biology* 33 (1968): 769–75.

14. Douglas E. Berg, David A. Jackson, and Janet E. Mertz, "Isolation of a λ*dv* Plasmid Carrying the Bacterial *gal* Operon," *Journal of Virology* 14 (1974): 1063–69

15. Janet Mertz, "Getting λ*dvgal* (λ*dv*120) back into *E. coli*," in "Calcium Transfection Experiments," Laboratory Notebook, May 1971, JEM.

16. Ibid.

17. Paul Berg, Robert Symons, and David Jackson used Mertz's λ*dvgal*-120 for their artificial synthesis of recombinant DNA. See Jackson, Symons, and Berg, "Biochemical Method for Inserting New Genetic Information into DNA of Simian Virus 40."

18. For an analysis of communal mode of experimental life, see Robert E. Kohler, *Lords of the Fly:* Drosophila *Genetics and the Experimental Life* (Chicago: University of Chicago Press, 1994), esp. chapter 4, "The Fly People," 91–132.

19. Susan Wright, *Molecular Politics: Developing American and British Regulatory Policy for Genetic Engineering, 1972–1982* (Chicago: University of Chicago Press, 1994).

20. Joshua Lederberg served as an adviser to the arms control administration that was negotiating the Biological Weapons Convention in the early 1970s. See Joshua Lederberg, "Dr. Lederberg Speaks Out on Biological Warfare Hazards," 112 Cong. Rec. 31395–96 (September 11, 1970).

21. Paul Berg to Morton Mandel, 11 September 1973, Paul Berg Papers, Stanford University Archives, SC 358 (hereafter PB), box 3, folder: Correspondence, 1973.

22. For a history of biosafety regulations regarding recombinant DNA experiments, see Wright, *Molecular Politics*.

23. Carl R. Merril, Mark R. Geier, and John C. Petricciani, "Bacterial Virus Gene Expression in Human Cells," *Nature* 233 (1971): 398–400.

24. Janet Mertz, "Putting λplac [λ DNAs] into Mammalian Cells," Laboratory Notebook, October 1971, JEM.

25. Paul Berg, interview with the author, October 6, 2005.

26. Paul Berg, "Viral Oncogenesis and Other Problems of Regulation," 1970, p. 12, PB, box 16, folder: ACS Grant.

27. Hamilton O. Smith, and Kent W. Wilcox, "A Restriction Enzyme from *He-*

mophilus influenzae. I: Purification and General Properties," *Journal of Molecular Biology* 51 (1970): 379–91; and Thomas J. Kelly, Jr., and Hamilton O. Smith, "A Restriction Enzyme from *Hemophilus Influenzae*. II.," *Journal of Molecular Biology* 51 (1970): 393–409.

28. Herbert W. Boyer, "DNA Restriction and Modification Mechanisms in Bacteria," *Annual Review of Microbiology* 25 (1971): 153–76.

29. Kathleen Danna and Daniel Nathans, "Specific Cleavage of Simian Virus 40 DNA by Restriction Endonuclease of *Hemophilus Influenzae*," *Proceedings of the National Academy of Sciences, USA* 68 (1971): 2913–17.

30. John F. Morrow and Paul Berg, "Cleavage of Simian Virus 40 DNA at a Unique Site by a Bacterial Restriction Enzyme," *Proceedings of the National Academy of Sciences, USA* 69 (1972): 3365–69.

31. Robert L. Sinsheimer at Caltech began to publish a series of seminar articles that examined the infectivity of DNAs from φX phage. Sinsheimer discovered that a single-stranded DNA in the φX virus can infect the cell. See Mehran Goulian, Arthur Kornberg, and Robert L. Sinsheimer, "Enzymatic Synthesis of DNA, XXVI. Synthesis of Infectious Phage φX174 DNA," *Proceedings of the National Academy of Sciences, USA* 58 (1967): 2321–28; and Robert L. Sinsheimer, *Interview with Robert L. Sinsheimer* (Pasadena: California Institute of Technology Archives, 1992).

32. Janet Mertz, "Infectivity of Defective, Cont[inuous], and R₁ Treated SV40 DNA," Laboratory Notebook, March 1972, JEM; and "Infectivity of λdvgal (λdv120) before and after R₁ RTF Restriction Enzyme," Laboratory Notebook, May 1972, JEM.

33. Janet E. Mertz and Ronald W. Davis, "Cleavage of DNA by RI Restriction Endonuclease Generates Cohesive Ends," *Proceedings of the National Academy of Sciences, USA* 69 (1972): 3370–74. Mertz and Davis's discovery was confirmed by Vittorio Sgaramella, a postdoctoral fellow in the Stanford Genetics Department down the hall from the Biochemistry Department, who showed that *Eco*RI-cut SV40 ends could not be ligated to phage T4 ends, which were known to be blunt. Vittorio Sgaramella, J. H. van de Sande, and H. Gobind Khorana, "Studies on Polynucleotides, C. A Novel Joining Reaction Catalyzed by the T4-Polynucleotide Ligase," *Proceedings of the National Academy of Sciences, USA* 69 (1972): 1468–75. Boyer and his colleagues subsequently determined the actual DNA sequence of these *Eco*RI-generated sticky ends. See Joe Hedgpeth, Howard M. Goodman, and Herbert W. Boyer, "DNA Nucleotide Sequence Restricted by the RI Endonuclease," *Proceedings of the National Academy of Sciences, USA* 69 (1972): 3448–52.

34. Janet Mertz, "Making λdvgal (λdv120)-SV40 Hybrids," Laboratory Notebook, May 1972, JEM.

35. Mertz and Davis. "Cleavage of DNA by RI Restriction Endonuclease Generates Cohesive Ends," 3371.

36. Janet E. Mertz, Ronald W. Davis, and Paul Berg, "Characterization of the Cleavage Site of the R1 Restriction Enzyme," abstract presented at the Fourth Annual Tumor Virus Meeting (arranged by Philip A. Sharp), August 16–19, 1972, Cold Spring Harbor Laboratory, Cold Spring Harbor, New York.

37. Janet Mertz, interview with the author, August 2009.

38. Alfred Hellman, M. N. Oxman, and Robert Pollack, eds., *Biohazards in Biological Research* (Cold Spring Harbor: Cold Spring Harbor Laboratory Press, 1973).

39. The Asilomar Conference Center in Monterey County had been used to host faculty members and graduate students of the Stanford Biochemistry Department for its annual department meetings. See Wright, *Molecular Politics*; James D. Watson and John Tooze, *The DNA Story: A Documentary History of Gene Cloning* (San Francisco: W. H. Freeman and Co., 1981); and John Lear, *Recombinant DNA: The Untold Story* (New York: Crown Publishers, 1978).

40. Hellman, Oxman, and Pollack, *Biohazards in Biological Research*, 354.

41. Ibid.

42. Interview with Janet Mertz, 9 March 1977, Oral History Programs, Recombinant DNA Controversy Collection, MC 100, MIT Institute Archive, esp. 16–31.

43. Paul Berg to Michael Stoker, 19 June 1972, PB, box 3.

44. Fourth Annual Tumor Virus Meeting (arranged by Philip A. Sharp), Abstracts, August 16–19, 1972, Cold Spring Harbor Laboratory, Cold Spring Harbor, New York.

45. Paul Berg, PhD, "A Stanford Professor's Career in Biochemistry, Science Politics, and the Biotechnology Industry," an oral history conducted in 1997 by Sally Smith Hughes, PhD, Regional Oral History Office, Bancroft Library, University of California, Berkeley, 2000, 114: "Mandel had discovered that if you took *E. coli* and exposed them to elevated calcium and gave them a shock, DNA entered these cells much, much more frequently and efficiently than if you didn't do that. And so, when Stan heard that, he came around and wanted to learn how to do this. Now, Janet Mertz, who was in my lab, was already doing that. And so she instructed Stan on how to do this, and they actually worked together." Mertz confirmed this in her interview with the author in February 2007. Stanley Cohen insists he independently developed his own method with occasional help of Peter Lobban. Stanley Cohen, interview with the author, March 2010.

46. Berg to Francis Crick, 5 November 1973, PB, box 3, folder: Correspondence, 1973. Crick indicated that he would like to have photos of SV40 chromatin in order to work on the details of protein-DNA interactions with Roger Kornberg (Arthur Kornberg's son, who went to Cambridge to work with Crick).

47. Arthur Kornberg, MD, "Biochemistry at Stanford, Biotechnology at DNAX," an oral history conducted in 1997 by Sally Smith Hughes, PhD, Regional Oral History Office, Bancroft Library, University of California, Berkeley, 1998.

48. A. Dale Kaiser, interview with the author, October 10, 2005. Jerald Hurwitz

did not join the Stanford Biochemistry Department when Kornberg established it in 1959.

49. For a historical analysis of the experimental life of plasmids from objects of research to experimental tools, see Angela N. H. Creager, "Adaptation or Selection? Old Issues and New Stakes in the Postwar Debates over Bacterial Drug Resistance," *Studies in History and Philosophy of Biological and Biomedical Sciences* 38 (2007): 159–90; and Mathias Grote, "Hybridizing Bacteria, Crossing Methods, Cross-checking Arguments: The Transition from Episomes to Plasmids (1961–1969)," *History and Philosophy of the Life Sciences* 30 (2008): 407–30.

50. Stanley N. Cohen, "Genetic Control of Episomal Drug Resistance Factors," National Institutes of Health, grant application, February 1967, Stanley N. Cohen Papers, Personal Collections, Stanford University (hereafter SNC).

51. Mertz later published her protocol; see Berg, Jackson, and Mertz, "Isolation of a Lambda dv Plasmid Carrying the Bacterial Gal Operon." On page 1067, they described the "establishment of λdvgal-120 120 in new bacterial hosts by transfection."

52. Stanley N. Cohen, Annie C. Chang, and Leslie Hsu, "Nonchromosomal Antibiotic Resistance in Bacteria: Genetic Transformation of *Escherichia coli* by R-Factor DNA," *Proceedings of the National Academy of Sciences, USA* 69 (1972): 2110–14. This article was communicated by Stanford biochemist A. Dale Kaiser in early 1972, and Cohen acknowledged Lobban's help in the adoption of the calcium chloride technique (or transformation assay).

53. Stanley Falkow, Declaration under Rule 132, in *Re: Application of Stanley N. Cohen et al.*, serial no. 959,288 for Biologically Functional Molecular Chimeras, September 27, 1982, exhibit 15, appendix to Finnegan, Henderson, Farabow, Garrett & Dunner, LLP, *Opinion Regarding Validity, Enforceability and Infringement Issues Presented by the Cohen and Boyer Patents*.

54. Stanley Falkow, "I'll Have the Chopped Liver Please, or How I Learned to Love the Clone," *American Society for Microbiology News* 67:11 (2001): 558.

55. Stanley N. Cohen, Annie Chang, Herbert W. Boyer, and Robert B. Helling, "Construction of Biologically Functional Bacterial Plasmids *In Vitro*," *Proceedings of the National Academy of Sciences, USA* 70 (1973): 3240–44.

56. Maxine Singer and Dieter Söll, "Guidelines for DNA Hybrid Molecules," *Science* 181 (1973): 1114.

57. Stanley N. Cohen and Annie C. Y. Chang, "Recircularization and Autonomous Replication of a Sheared R-Factor DNA Segment in *Escherichia coli* Transformants," *Proceedings of the National Academy of Sciences, USA* 70 (1973): 1293–97. It later turns out that pSC101 was not derived from R6–5 plasmid. See Stanley N. Cohen and Annie C. Y. Chang, "Revised Interpretation of the Origins of the pSC101 Plasmid," *Journal of Bacteriology* 132 (1977): 734–37.

58. Recent National Institutes of Health guidelines have very strict rules regarding cloning genes from *Staphylococcus aureus*.

59. Quoted in Wright, *Molecular Politics*, 246; and Stanley N. Cohen, "Recombinant DNA: Fact and Fiction," Science 195 (1977): 654.

60. Bruno Latour, *Science in Action: How to Follow Scientists and Engineers through Society* (Cambridge: Harvard University Press, 1988).

61. Donald Brown, interview with the author, February 2006. At that time, Morrow was writing his dissertation and planned to join Brown's laboratory as a postdoctoral fellow. Donald D. Brown, "Some Genes were Isolated and Their Structure Studied before the Recombinant DNA Era," *BioEssays* 16 (1994): 139–43. The ribosomal genes of *Xenopus laevis* were first isolated in 1966 by Max L. Birnstiel and H. Wallace as a way to study differential gene action in embryos. See Max L. Birnstiel, H. Wallace, J. L. Hirlin, and M. Fischberg, "Localization of the Ribosomal DNA Complements in the Nucleolar Organizer Region of *Xenopus Laevis*," *National Cancer Institute Monograph* 23 (1966): 431–44.

62. Morrow et al., "Replication and Transcription of Eukaryotic DNA in *Escherichia coli*."

63. Stanley N. Cohen, "Extraction of *Staphylococcus* DNAs," 27 June 1973, Laboratory Notebook, SNC.

64. Chang and Cohen, "Genome Construction between Bacterial Species *in vitro*."

65. John F. Morrow, correspondence with the author, March 2010.

66. For an analysis of the so-called Asilomar II conference and subsequent developments in the regulation of recombinant DNA experiments, see Wright, *Molecular Politics*.

67. Scholars in science studies have focused on systems of exchange. For more, see Robert Kohler, *Lords of the Fly*; and Creager and Morgan, "After the Double Helix."

68. Kohler, *Lords of the Fly*, 103.

69. Ibid.

70. Berg, "A Stanford Professor's Career in Biochemistry, Science Politics, and the Biotechnology Industry," 56.

Chapter Four

1. Stephen S. Hall, *Invisible Frontiers: The Race to Synthesize a Human Gene*, 1st ed. (New York: Atlantic Monthly Press, 1987). As historian Robert Bud points out, there had been earlier enthusiasm and recognition of the commercial promise of the industrial uses of life before the advent of recombinant DNA technology. See Robert Bud, *The Uses of Life: A History of Biotechnology* (Cambridge: Cambridge University Press, 1993).

2. Richard Roblin to Paul Berg, 7 May 1970, Paul Berg Papers, Stanford University Archives, SC 215 (hereafter PB), box 2, folder: General Correspondence,

1970 (P–Z). Roblin and Theodore Friedmann published their speculation on human gene therapy in 1972 and became one of the earliest proponents of human gene therapy using gene modification technologies. See Theodore Friedmann and Richard Roblin, "Gene Therapy for Human Genetic Disease?," *Science* 175 (1972): 949–55.

3. Robert E. Kohler, *Lords of the Fly:* Drosophila *Genetics and the Experimental Life* (Chicago: University of Chicago Press, 1994).

4. Evelyn Fox Keller, *The Century of the Gene* (Cambridge: Harvard University Press, 2002).

5. François Jacob and Jacques Monod, "Genetic Regulatory Mechanisms in the Synthesis of Proteins," *Journal of Molecular Biology* 3 (1961): 354.

6. Jacques Monod and François Jacob, "General Conclusions: Teleonomic Mechanisms in Cellular Metabolism, Growth, and Differentiation," *Cold Spring Harbor Symposia on Quantitative Biology* 26 (1961): 397.

7. Scott Gilbert, "Entrance of Molecular Biology into Embryology," in Sahotra Sarkar, ed., *The Philosophy and History of Molecular Biology: New Perspectives* (Dordrecht: Kluwer Academic, 1996), 101–23.

8. Sahotra Sarkar, "From Genes as Determinants to DNA as Resource: Historical Notes on Development and Genetics," in Eva M. Neumann-Held and Christoph Rehmann-Sutter, eds., *Genes in Development: Re-reading the Molecular Paradigm* (Durham: Duke University Press, 2006), 77–95. For Edward B. Lewis's adoption of the operon model, see Edward B. Lewis, "Genes and Developmental Pathways," *American Zoologist* 3 (1963): 33–56. Walter Gehring later recalls that "inspired by the famous work of François Jacob and Jacques Monod on gene regulation in bacteria, [I] contemplated the isolation of *Drosophila*-analogs to DNA-binding proteins such as the *E. coli lac* repressor in the 1960s." Quoted in Marcel Weber, "Walking on the Chromosome: *Drosophila* and the Molecularization of Development," in Jean-Paul Gaudillière and Hans-Jörg Rheinberger, eds., *From Molecular Genetics to Genomics: Mapping Cultures of Twentieth Century Genetics* (London: Routledge, 2004), 73.

9. For example, see James Bonner, *The Molecular Biology of Development* (Oxford: Oxford University Press, 1965); and Evelyn Fox Keller, "Decoding the Genetic Program: Or, Some Circular Logic in the Logic of Circularity," in Peter J. Beurton, Raphael Falk, and Hans-Jörg Rheinberger, eds., *The Concept of the Gene in Development and Evolution: Historical and Epistemological Perspectives* (Cambridge: Cambridge University Press, 2000), 159–77. On Sydney Brenner's developmental biology work, see Soraya de Chadarevian, "Of Worms and Programmes: *Caenorhabditis elegans* and the Study of Development," *Studies in History and Philosophy of Biological and Biomedical Sciences* 29 (1998): 81–105. For Jacob's move into developmental biology, see Michel Morange, "François Jacob's Lab in the Seventies: The T-Complex and the Mouse Developmental Genetic Program," *History and Philosophy of the Life Sciences* 22 (2000): 397–411.

10. Michel Morange, "The Transformation of Molecular Biology on Contact

with Higher Organisms, 1960–1980: From a Molecular Description to a Molecular Explanation," *History and Philosophy of the Life Sciences* 19 (1997): 369–93. As early as the third anniversary of the operon model in 1964, molecular biologist Gunter Stent critically assessed its implications for developmental biology. See Gunter Stent, "The Operon: On Its Third Anniversary," *Science* 144 (1964): 816–20.

11. Morange, "Transformation of Molecular Biology."

12. Monod and Jacob, "General Conclusions," 363.

13. C. A. Thomas, Jr., "The Genetic Organization of Chromosomes," *Annual Review of Genetics* 5 (1971): 237 (emphasis added).

14. For the productivity of *Drosophila* for genetic research, see Kohler, *Lords of the Fly*.

15. On the one-gene, one-band hypothesis, see Wolfgang Beermann, "Chromomerenkonstanz und spezifische Modifikationen der Chromosomenstruktur in der Entwicklung und Organdifferenzierung von *Chironomus tentans*," *Chromosoma* 5 (1952): 139–98.

16. For example, the molecular biologist Francis Crick assumed that most DNA segments in animal chromosomes did not code for proteins. Rather, in a globular form, these noncoding DNA segments would function as a controlling element in the expression of genes including differentiation and development. See Francis Crick, "General Model for the Chromosomes of Higher Organisms," *Nature* 234 (1971): 25–27. Eric Davidson began to claim that the high-level concentration of nuclear RNA sequences might be a product of repetitive DNA sequences. See Roy J. Britten and Eric H. Davidson, "Gene Regulation for Higher Cells: A Theory," *Science* 165 (1969): 349–57. On the discovery of repetitive sequences in DNA, see Roy J. Britten and David Kohne, "Repeated Sequences in DNA," *Science* 161 (1968): 529–40.

17. J. Barry Egan and David S. Hogness, "The Topography of Lambda DNA: Isolation of Ordered Fragments and the Physical Mapping of Point Mutations," *Journal of Molecular Biology* 71 (1972): 363–81.

18. David S. Hogness, "The Structure and Function of Animal Chromosomes," National Science Foundation Grant Proposal, 1970, p. 2, David S. Hogness Papers, Personal Collections, Stanford University (hereafter DSH). Because it was not clear whether the mode of gene regulation in bacteria could be applied to higher organisms, Hogness used the more general term "transcripton for a unit of transcription. It does not imply any particular mode of regulation as does the term operon; operons are then one class of transcriptions."

19. David S. Hogness, "The Arrangement and Function of DNA Sequences in Animal Chromosomes," National Institutes of Health Grant Proposal, 1972, DSH.

20. Ibid., 1.

21. Pieter C. Wensink, David Finnegan, John E. Donelson, and David S. Hogness, "A System for Mapping DNA Sequences in the Chromosome of *Drosophila melanogaster*," *Cell* 3 (1974): 315–25.

22. Ibid.

23. David S. Hogness, declaration before the United States Patent and Trademark Office, for "Biologically Functional Molecular Chimeras," signed on 27 October 1982, Stanford University Office of Technology and Licensing Archives, Stanford University (hereafter SUOTL).

24. Janet Mertz to Mary Lou Pardue, 14 November 1973, Janet E. Mertz Papers, Personal Collections, University of Wisconsin, Madison.

25. Stanley N. Cohen and Annie C. Y. Chang, "Recircularization and Autonomous Replication of a Sheared R-Factor DNA Segment in *Escherichia coli* Transformants," *Proceedings of the National Academy of Sciences, USA* 70 (1973): 1293–97 (published May 1973).

26. For Cohen's proposal for collaboration with David Hogness, see Stanley N. Cohen, MD, "Science, Biotechnology, and Recombinant DNA: A Personal History," an oral history conducted in 1995 by Sally Smith Hughes, PhD, Regional Oral History Office, Bancroft Library, University of California, Berkeley, 2009, 66.

27. Ibid., 64. His first intraspecies cloning paper was in fact published in November 1973.

28. Ibid., 64–65.

29. Paul Berg (Principal Investigator), with Douglas Brutlag, Ronald Davis, David Hogness, and George Stark, "Biochemical, Genetic, and Electron Microscopic Studies of Chromosome Structure and Expression in Eukaryotic Organisms," National Science Foundation Grant Proposal, November 1973, PB, box 13, folder: NSF Grant Application.

30. For an illustration of the competition in molecular biology, see James D. Watson, *The Double Helix: A Personal Account of the Discovery of the Structure of DNA* (New York: Atheneum, 1968).

31. Cohen, "Science, Biotechnology, and Recombinant DNA," 66.

32. Ibid., 17.

33. Ibid., 19.

34. Ibid., 64–65.

35. Ibid., 65–66; and Paul Berg, PhD, "A Stanford Professor's Career in Biochemistry, Science Politics, and the Biotechnology Industry," an oral history conducted in 1997 by Sally Smith Hughes, PhD, Regional Oral History Office, Bancroft Library, University of California, Berkeley, 2000, 112.

36. Stanley Cohen's interspecies cloning paper (*Staph* cloning) was submitted to the *Proceedings of the National Academy of Sciences, USA (PNAS)* in November 1973 (published April 1974) and Morrow's *Xenopus* cloning paper was submitted to *PNAS* in January 1974 (published May 1974). Hogness's cloning of *Drosophila* DNAs was submitted to *PNAS* in August 1974 (published December 1974). Regardless of Cohen's concerns, it would be unlikely for Hogness to publish the cloning of random DNAs from the entire fly genome before the publication of Cohen's cloning of *Staph* and *Xenopus* genes. Hogness would need a substantial amount of time to characterize whatever clones he obtained from the fly genome,

whereas, for Cohen, both the *Staph* and *Xenopus* genes were well defined and easy to prove that they were actually cloned.

37. Michael Grunstein and David Hogness, "Colony Hybridization: A Method for the Isolation of Cloned DNAs that Contain a Specific Gene," *Cell* 72 (1975): 3961–65; Welcome Bender, Pierre Spierer, and David Hogness, "Chromosomal Walking and Jumping to Isolate DNA from the *Ace* and *rosy* Loci and the Bithorax Complex in *Drosophila melanogaster*," *Journal of Molecular Biology* 168 (1983): 17–33; and Marcel Weber, "Redesigning the Fruit Fly: The Molecularization of *Drosophila*," in Angela N. H. Creager, Elizabeth Lunbeck, and M. Norton Wise (eds.), *Science without Laws: Model Systems, Cases, Exemplary Narratives* (Durham: Duke University Press, 2008), 23–45.

38. Evenly Fox Keller, "*Drosophila* Embryos as Transitional Objects: The Work of Donald Poulson and Christiane Nüsslein-Volhard," *Historical Studies in the Physical and Biological Sciences* 26 (1996): 313–46; and Keller, "Developmental Biology as a Feminist Cause?," *Osiris* 12 (1997): 16–28.

39. On the application of recombinant DNA technology for *Drosophila* developmental biology, see Marcel Weber, *Philosophy of Experimental Biology* (Cambridge: Cambridge University Press, 2004), 159–64. Through a series of cloning developmentally significant genes in the laboratories of Gehring at the University of Basel and Thomas Kaufman at Indiana University, scientists found a homologous sequence among the *Antp, Ftz,* and *Ultrabithorax* complex of genes in *Drosophila*, namely, the homeobox. More surprisingly a few years after the initial discovery of the homeobox, it turned out to be highly conserved in other organisms, such as the frog *Xenopus laevis,* a mouse species, and other metazoans. The homeobox genes, or, more generally, the genetic tool kit for development, presented a striking picture of a regulatory circuit, in which genetic and epigenetic factors coordinate gene activation for cellular differentiation and development. Moreover, the highly conserved regulatory tool kits across different species barriers prompted a series of new questions about the relationship between development and evolution, in which genetic tool kits are regarded as one of the most significant engine for evolutionary changes. For an accessible general introduction, see Sean Carroll, *Endless Forms Most Beautiful: The New Science of Evo Devo and the Making of the Animal Kingdom* (New York: W. W. Norton, 2005).

40. Lobban, correspondence with the author, March 7, 2007. According to Cohen's recollection, he did not specifically recall Lobban's speculations about the industrial application of recombinant DNA technology, though he acknowledges that he read Lobban's 1972 dissertation in 1975. See Cohen, "Science, Biotechnology, and Recombinant DNA," 53.

41. Hans-Jörg Rheinberger, "Beyond Nature and Culture: A Note on Medicine in the Age of Molecular Biology," *Science in Context* 8 (1995): 252.

42. Victor K. McElheny, "Animal Gene Shifted to Bacteria; Aid Seen to Medicine and Farm," *New York Times*, May 20, 1974, 1.

43. Bertram Rowland, "Bertram Rowland and the Cohen/Boyer Cloning Patent," available at http://www.law.gwu.edu/Academics/FocusAreas/IP/Pages/Cloning.aspx, accessed October 2011.

44. Cohen, "Science, Biotechnology, and Recombinant DNA," 151.

45. Sally S. Hughes, "Making Dollars out of DNA: The First Major Patent in Biotechnology and the Commercialization of Molecular Biology, 1974–1980," *Isis* 92 (2001): 541–75. Her article, however, does not include a set of controversies on the validity and scope of the Cohen-Boyer patent applications.

46. Cohen, "Science, Biotechnology, and Recombinant DNA," 157.

47. Dennis Flanagan to Paul Berg, 17 January 1974, PB, SC358, box 4, folder: *Scientific America.*

48. Dennis Flanagan to Paul Berg, 13 September 1974, PB, SC358, box 4, folder: *Scientific America.*

49. Paul Berg to Dennis Flanagan, 24 September 1974, PB, SC358, box 4, folder: *Scientific America.*

50. John Morrow to Stanley Cohen and Herbert Boyer, 23 January 1975, SUOTL.

51. John Poitras, "Meeting on DNA Patent," 17 May 1976; attendees were Clayton Rich, Joshua Lederberg, William Massy, William Baxter, Robert Rosenzweig, Peter Carpenter, Paul Berg, John Poitras, and Stanley Cohen, SUOTL.

52. Anonymous patent reviewer, "In Confidence, Re: Process and Composition for Biologically Functional DNA Chimeras," 1 July 1975, SUOTL.

53. Notes on Continuous Application, serial no. 687, 430, filed on 17 May 1976, SUOTL.

54. Alvin E. Tanenholtz, October 1977, SUOTL; and Janet E. Mertz and Ronald W. Davis, "Cleavage of DNA by RI Restriction Endonuclease Generates Cohesive Ends," *Proceedings of the National Academy of Sciences, USA* 69 (1972): 3370–74. For a more technical analysis, see chapter 3.

55. Stanley Cohen, "Our Discussions about the Origin of the 'Recombinant DNA Technique,'" memorandum to Joshua Lederberg, 10 July 1978, in Cohen, "Science, Biotechnology, and Recombinant DNA," appendix, 2.

56. Tsunehiro Mukai, Kenichi Matsubara, and Yasuyuki Takagi, "Cloning Bacterial Genes with Plasmid Lambda-dv," *Molecular and General Genetics* 146 (1976): 269–74.

57. Bertram Rowland to Stanford Office of Technology Licensing, SUOTL.

58. Cetus's patent lawyer, Albert P. Halluin, raised objections to the Cohen-Boyer patenting on a set of legal grounds. See Albert P. Halluin, "Patenting the Results of Genetic Engineering Research: An Overview," in David W. Plant, Neils J. Reimers, and Norton D. Zinder, eds., *Patenting of Life Forms* (Cold Spring Harbor: Cold Spring Harbor Laboratory Press, 1982), 67–126. Halluin's position regarding the Cohen-Boyer patent is summarized succinctly in a letter from D. R. Dunner to SUOTL, 6 May 1985, SUOTL.

59. On the tensions arising from an effort for the demarcation between intellectual commons and intellectual property, see Arti K. Rai, "Regulating Scientific Research: Intellectual Property Rights and the Norms of Science," *Northwestern University Law Review* 94 (1999): 77–152. For issues on the control and ownership of intellectual property in academic research, see Daniel J. Kevles, "Principles, Property Rights, and Profits: Historical Reflections on University/Industry Tensions," *Accountability in Research* 8 (2001): 12–26. The fuller discussion of the prosecution history with its complete legal technicalities would require access to the PTO's prosecution file on the Cohen-Boyer patents, whose access was briefly open to the public between June and December 1982 but closed by Stanford's request amidst the rejection of its second patent application. "Stanford Shuts Open Door to Cohen-Boyer Patent File as Application Hangs Fire," *Biotechnology Newswatch* 2:24 (1982): 1.

60. Anonymous patent reviewer, "In Confidence, Re: Process and Composition for Biologically Functional DNA Chimeras," Office of Technology Licensing Correspondence, 1974–79, 1 July 1975, SUOTL.

61. Joshua Lederberg to Senator Gaylord Nelson, 22 May 1978, Joshua Lederberg Papers, National Library of Medicine, box 35, folder: Correspondence.

62. On the modes and meanings of gift exchange, see Natalie Zemon Davis, *The Gift in Sixteenth-Century France* (Madison: University of Wisconsin Press, 2000).

63. See Niels Reimers, "Stanford's Office of Technology Licensing and the Cohen/Boyer Cloning Patents," an oral history conducted in 1997 by Sally Smith Hughes, PhD, Regional Oral History Office, Bancroft Library, University of California, Berkeley, 1998, 13–14.

64. Watson, *The Double Helix*; and Steven Shapin, *The Scientific Life: A Moral History of a Late Modern Vocation* (Chicago: University of Chicago Press, 2008), 218.

Chapter Five

1. The Stanford and University of California patent application cited three papers as crucial evidence for Cohen and Boyer's invention of recombinant DNA technology: Stanley N. Cohen, Annie Chang, Herbert W. Boyer, and Robert B. Helling, "Construction of Biologically Functional Bacterial Plasmids *In Vitro*," *Proceedings of the National Academy of Sciences, USA (PNAS)* 70 (1973): 3240–44; Annie C. Chang and Stanley N. Cohen, "Genome Construction between Bacterial Species *in vitro*: Replication and Expression of Staphylococcus Plasmid Genes in *Escherichia coli*," *Proceedings of the National Academy of Sciences, USA* 71 (1974): 1030–34; and John F. Morrow, Stanley N. Cohen, Annie Chang, Herbert W. Boyer, Howard M. Goodman, and Robert B. Helling, "Replication and Transcription of Eukaryotic DNA in *Escherichia coli*," *Proceedings of the National Academy of Sciences, USA* 71 (1974): 1743–47.

2. For a historical analysis of the rise of recombinant DNA technology out of the vibrant circulation of materials and technologies within the dense research network formed around Stanford biochemists, see Doogab Yi, "Cancer, Viruses, and Mass Migration: Paul Berg's Venture into Eukaryotic Biology and the Advent of Recombinant DNA Research and Technology, 1967–1974," *Journal of the History of Biology* 41 (2008): 589–636. For Cohen's contribution to the transformation of plasmid from an object of research into a technology for genetic engineering, see Angela N. H. Creager, "Adaptation or Selection? Old Issues and New Stakes in the Postwar Debates over Bacterial Drug Resistance," *Studies in History and Philosophy of Biological and Biomedical Sciences* 38 (2007): 159–90.

3. For the evolution of regulatory regimes for recombinant DNA research and technology, see Susan Wright, *Molecular Politics: Developing American and British Regulatory Policy for Genetic Engineering, 1972–1982* (Chicago: University of Chicago Press, 1994); and Sheldon Krimsky, *Genetic Alchemy: The Social History of the Recombinant DNA Controversy* (Cambridge: MIT Press, 1985).

4. Robert M. Rosenzweig to Donald S. Fredrickson, 18 June 1976, folder: Recombinant DNA Patent, Stanford University Office of Technology Licensing Archives, Stanford University (hereafter SUOTL).

5. "Institutional Patent Agreement Governing Grants and Awards from the Department of Health, Education, and Welfare," August 1968, Norman J. Latker Papers, Personal Collection, Bethesda, MD (hereafter NJL).

6. Barry Leshowitz, "The Demise of Technology Transfer in DHEW," 1978, NJL.

7. Stanley N. Cohen and Herbert W. Boyer, Process for Producing Biologically Functional Molecular Chimeras, US Patent 4,237,224, filed November 4, 1974, issued December 2, 1980.

8. Historians have instead attributed the prolonged prosecution of recombinant DNA technology patent applications to heated public health discussions about the safety of recombinant DNA technology and the lack of legal means for patenting organisms in the 1970s. See Sally S. Hughes, "Making Dollars out of DNA: The First Major Patent in Biotechnology and the Commercialization of Molecular Biology, 1974–1980," *Isis* 92 (2001): 541–75. For a history of the expansion of intellectual property rights to biological organisms, see Daniel J. Kevles, "Ananda Chakrabarty Wins a Patent: Biotechnology, Law and Society, 1972–1980," *Historical Studies in the Physical and Biological Sciences* 25 (1994): 111–35, and Kevles, "Of Mice & Money: The Story of the World's First Animal Patent," *Daedalus* 131 (2002): 78–88. For a broader overview of the commercialization of science, with an emphasis on shifts in funding and in the organization of science in the latter half of the twentieth century, see Philip Mirowski and Esther-Mirjam Sent, "The Commercialization of Science, and the Response of STS," in Edward J. Hackett, Olga Amsterdamska, Michael Lynch, and Judy Wajcman, eds., *Handbook of Science, Technology and Society Studies* (Cambridge: MIT Press, 2007), 635–89.

9. My analysis builds on the work of social scientists who have examined the

broader political and economic context of American research universities' involvement in patent management. See David C. Mowery, Richard R. Nelson, Bhaven N.
Sampt, and Arvids A. Ziedonis, eds., *Ivory Tower and Industrial Innovation* (Stanford: Stanford University Press, 2004). For an insightful analysis of the enactment
of the Bayh-Dole Act in terms of institutional building, see Elizabeth P. Berman,
"Why Did Universities Start Patenting? Institution-Building and the Road to the
Bayh-Dole Act," *Social Studies of Science* 38 (2008): 835–71.

10. For a historical overview of the emergence of biotechnology, see Martin
Kenney, *Biotechnology: The University-Industrial Complex* (New Haven: Yale University Press, 1986); and Sheldon Krimsky, *Biotechnics and Society: The Rise of Industrial Genetics* (New York: Praeger, 1991). For a pioneering attempt to histori-
cize the emergence of biotechnology, see essays in Arnold Thackray, ed., *Private
Science: Biotechnology and the Rise of Molecular Sciences* (Philadelphia: University of Pennsylvania Press, 1998).

11. The origins of this standard history can be traced to Stanley N. Cohen, "The
Manipulation of Genes," *Scientific American* 233 (1975): 24–33.

12. For an analysis of the transformation of recombinant DNA technology for
a research technique to a commercial biotechnology, see Susan Wright, "Recombinant DNA Technology and Its Social Transformation, 1972–1982," *Osiris* 2 (1986):
303–60.

13. For a historical account of the rise of biotech companies in the 1970s and
1980s, see Stephen S. Hall, *Invisible Frontiers: The Race to Synthesize a Human
Gene* (New York: Atlantic Monthly Press, 1987); and Cynthia Robbins-Roth, *From
Alchemy to IPO: The Business of Biotechnology* (Cambridge: Perseus Publishing,
2000).

14. Kevles, "Ananda Chakrabarty Wins a Patent"; Hughes, "Making Dollars out
of DNA"; Mowery et al., *Ivory Tower and Industrial Innovation*; and Roger L. Geiger, *Knowledge and Money: Research Universities and the Paradox of the Marketplace* (Stanford: Stanford University Press, 2004).

15. A wave of economic analyses pointed out the slower growth of productivity in the 1970s, esp. a substantial drop in the growth of labor productivity between
1973 and 1976. See Edward F. Denison, *Accounting for Slower Economic Growth:
The United States in the 1970's* (Washington, DC: Brookings Institution, 1979); and
William J. Baumol, Sue Anne Batey Blackman, and Edward N. Wolff, *Productivity
and American Leadership: The Long View* (Cambridge: MIT Press, 1989).

16. For a comprehensive legal history of government patent policy see Rebecca S. Eisenberg, "Public Research and Private Development: Patents and Technology Transfer in Government-Sponsored Research," *Virginia Law Review* 82
(1996): 1663–727.

17. On the history of the managers of research, see Robert E. Kohler, "The
Management of Science: The Experience of Warren Weaver and the Rockefeller
Foundation Programme in Molecular Biology," *Minerva* 14 (1976): 279–306. Crit-

ics of the commercialization of academic research have characterized this framework as emblematic of the "corruption" of academic scientists. See Jennifer Washburn, *University, Inc.: The Corporate Corruption of American Higher Education* (New York: Basic Books, 2005).

18. For an overview of the law and economics perspective on intellectual property rights, see Edmund W. Kitch, "The Nature and Function of the Patent System," *Journal of Law and Economics* 20 (1977): 265–90. For a historical analysis of the rise of the law and economics movement, see Steven M. Teles, *The Rise of the Conservative Legal Movement: The Battle for Control of the Law* (Princeton: Princeton University Press, 2008).

19. For an insightful analysis of the scientific life in the age of commercial biotechnology, see Steven Shapin, *The Scientific Life: A Moral History of a Late Modern Vocation* (Chicago: University of Chicago Press, 2008).

20. Berman, "Why Did Universities Start Patenting?"

21. On the central role of law in reshaping the relationship between technoscience and public policy, see Sheila Jasanoff, *Science at the Bar: Law, Science, and Technology in America* (Cambridge: Harvard University Press, 1995). For a historical analysis of the role of patents in the creation and regulation of new markets, see essays in the special issue of *History and Technology* 24:2 (2008), esp. Jean-Paul Gaudillière, "How Pharmaceuticals became Patentable: the Production and Appropriation of Drugs in the Twentieth Century," *History and Technology* 24 (2008): 99–106.

22. Roger L. Geiger, *Research and Relevant Knowledge: American Research Universities since World War II* (New York: Oxford University Press, 1993); and Nathan Reingold, "Science and Government in the United-States since 1945," *History of Science* 32 (1994): 361–86. For a historical analysis of ideological cultural valences between science and democracy in the post–World War II years, see David A. Hollinger, "Science as a Weapon in *Kulturkämpfe* in the United States during and after World War II," *Isis* 86 (1995): 440–54.

23. Stuart W. Leslie, *The Cold War and American Science: The Military-Industrial-Academic Complex at MIT and Stanford* (New York: Columbia University Press, 1993); Peter Galison and Bruce Hevly, eds., *Big Science: The Growth of Large-Scale Research* (Stanford: Stanford University Press, 1992); and Rebecca S. Lowen, *Creating the Cold War University: The Transformation of Stanford* (Berkeley: University of California Press, 1997). For the post–World War II growth of federal support in biomedical research, see Angela N. H. Creager, *The Life of a Virus: Tobacco Mosaic Virus as an Experimental Model, 1930–1965* (Chicago: University of Chicago Press, 2002), esp. chapter 5, "The War against Polio," 141–84. On the public funding for science before World War II, see A. Hunter Dupree, *Science and the Federal Government: A History of Policies and Activities*, rev. ed. (1957; Baltimore: Johns Hopkins University Press, 1986); and Charles E. Rosenberg, *No Other Gods: On Science and American Social Thought* (Baltimore: Johns Hopkins University Press, 1976).

24. Clark Kerr, *The Uses of the University* (Cambridge: Harvard University Press, 1963); and Daniel J. Kevles, "The National Science Foundation and the Debate over Postwar Research Policy, 1942–1945: A Political Interpretation of Science-the Endless Frontier," *Isis* 68 (1977): 5–26.

25. Kerr relied on Fritz Machlup's *The Production and Distribution of Knowledge in the United States* (Princeton: Princeton University Press, 1962) in his discussion of the emergence of the knowledge economy. According to Machlup, "the production, distribution, and consumption of 'knowledge' in all its forms is said to accounted for 29 percent of gross national product . . . ; and 'knowledge production' is growing at about twice the rate of the rest of the economy," quoted in Kerr, *Uses of the University*, 66. See also Paddy Riley, "Clark Kerr: From the Industrial to the Knowledge Economy," in Nelson Lichtenstein, ed., *American Capitalism: Social Thought and Political Economy in the Twentieth Century* (Philadelphia: University of Pennsylvania Press, 2006), 71–87.

26. For a historical analysis of the rising call for relevance in research in the 1960s and 1970s, see Daniel J. Kevles, *The Physicists: The History of a Scientific Community in Modern America* (New York: Knopf, 1977), 413–16; Eric J. Vettel, *Biotech: The Countercultural Origins of an Industry* (Philadelphia: University of Pennsylvania Press, 2006), esp. chapter 6, "A Season of Policy Reform," 129–56; and Matthew Wisnioski, *Engineers for Change: Competing Visions of Technology in 1960s America* (Cambridge: MIT Press, 2012).

27. A wave of economic studies in the 1970s questioned the efficiency of government investment in scientific and technological enterprises. See Bruce Williams, "The Economic Impact of Science and Technology in Historical Perspective," *Minerva* 20 (1982): 301–12. For a critical reassessment of the history of the linear model of science and technology, see Donald Stokes, *Pasteur's Quadrant: Basic Science and Technological Innovation* (Washington, DC: Brookings Institution, 1997); and David Edgerton, "The Linear Model Did Not Exist: Reflections on the History and Historiography of Science and Research in Industry in the Twentieth Century," in Karl Grandin, Nina Wormbs, and Sven Widmalm, eds., *The Science-Industry Nexus: History, Policy, Implications* (Sagamore Beach: Science History Publications, 2004), 31–57.

28. Chalmers W. Sherwin and Raymond S. Isenson, "Project Hindsight: A Defense Department Study of the Utility of Research," *Science* 156 (1967): 1571. For a historical analysis of Project Hindsight, see Glen R. Asner, "U.S. Department of Defense, and the Golden Age of Industrial Research," in Grandin et al., *Science-Industry Nexus*, 3–30.

29. On the role of economists at the RAND corporation in the articulation of the economics of science and innovation, see David A. Hounshell, "The Medium is the Message, or How Context Matters: The RAND Corporation Builds an Economics of Innovation, 1946–1962," in Thomas P. Hughes and Agatha C. Hughes, eds., *Systems, Experts, and Computers: The System Approach in Management and Engineering, World War II and After* (Cambridge: MIT Press, 2000), 255–310.

30. Kenneth J. Arrow, "Economic Welfare and the Allocation of Resources for Invention," RAND Corporation, P-1856-RC, 15 December 1959, p. 20, reprinted in Philip Mirowski and Esther-Mirjam Sent, eds., *Science Bought and Sold: Essays in the Economics of Science* (Chicago: University of Chicago Press, 2002), 165–80.

31. Kevles, "National Science Foundation and the Debate over Postwar Research Policy, 1942–1945." On the growth of federal research economy through institutional pluralism, see Geiger, *Research and Relevant Knowledge.*

32. Kenneth S. Pitzer to Robert H. Finch (secretary of the DHEW), 9 June 1969, Richard W. Lyman Papers, box 8, folder: Federal Government, Stanford University Archives, SC 215.

33. Ibid.

34. Ibid.

35. White House, *Biomedical Science and Its Administration: A Study of the National Institutes of Health* (Washington, DC: US Government Printing Office, 1965), xv.

36. Stephen P. Strickland, *Politics, Science, and Dread Disease: A Short History of United States Medical Research Policy* (Cambridge: Harvard University Press, 1972), 207.

37. Jean-Paul Gaudillière, "The Molecularization of Cancer Etiology in the Postwar United States: Instruments, Politics and Management," in Soraya de Chadarevian and Marmke Kamminga, eds., *Molecularizing Biology and Medicine: New Practices and Alliances, 1910–1970s* (Amsterdam: Harwood Academic Press, 1998), 139–70; and Daniel J. Kevles, "Pursuing the Unpopular: A History of Courage, Viruses and Cancer," in Robert B. Silvers, ed., *Hidden Histories of Science* (New York: New York Review of Books, 1995), 69–114.

38. Vettel, *Biotech*, 136.

39. See, e.g., Philip M. Boffey, "Federal Research Funds: Science Gets Caught in a Budget Squeeze," *Science* 158 (1967): 1286–88; and Earl Frank Cheit, *The New Depression in Higher Education: A Study of Financial Conditions at 41 Colleges and Universities* (New York: McGraw-Hill, 1971). For a useful set of data on the federal support for academic research and development in the United States from the early 1970s to the late 1990s, see Daniel S. Greenberg, *Science, Money, and Politics: Political Triumph and Ethical Erosion* (Chicago: University of Chicago Press, 2001), appendix.

40. National Cancer Institute, "A New Attack upon Disease Research Problems," 15 July 1964, p. 3, Carl G. Baker Papers, Office of NIH History Collections, National Institutes of Health, Bethesda, MD (hereafter CGB), box 2, folder: 1964.

41. Memorandum, Kenneth Endicott to National Institutes of Health (NIH) director James Shannon, "Applied Developmental Research and Research Services: Need for Legislation," 28 February 1963, p. 2, CGB, box 2, folder: 1963.

42. Leo B. Slater, *War and Disease Biomedical Research on Malaria in the Twentieth Century* (New Brunswick: Rutgers University Press, 2009).

43. James Shannon, NIH director, to the Surgeon General, Public Health Service, "Need for Changes in Department Patent Policy to Permit Effective Collaboration with Industry," 14 August 1964, NJL.

44. US General Accounting Office, *Problem Areas Affecting Usefulness of Results of Government-Sponsored Research in Medical Chemistry: A Report to the Congress* (Washington, DC: US Government Printing Office, 1968); and Harbridge House, Inc., *Government Patent Policy Study: Final Report*, prepared for the Federal Council for Science and Technology Committee on Government Patent Policy (Washington, DC: US Government Printing Office, 1968).

45. Eisenberg, "Public Research and Private Development."

46. As one referee of this book pointed out, some scholars view the DHEW's commitment to the public availability of the results of biomedical research as a consequence of the failed attempt of the Truman administration to implement universal health care. See Strickland, *Politics, Science, and Dread Disease.* The Atomic Energy Commission instituted a title policy as a way for the US government to monopolize inventions with national security implications. See Alex Wellerstein, "Patenting the Bomb: Nuclear Weapons, Intellectual Property, and Technological Control," *Isis* 99 (2008): 57–87.

47. On Latker's central role in the enactment of the Bayh-Dole Act, see Washburn, *University, Inc.*; and Berman, "Why Did Universities Start Patenting?"

48. Norman Latker, *Patent Council, Department of Health, Education, and Welfare, US Congressional Hearing, Committee on Science and Technology, Government Patent Policy: The Ownership of Inventions Resulting from Federally Funded Research and Development*, 94th Cong., 2nd sess. (Washington, DC: US Government Printing Office, 1976), 723.

49. Berman, "Why Did Universities Start Patenting?"; and Jon Sandelin, "A History of the Association of University Technology Managers," unpublished manuscript, document courtesy of Jon Sandelin (October 2005).

50. David Mowery and Bhaven N. Sampat, "University Patents and Patent Policy Debates in the USA, 1925–1980," *Industrial and Corporate Change* 10 (2001): 781–814.

51. On a few occasions, universities and academic researchers used patenting as a means to control beneficial medical inventions and therapeutics for the public benefit. See Angela N. H. Creager, "Biotechnology and Blood: Edwin Cohn's Plasma Fractionation Project, 1940–1953," in Thackray, *Private Science,* 39–62; and Maurice Cassier and Christiane Sinding, "'Patenting in the Public Interest': Administration of Insulin Patents by the University of Toronto," *History and Technology* 24 (2008): 153–71.

52. The Research Corporation was established in 1912 by the Berkeley chemist Frederick G. Cottrell as a nonprofit organization for managing academic patents. See David Mowery and Bhaven N. Sampat, "Patenting and Licensing University Inventions: Lessons from the History of the Research Corporation," *Industrial and*

Corporate Change 10 (2001): 317–55. The Wisconsin Alumni Research Foundation was founded as a separate corporation in the 1920s to handle the University of Wisconsin's patents, most notably those related to vitamin D. See Rima D. Apple, "Patenting University Research: Harry Steenbock and the Wisconsin Alumni Research Foundation," *Isis* 80 (1989): 375–94.

53. Eisenberg, "Public Research and Private Development."

54. Ibid., 1700.

55. Mowery, "University Patents and Patent Policy Debates in the USA, 1925–1980," esp. 338–41.

56. For a history of Stanford's Office of Technology, see Hughes, "Making Dollars out of DNA."

57. Niels Reimers, "Tiger by the Tail," *Chemtech* 17 (1987): 464–71.

58. Lowen, *Creating the Cold War University*.

59. David L. Webster, 15 December 1939, David L. Webster Papers, addendum, box I, Stanford University Archives, quoted in Lowen, *Creating the Cold War University*, 42.

60. C. Stewart Gillmor, *Fred Terman at Stanford: Building a Discipline, a University, and Silicon Valley* (Stanford: Stanford University Press, 2004), 152–54.

61. Richard W. Lyman (Stanford provost) to department chairmen and principal investigators, "Patent Licensing," 16 May 1969, Paul Berg Papers, Stanford University Archives, SC 358 (hereafter PB), box 17, folder: Faculty Meeting, 1968–69. Lyman served as president of Stanford from 1970 to 1980.

62. William F. Miller (vice president for research), "University Patent Policy and Patent Licensing Program, Guide Memo: 75, Patents," 19 June 1970, Joshua Lederberg Papers, Stanford University Archives, SC 186, box 22(B), folder: Patents, SUA.

63. Niels Reimers, "A Personal History of the Stanford University Office of Technology Licensing," unpublished manuscript as of 2003, document courtesy of Stanford Office of Technology Licensing (OTL). The RC, whose royalty income had largely depended on a few "home-run" patents in chemical, pharmaceutical, and agricultural inventions, such as vitamin B, cortisone, reserpine, and hybrid seed corn, was also affected by this industry-specific disparity in academic patenting.

64. Stanford OTL's first patent was granted to Johnson. William S. Johnson, Intermediates in Synthesis of 16-Dehydroprogesterone, US Patent 3,598,845, filed December 30, 1968, issued August 10, 1971.

65. Latker, Hearings before the Committee on Science and Technology, *Government Patent Policy*, 723.

66. Niels Reimers to NIH, 24 May 1976, SUOTL.

67. Niels Reimers, "Stanford's Office of Technology Licensing and the Cohen/Boyer Cloning Patents," an oral history conducted in 1997 by Sally Smith Hughes, PhD, Regional Oral History Office, Bancroft Library, University of California, Berkeley, 1998.

68. Joshua Lederberg, News Release, *Stanford University News Service*, May 20, 1974, quoted in Hughes, "Making Dollars out of DNA," 545.

69. Ibid., 548–49. Cohen already held a patent on a laboratory instrument. Stanley Cohen and Myron Tannenbaum, Filtration apparatus, US Patent, 3,730,352, filed December 6, 1971, and issued May 1, 1973.

70. David A. Jackson, Robert H. Symons, and Paul Berg, "Biochemical Method for Inserting New Genetic Information into DNA of Simian Virus 40: Circular SV40 DNA Molecules Containing Lambda Phage Genes and Galactose Operon of *Escherichia coli*," *Proceedings of the National Academy of Sciences, USA* 69 (1972): 2904–9; and Peter E. Lobban, *An Enzymatic Method for End-to-End Joining of DNA Molecules* (PhD dissertation, Stanford University, 1972); and Janet E. Mertz and Ronald W. Davis, "Cleavage of DNA by RI Restriction Endonuclease Generates Cohesive Ends," *PNAS* 69 (1972): 3370–74.

71. Stanley N. Cohen, MD, "Science, Biotechnology, and Recombinant DNA: A Personal History," an oral history conducted in 1995 by Sally Smith Hughes, PhD, Regional Oral History Office, Bancroft Library, University of California, Berkeley, 2009, 150.

72. William Carpenter to Niels Reimers, *S74–43 Gene Transplantation*, 18 October 1974. SUOTL.

73. Ibid.

74. Cohen et al., "Construction of Biologically Functional Bacterial Plasmids *In Vitro*."

75. Stanley N. Cohen, "The Stanford DNA Cloning Patent," in W. J. Whelan and Sandra Black, eds., *From Genetic Engineering to Biotechnology: The Critical Transition* (New York: Wiley, 1982), 213–16; and John Lear, *Recombinant DNA: The Untold Story* (New York: Crown Publishers, 1978), 59–60.

76. Stanford biochemists' contribution to the development of recombinant DNA–related technologies became a subject of intense discussion between Stanford scientists and PTO examiners. See Yi, "Cancer, Viruses, and Mass Migration."

77. John Morrow to Stanley Cohen and Herbert Boyer, 23 January 1975, SUOTL.

78. Cohen to Bertram Rowland, 22 January 1975, SUOTL. Cohen wrote this letter immediately after his phone conversation with Morrow and Helling.

79. Ibid.

80. "In Confidence, Re: Process and Composition for Biologically Functional DNA Chimeras," Office of Technology Licensing Correspondence, 1974–79, 1 July 1975, SUOTL.

81. "Meeting with Bill Massy, Stanley Cohen, Charles Yanofsky, Paul Berg, Ronald Davis, David Hogness, and Niels Reimer at the Stanford Medical School," 9 April 1975, SUOTL, file: "Gene Transplant." Cohen reneged on his promise not to benefit personally from the recombinant DNA patent sometime in 1983, after Stanford first began to distribute royalty incomes from the first recombinant DNA

patent to the inventors in the fall of 1982 (the university began to receive licensing fees in December 1981, and by August 1982 the royalty income from the recombinant DNA patent was $1.4 million). See Reimers, "Stanford's Office of Technology Licensing and the Cohen/Boyer Cloning Patents," 13–14.

82. Kevles, "Ananda Chakrabarty Wins a Patent."

83. Finnegan, Henderson, Farabow, Garrett & Dunner, LLP, *Opinion Regarding Validity, Enforceability and Infringement Issues Presented by the Cohen and Boyer Patents*, prepared for Leland Stanford Junior University (confidential opinion of counsel, August 9, 1985), "VI. Prosecution History of the Cohen and Boyer Patents," 35–55.

84. For a history of Genentech, see Hughes, "Making Dollars out of DNA"; and Hall, *Invisible Frontiers.*

85. Niels Reimers, "Licensing Plan (Original Draft, May 14, 1976)," 13 July 1976, SUOTL. For the enthusiasm on the industrial uses of life, see Robert Bud, *The Uses of Life: A History of Biotechnology* (Cambridge: Cambridge University Press, 1993).

86. John Poitras, "Meeting on DNA Patent: May 17, 1976," SUOTL.

87. For the remarks of Berg and Lederberg, see Poitras, "Meeting on DNA Patent: May 17, 1976," SUOTL.

88. Tanenholtz ultimately acquiesced to some of Rowland's counterarguments, granting the first Cohen-Boyer patent in 1980 (the other two Cohen-Boyer patents were granted in 1984 and 1986, respectively). For a brief prosecution history of recombinant DNA technology patent applications, see Yi, "Cancer, Viruses, and Mass Migration," esp. 622–27; and Albert P. Halluin, "Patenting the Results of Genetic Engineering Research: An Overview," in David W. Plant, Niels J. Reimers, and Norton D. Zinder, eds., *Patenting of Life Forms* (Cold Spring Harbor: Cold Spring Harbor Laboratory Press, 1982), 67–126.

89. Letter to Paul Berg, 1977, PB, Box 5, Personal Correspondence, SUA.

90. Paul Berg (chairman), David Baltimore, Herbert Boyer, Stanley Cohen, Ronald Davis, David Hogness, Daniel Nathans, Richard Roblin, James Watson, Sherman Weissman, and Norton Zinder, "Potential Biohazards of Recombinant DNA Molecules," *Science* 185 (1974): 303.

91. Paul Berg, personal correspondence with the author, September 2009.

92. Poitras, "Meeting on DNA Patent: May 17, 1976," SUOTL. For useful background information see Charles Weiner, "Patenting and Academic Research: Historical Case Studies," in Vivian Weil and John W. Snapper, eds., *Owning Scientific and Technical Information: Value and Ethical Issues* (New Brunswick: Rutgers University Press, 1989), 87–109.

93. Poitras, "Meeting on DNA Patent: May 17, 1976," SUOTL.

94. Niels Reimers, "Memo: Meeting on Gene Transplant at the Medical School," 7 April 1975, SUOTL.

95. Poitras, "Meeting on DNA Patent: May 17, 1976," SUOTL.

96. Cohen, "Science, Biotechnology, and Recombinant DNA," 152; and "News: Genetic Manipulation to be Patented?," *Nature* 261 (1976): 624.

97. Stanley Cohen to Niels Reimers, 14 June 1976, SUOTL.

98. Niels Reimers, "Meeting with Norman Latker and David Eden, 24 May 1976, SUOTL.

99. Here my interpretation differs from Hughes's view of Rosenzweig's letter in "Making Dollars out of DNA." Though his letter exhibits Stanford's prudence, the political situation of recombinant DNA technology at the NIH, esp. Latker's temporary resignation in 1978, underlined the contentiousness of NIH's patent waiver in this particular case. See William J. Broad, "Patent Bill Returns Bright Idea to Inventor," *Science* 205 (1979): 473–76; and Mowery et al., *Ivory Tower and Industrial Innovation*, esp. chapter 5, "A Political History of the Bayh-Dole Act of 1980," 85–98.

100. From Robert M. Rosenzweig, office memorandum, 30 June 1976; Subject: Telephone Conversation with Joseph Perpich, Associate Director for Program Planning, NIH; Distribution: R. Augsburger, W. F. Massy, P. Berg, S. Cohen, R. W. Lyman, N. Reimers, C. Rich, and J. Siena, PB, box 19, folder: Recombinant DNA Patent Business.

101. Ibid.

102. Under the terms of the IPA, the DHEW reserved its right to revoke the entire title if the grantee failed to comply with the IPA obligations. See Appendix II, "Institutional Patent Agreement Governing Grants and Awards from the Department of Health, Education, and Welfare," in the Office of Director, NIH, *Recombinant DNA Research*, vol. 2. Documents Relating to "NIH Guidelines for Research Involving Recombinant DNA Molecules," June 1976–November 1977, prepared by the Office of the Director, NIH, Department of Health, Education, and Welfare Publication No. (NIH) 78–1139 (Washington, DC: U.S. Government Printing Office, 1978). It should be noted that the DHEW had never exercised this stipulation. See Washburn, *University, Inc.*

103. Hearings before the Committee on Science and Technology, *Government Patent Policy*, 1260.

104. Niels Reimers, "Mechanisms for Technology Transfer: Marketing University Technology," in *Technology Transfer: University Opportunities and Responsibilities: A Report on the Proceedings of a National Conference on the Management of University Technology Resources* (Cleveland: Case Western Reserve University, 1974), 100–101.

105. Preface, *Technology Transfer*, 8.

106. On the establishment of the Society of University Patent Administrators, see Jon Sandelin, "A History of the Association of University Technology Managers" (unpublished manuscript, October 2005). Document courtesy of Jon Sandelin.

107. On the ascendancy of market rationales in public policy debates, see

Mark A. Smith, *The Right Talk: How Conservatives Transformed the Great Society into the Economic Society* (Princeton: Princeton University Press, 2007).

108. Niels Reimers, "Mechanisms for Technology Transfer," in *Technology Transfer,* 100–101.

109. Ibid.

110. Donald S. Fredrickson, *The Recombinant DNA Controversy: A Memoir—Science, Politics, and the Public Interest, 1974–1981* (Washington, DC: ASM Press, 2001), 94–95.

111. Fredrickson, Letter, 8 September 1976, in Office of the Director, NIH, *Recombinant DNA Research,* 2:21–24.

112. Fredrickson, *Recombinant DNA Controversy,* 95.

113. David Baltimore to Donald S. Fredrickson, 20 September 1976, in Office of the Director, NIH, *Recombinant DNA Research,* 2:80.

114. Garret M. Ihler, Department of Biochemistry, University of Pittsburgh, to Fredrickson, 30 June 1976, in Office of the Director, NIH, *Recombinant DNA Research,* 2:62–63.

115. Paul Berg to Fredrickson, 27 September 1976, in Office of the Director, NIH, *Recombinant DNA Research,* 2:90–92.

116. Ronald Cape to Fredrickson, 28 September 1976, in Office of the Director, NIH, *Recombinant DNA Research,* 2:94–97. It should be acknowledged that Cetus had commercial interests in recombinant DNA technology, and if the patent were upheld, Cetus would have to license it from Stanford and University of California.

117. Ibid.

118. Jerome Birnbaum to Fredrickson, 30 September 1976, in Office of the Director, NIH, *Recombinant DNA Research,* 2:129–30.

119. C. Joseph Stetler to Fredrickson, 29 September 1976, in Office of the Director, NIH, *Recombinant DNA Research,* 2:106–8.

120. Latker, Hearings before the Committee on Science and Technology, *Government Patent Policy,* 705.

121. Ibid., 724.

122. Ibid.

123. Niels Reimers, "Meeting with Genentech," 2 June 1976, SUOTL.

124. Niels Reimers, "Various Telephone Conversations and Meetings," 2 August 1976, files S74–43: Recombinant DNA, SUOTL.

125. Ibid.

126. Ibid.

127. Marshall Dann, commissioner of the Patent and Trademark Office, "Patent and Trademark Office: Recombinant DNA Accelerated Processing of Patent Applications for Inventions," *Federal Register* 42 (1977): 2712–13. Betsy Ancker-Johnson approved this new policy on January 7, 1977.

128. The term "tragedy of the commons" was popularized by the biologist Garrett Hardin, who was concerned about market failures in areas of common goods

such as the environment and natural resources. See Garrett Hardin, "The Tragedy of the Commons," *Science* 162 (1968): 1243–48. In the 1970s the phrase gained a life of its own among scholars of law and economics, who argued that market failures in public goods could be addressed by "privatizing" the commons. See James Boyle, *The Public Domain: Enclosing the Commons of the Mind* (New Haven: Yale University Press, 2008).

129. For a historical overview of the Chicago school of law and economics, see Johan Van Overtveldt, *The Chicago School: How the University of Chicago Assembled the Thinkers Who Revolutionized Economics and Business* (Chicago: B2 Books, 2007).

130. For example, see Harold Demsetz, "Toward a Theory of Property Rights," *American Economic Review* 57 (1967): 347–59; and Richard A. Posner, *Economic Analysis of the Law* (Boston: Little Brown, 1973).

131. Kitch, "Nature and Function of the Patent System." Machlup's warning had become a legal basis for the public ownership of common resources and goods in the 1950s and 1960s. See Fritz Machlup, *An Economic Review of the Patent System*, Study #15 of the Subcommittee on Patents, Trademarks, and Copyrights of the Committee on the Judiciary, US Senate, 85th Congress (Washington, DC: US Government Printing Office, 1958).

132. Kitch, "Nature and Function of the Patent System," 287.

133. Betsy Ancker-Johnson and David B. Change, *U.S. Technology Policy: A Draft Study* (Office of the Assistant Secretary for Science and Technology, US Department of Commerce, National Technical Information Service, PB-263 806, March 1977), 72.

134. Ibid., 26.

135. Joseph A. Califano, Jr., to Juanita M. Kreps, secretary of commerce, 8 February 1977, in Office of the Director, NIH, *Recombinant DNA Research*, 2:11.

136. Joseph A. Hill, Department of Justice, memorandum, "Patenting of Recombinant DNA Research Inventions," 5 May 1977, MS.C 526, Donald S. Fredrickson Papers, box 15, folder 8: Patents, 1977–1981, National Library of Medicine, Bethesda, MD.

137. Leshowitz, "Demise of Technology Transfer in DHEW."

138. Washburn, *University, Inc.*; and Berman, "Why Did Universities Start Patenting?"

139. *Government Patent Policies: Institutional Patent Agreements: Hearings on S., Before the Subcommittee on Monopoly and Anticompetitive Activities of the Select Committee on Small Business,* 95th Cong., 1st session, vol. 1 (December 1977), and 2nd session, vol. 2 (May and June 1978) (Washington, DC: US Government Printing Office, 1978).

140. Nancy K. Eskridge, "Dole Blasts HEW for 'Stonewalling' Patent Applications," *BioScience* 28 (1978): 605; and Washburn, *University, Inc.*, 66–67.

141. Fredrickson to Rosenzweig, March 1978, SUOTL.

142. *The University and Small Business Patent Procedures Act: Hearings on S. 414, Before the Committee on the Judiciary*, 96th Cong., 1st sess. (May 16, and June 6, 1979), serial no. 96–11 (Washington, DC: US Government Printing Office, 1979). The bill became Pub. L. No. 96–517 when President Jimmy Carter signed it into law on December 12, 1980. US Patent and Trademark Office issued the first recombinant DNA patent in December 2, 1980, Patent 4,237,224, "Process for Producing Biologically Functional Molecular Chimeras."

143. Kevles, "Ananda Chakrabarty Wins a Patent"; Masao Miyoshi, "Ivory Tower in Escrow," *Boundary 2* 27 (2000): 7–50; Hughes, "Making Dollars out of DNA"; and Geiger, *Knowledge and Money*.

144. "Biotechnology Patent Streamlining Mechanism: Study Proposal," 13 August 1982, prepared for Stanford University, Arthur Kornberg Papers, box 5, Stanford University Archives, SC 359.

145. Earned royalty rates for end products were graduated according to sales. For the first five million dollar sales, it was 1 percent, then reduced to 0.75 percent for the next five million dollars, and then reduced again to 0.5 percent for over ten million dollars.

146. Recombinant DNA product patents were granted in 1984 (for prokaryotes) and 1986 (for eukaryotes). Stanley N. Cohen and Herbert W. Boyer, Biologically Functional Molecular Chimeras, product patent (prokaryotic), US Patent 4,468,464, issued August 28, 1984, filed initially November 4, 1974, abandoned May 17, 1976, and refiled November 9, 1978; and Cohen and Boyer, Biologically Functional Molecular Chimeras, product patent (eukaryotic), US Patent 4,740,470, issued April 26, 1988, filed initially November 4, 1974, abandoned May 17, 1976, and refiled April 20, 1984.

147. Maryann Feldman, Alessandra Colaianni, and Kang Liu, "Commercializing Cohen-Boyer 1980–1997," no. 05–21, 2005, DRUID Working Papers, DRUID, Copenhagen Business School, Department of Industrial Economics and Strategy/ Aalborg University, Department of Business Studies.

148. Adrian Arima to Kennedy, 8 December 1984, Recombinant DNA Files (compiled by Katharine Ku), SUOTL.

149. Katherine Ku, memorandum to Kennedy, "Licensing: Background," 31 January 1986, Recombinant DNA Files, SUOTL.

150. The Reagan administration abolished this restriction without any debate in 1984. This revision allowed large business organizations to take ownership of government-sponsored research results, extending patent waive to all contractors. See Eisenberg, "Public Research and Private Development," 1704.

151. Niels Reimers, "Stanford's Office of Technology Licensing and the Cohen/ Boyer Cloning Patents," an oral history conducted in 1997 by Sally Smith Hughes, PhD, Regional Oral History Office, Bancroft Library, University of California, Berkeley, 1998, 44.

152. On the increasing calls for accountability, see Vettel, *Biotech*.

153. For the role of lay activists in the formation of the nation's biomedical research policy in the post–World War II era, see Angela N. H. Creager, "Mobilizing Biomedicine: Virus Research between Lay Health Organizations and the U.S. Federal Government, 1935–1955," in Caroline Hannaway, ed. *Biomedicine in the Twentieth Century: Practices, Policies, and Politics* (Amsterdam: IOS Press, 2008), 171–201.

154. For federal budget data for academic research and on the broader implications of federal budget trends in the 1970s for the commercialization of science, see Greenberg, *Science, Money, and Politics*; and Mirowski and Sent, "Commercialization of Science, and the Response of STS."

155. Jean-Paul Gaudillière and Ilana Löwy, eds., *The Invisible Industrialist: Manufactures and the Production of Scientific Knowledge* (London: Routledge, 1998); and P. Roy Vagelos and Louis Galambos, *Medicine, Science, and Merck* (Cambridge: Cambridge University Press, 2004).

156. Gaudillière, "How Pharmaceuticals became Patentable."

157. On the rise of American conservatism and the ascendency of market ideology, see George H. Nash, *Conservative Intellectual Movement in America since 1945* (New York: Basic Books, 1976). For the reconfiguration of public policy in terms of economic rationales in the 1970s, see Smith, *Right Talk*. For a historical analysis of the broader political and economic context of the commercialization of biomedical materials, see Doogab Yi, "The Scientific Commons in the Marketplace: The Industrialization of Biomedical Materials at the New England Enzyme Center, 1963–1980," *History and Technology* 25 (2009): 69–87.

158. Edward Yoxen, "Life as a Productive Force: Capitalizing the Science and Technology of Molecular Biology," in Les Levidow and Robert M. Young, eds., *Science, Technology, and the Labor Process: Marxist Studies* (London: Blackrose Press, 1981), 66–122. For the different institutional path toward commercial biotechnology without patenting that predominated in Britain (at least in its first stage in the 1970s), see Michael Mackenzie, Alberto Cambrosio, and Peter Keating, "The Commercial Application of a Scientific Discovery: The Case of the Hybridoma Technique," *Research Policy* 17 (1988): 155–70; and Soraya de Chadarevian, *Designs for Life: Molecular Biology after World War II* (Cambridge: Cambridge University Press, 2002), 336–62.

159. The rise of a discourse of the tragedy of the commons in intellectual property exhibited the shifting nexus among intellectual property, open science, and economic innovation forged in the early twentieth century. For the coproduction of open science and intellectual property in the 1930s, see Adrian Johns, "Intellectual Property and the Nature of Science," *Cultural Studies* 20 (March–May 2006): 145–64, and Johns, *Piracy: The Intellectual Property Wars from Gutenberg to Gates* (Chicago: University of Chicago Press, 2009).

160. "Cumulative Cohen-Boyer Royalties," folder: Recombinant DNA Patent Files, SUOTL.

161. For a contemporary discussion on the commercialization of biomedicine, see Sheldon Krimsky, *Science in the Private Interest: Has the Lure of Profits Corrupted Biomedical Research?* (Lanham: Rowman & Littlefield, 2003); and Philip Mirowski, *Science-Mart: Privatizing America Science* (Cambridge: Harvard University Press, 2011). For biologists' attempt to counter privatization in molecular biology through promoting open access to biomedical databases, see Bruno J. Strasser, "The Experimenter's Museum: GenBank, Natural History, and the Moral Economies of Biomedicine," *Isis* 102 (2011): 60–96. Jeremy Green's recent work provides a nuanced analysis of the role of the pharmaceutical industry in the advent of the current prescription drug regime. See Jeremy A. Greene, *Prescribing by Numbers: Drugs and the Definition of Disease* (Baltimore: Johns Hopkins University Press, 2007). For a historical analysis of ethical issues in clinical trials, see Gerald J. Kutcher, *Contested Medicine: Cancer Research and the Military* (Chicago: University of Chicago Press, 2009). For a discussion on the legal and ethical issues in gene patenting, see Myles W. Jackson, "Intellectual Property and Molecular Biology: Biomedicine, Commerce and the CCR5 Gene," Francis Bacon Lecture, May 5, 2011, Caltech. For a critical analysis of the theory of the tragedy of the commons in intellectual property rights, see James Boyle, *Shamans, Software, and Spleens: Law and the Construction of the Information Society* (Cambridge: Harvard University Press, 1996); and Michael A. Heller, "The Tragedy of the Anticommons: Property in the Transition from Marx to Markets," *Harvard Law Review* 111 (1998): 621–88.

Chapter Six

1. Andy Warhol, *The Philosophy of Andy Warhol: From A to B and Back Again* (New York: Harcourt Brace Jovanovich, 1975), 92.

2. Arthur Kornberg, *The Golden Helix: Inside Biotech Ventures* (Sausalito: University Science Books, 1995), 181. *The Golden Helix* provoked a heated debate among academic biologists on the proper relationship between academic research and biotechnology industry. See Robert S. Schwartz, "Review of *The Golden Helix: Inside Biotech Ventures*," *New England Journal of Medicine* 333:19 (1995): 1292–93. For Kornberg's and his colleagues' rebuttals on Schwartz's review, see Edgar Haber, Daniel E. Koshland, David Morris, Arthur Kornberg, David Hogness, I. Robert Lehman, Robert Baldwin, Paul Berg, A. Dale Kaiser, Philip Sharp, J. Michael Bishop, Robert J. Glaser, and Robert S. Schwartz's response, "Review of Kornberg's *The Golden Helix*," *New England Journal of Medicine* 334:15 (1996): 994–95.

3. "The Nobel Prize in Chemistry 1980," Nobelprize.org, http://nobelprize.org /nobel_prizes/chemistry/laureates/1980/

4. Paul Berg to Donald S. Fredrickson, 27 September 1976, Office of Director,

National Institutes of Health, *Recombinant DNA Research,* Vol. 2, Documents Relating to "NIH Guidelines for Research Involving Recombinant DNA Molecules," June 1976–November 1977, prepared by the Office of the Director, NIH, Department of Health, Education, and Welfare, publication no. NIH 78–1139 (Washington, DC: US Government Print Office, 1978), 90–92.

5. For the history of monoclonals and its commercialization, see Albert Cambrosio and Peter Keating, *Exquisite Specificity: The Monoclonal Antibody Revolution* (Oxford: Oxford University Press, 1995); Soraya de Chadarevian, "The Making of an Entrepreneurial Science," *Isis* 102 (2011): 601–33. For the history of Hybritech, see Mark P. Jones, "Biotech's Perfect Climate: The Hybritech Story" (PhD dissertation, University of California, San Diego, 2005).

6. A. Dale Kaiser, "Memo: Symposium Luncheon Tours of Biochemistry," an invitation to join the Biochemistry Department Industrial Affiliates Program, 24 March 1980, Arthur Kornberg Papers, Stanford University Archives, SC 359 (hereafter AK), box 3, folder: Industrial Affiliates Program.

7. David C. Mowery, Richard R. Nelson, Bhaven N. Sampt, and Arvids A. Ziedonis, eds., *Ivory Tower and Industrial Innovation: University-Industry Technology Transfer Before and After the Bayh-Dole Act in the United States* (Stanford: Stanford University Press, 2004).

8. For a comparative analysis of start-up biotech companies in the Bay Area and the organizational genealogies of their different business strategies and culture, see Simcha Jong, "How Organizational Structures in Science Shape Spin-off Firms: The Biochemistry Departments of Berkeley, Stanford, and UCSF and the Birth of the Biotech Industry," *Industrial and Corporate Change* 15 (2006): 251–83.

9. For an insightful analysis of a new moral economy of science in commercial biotechnology, see Paul Rabinow, *Making PCR: A Story of Biotechnology* (Chicago: University of Chicago Press, 1996).

10. See Eric J. Vettel, *Biotech: The Countercultural Origins of an Industry* (Philadelphia: University of Pennsylvania Press, 2006), for sociocultural explanations for the emergence of a new generation of biotech entrepreneurs. For a discussion on the managerial and technological dimensions of the industrialization of the life sciences, see Jean-Paul Gaudillière and Ilana Löwy, eds., *The Invisible Industrialist: Manufactures and the Production of Scientific Knowledge* (New York: Rutledge, 1998). Elizabeth Berman notes that legal shifts in venture capital investment, such as the passage of the Employee Retirement Income Security Act in 1974 and the passage of the Revenue Act in 1978, dramatically increased the availability of venture capital investment, providing a fertile ground for entrepreneurship in biotechnology. See Elizabeth P. Berman, *Creating the Market University: How Academic Science Became an Economic Engine* (Princeton: Princeton University Press, 2011).

11. For an insightful analysis of the emergence of scientist-entrepreneurs, see

Steven Shapin, *The Scientific Life: A Moral History of a Late Modern Vocation* (Chicago: University of Chicago Press, 2008).

12. Sally S. Hughes, *Genentech: The Beginnings of Biotech* (Chicago: University of Chicago Press, 2011), 72.

13. Clark Kerr, *The Uses of the University* (Cambridge: Harvard University Press, 1963).

14. Vettel, *Biotech*.

15. Daniel J. Kevles, *The Physicists: The History of a Scientific Community in Modern America* (New York: Knopf, 1977); and Angela N. H. Creager, "Mobilizing Biomedicine: Virus Research between Lay Health Organizations and the U.S. Federal Government, 1935–1955," in Caroline Hannaway, ed., *Biomedicine in the Twentieth Century: Practices, Policies, and Politics* (Amsterdam: IOS Press, 2008), 171–201.

16. President's Biomedical Panel, final draft, October 1975, Paul Berg Papers, Stanford University Archives, SC 358 (hereafter PB), box 4, p. 40.

17. Ibid.

18. Paul Berg to Arthur Kornberg, 3 April 1970, PB, box 2, folder: General Correspondence, 1970.

19. Porter E. Coggeshall and Prudence W. Brown, Committee on National Needs for Biomedical and Behavioral Research Personnel, Institute of Medicine, *The Career Achievements of NIH Predoctoral Trainees and Fellows* (Washington, DC: National Academy Press, 1984).

20. Ibid., 47.

21. Ibid., 50.

22. Clayton Rich to faculty, School of Medicine, 30 December 1974, AK, box 5, folder: 1971–77.

23. Arthur Kornberg, "Notes for the Discussion of an Affiliation with a Commercial Company," AK, box 5, folder: 1981.

24. Richard A. Knox, "Stanford Medical School Suffers Fiscal Ideological Crises," *Science* 203 (1979): 148.

25. On the impact of the Medicare Act on the medical school, see Kenneth M. Ludmerer, *Time to Heal: American Medical Education from the Turn of the Century to the Era of Managed Care* (Oxford: Oxford University Press, 1999).

26. Arthur Kornberg, "Meeting: Notes for Stanford," AK, box 5, folder: 1977.

27. Ibid.

28. Interview with Stanford University president Donald Kennedy, "Gene Splicing: Patents, Problems, and 'Enormous Potential Benefit,'" *SIPI Scope* 9 (1981): 1–9.

29. Annie C. Y. Chang, Jack H. Nunberg, Randal J. Kaufman, Henry A. Erlich, Robert T. Schimke, and Stanley N. Cohen, "Phenotypic Expression in *E. coli* of a DNA Sequence Coding for Mouse Dihydrofolate Reductase," *Nature* 275 (1978): 617–24.

30. Paul Berg to Robert Schimke, 5 February 1979, PB, box 5, folder: Correspondence, 1978–79, (S–Z). MUA (Memorandum of Understanding Agreement) had to be approved by the Biohazards Safety Committee prior to the start of research involving recombinant DNA.

31. Arthur Kornberg, "Notes," AK, box 5, folder: 1981.

32. Arthur Kornberg, "Departmental Finances, September 10, 1979," AK, box 29, folder: 1979 (emphasis added).

33. Ibid.

34. Minutes of the Faculty Meeting, July 1980, A. Dale Kaiser Papers, Stanford University Archives, SC 359 (hereafter ADK), box 11, folder: Faculty Minutes.

35. Philip H. Abelson, "Academic Science and Industry," *Science* 183 (1974): 1251.

36. Minutes of the Faculty Meeting, "Guidelines for the Industrial Affiliation Program," 17 October 1979, ADK, box 11, folder: Faculty Minutes.

37. Ibid.

38. A. Dale Kaiser, "Memo: Symposium Luncheon Tours of Biochemistry," an invitation to join the Biochemistry Department Industrial Affiliates Program, 24 March 1980, AK, box 3, folder: Industrial Affiliates Program.

39. A. Dale Kaiser, Faculty minutes, 28 October 1980, AK, box 11.

40. Timothy Lenoir, "Biochemistry at Stanford: A Case Study in the Formation of an Entrepreneurial Culture" (unpublished manuscript, April 21, 2002), Microsoft Word file.

41. Susan Wright, *Molecular Politics: Developing American and British Regulatory Policy for Genetic Engineering, 1972–1982* (Chicago: University of Chicago Press, 1994), 83; and Hughes, *Genentech.*

42. Cynthia Robbins-Roth, *From Alchemy to IPO: The Business of Biotechnology* (Cambridge: Perseus Publishing, 2000).

43. The initial six investors were Bendix, General Foods, Koppers, Mead, MacLaren Power and Paper, and Elf Technologies of Société Nationale Elf Aquitaine.

44. Martin Kenney, *Biotechnology: The University-Industrial Complex* (New Haven: Yale University Press, 1986), 48–49; and "Pure Knowledge Vs. Pure Profit," *Time*, September 28, 1981.

45. "Alza Stock Gift," PB, box 2, folder: Correspondence, 1969.

46. Kornberg, *Golden Helix*, 181.

47. Paul Berg, PhD, "A Stanford Professor's Career in Biochemistry, Science Politics, and the Biotechnology Industry," an oral history conducted in 1997 by Sally Smith Hughes, PhD, Regional Oral History Office, Bancroft Library, University of California, Berkeley, 2000, 117.

48. For Genentech's research on human growth hormone and its dispute with UCSF, see Hughes, *Genentech*, esp. chapter 5, "Human Growth Hormone: Shaping a Commercial Future," 107–35.

49. For a business history perspective on this early trend in the biotech indus-

try, see Gary P. Pisano, *Science Business: The Promise, the Reality, and the Future of Biotech* (Cambridge: Harvard Business School Press, 2006).

50. On the criticisms on the monetization of intellectual property as a business strategy in the biotech industry, see Pisano, *Science Business*.

51. DNAX, Ltd., "Business Plan: Confidential," 18 November 1980 (draft), 20 May 1981 (revision), pp. 11–12, DNAX Papers, Arthur Kornberg Personal Collections, Stanford University (hereafter DNAX).

52. The relationship of the DNAX Research Institute may be compared to that of the Roche Institute of Molecular Biology and/or the Basel Institute of Immunology to the Hoffman–La Roche Company.

53. "DNAX Business Plan," 20 May 1981, p. 15, DNAX.

54. "DNAX Mission Statement," 1980, DNAX.

55. "DNAX Business Plan," 20 May 1981, p. 12 (emphasis added), DNAX.

56. "Minutes of the Faculty Meeting," 12 April 1979, ADK, box 11, folder: Faculty Meeting.

57. Ken-ichi Arai, "Interview: Biological Research in Japan and Asia," *EMBO Reports* 2 (2001): 549–51.

58. Kornberg, *Golden Helix*, 152–53.

59. "DNAX Proposal for Stanford University Concerning Henry Kaplan Cell Lines and Antibodies," ca. October/November 1981, DNAX.

60. Paul Berg to Alejandro Zaffaroni, 13 November 1980, DNAX.

61. Ibid.

62. Robert Bud, *The Uses of Life: A History of Biotechnology* (Cambridge: Cambridge University Press, 1993).

63. On the history of hybridoma technology and its commercialization, see Michael Mackenzie, Alberto Cambrosio and Peter Keating, "The Commercial Application of a Scientific Discovery: The Case of the Hybridoma Technique," *Research Policy* 17 (1988): 155–70; Alberto Cambrosio and Peter Keating, "Between Fact and Technique: The Beginnings of Hybridoma Technology," *Journal of the History of Biology* 25 (1992): 175–230; and Soraya de Chadarevian, *Designs for Life: Molecular Biology after World War II* (Cambridge: Cambridge University Press, 2002).

64. "DENAX/DNAX: Business Plan," draft, 18 November 1980, DNAX.

65. Arthur Kornberg, "DNAX, 1980: Memo," folder: DNAX Misc, 1980–81, DNAX.

66. "DNAX Research Program," October 1981, p. 8, DNAX.

67. The DNAX's scientific advisory board consisted of the following scientists, as of 1981: Paul Berg (Stanford), Harvey Cantor (Harvard), William Dreyer (Caltech), Judah Folkman (Harvard), Avram Goldstein (Stanford), Edgar Haber (Harvard), Leroy Hood (Caltech), Michael Hunkapiller (Caltech), Kurt Isselbacher (Harvard), Arthur Kornberg (Stanford), Roger Kornberg (Stanford), Thomas Kornberg (UCSF), Ronald Levy (Stanford), Harden McConnell (Stanford), George Palade (Yale), Samuel Strober (Stanford), Irving Weissman (Stanford), and Charles Yanofsky (Stanford).

68. "DNAX Scientific Advisory Board Meeting," Palo Alto, 7 July 1981, p. 12, DNAX.

69. Kornberg, *Golden Helix*, 47–57. With Christian B. Anfinsen, Haber demonstrated that the enzyme ribonuclease could be refolded from the linear sequence of its amino acid chains. Anfinsen received the Nobel Prize in 1972 for this demonstration of the DNA sequence hypothesis. For more, see Angela N. H. Creager, "Christian B. Anfinsen," in Noretta Koertge, ed., *New Dictionary of Scientific Biography* (Charles Scribner's Sons/Thomson Gale, 2008), 1:76–82; and Bruno J. Strasser, "World in One Dimension: Linus Pauling, Francis Crick and the Central Dogma of Molecular Biology," *History and Philosophy of the Life Sciences* 28 (2006): 491–512.

70. "DNAX's Immunobiology Strategy: Confidential," 18 February 1981, DNAX.

71. "DNAX Scientific Advisory Board Meeting," 7 July 1981, Palo Alto, pp. 13–14, DNAX.

72. "DNAX's Immunobiology Strategy: Confidential," 18 February 1981, DNAX.

73. "DNAX Business Plan," 20 May 1981, p. 15, DNAX.

74. *Newswatch*, "What Japanese Industry is Doing in Biotechnology," 6 July 1981, p. 3.

75. Ken-ichi Arai, "Memo on Drs. Noguchi and Miura," 29 November 1981, DNAX.

76. Ibid.

77. Kornberg, *Golden Helix*, 99.

78. Ken-ichi Arai to Kornberg, 25 November 1982, DNAX.

79. Kenney, *Biotechnology*, 91.

80. "Schering-Plough to Acquire DNAX Limited," 7 April 1982, DNAX.

81. Ken-ichi Arai to Kornberg, 5 December 1982, DNAX.

82. "Faculty Minutes," 4 December 1981, PB, box 18, folder: Faculty Minutes.

83. Ibid.

84. Douglas Brutlag to A. Dale Kaiser, "Arrangement Proposed by SKF," 30 November 1981, PB, box 18, folder: Faculty Minutes.

85. A. Dale Kaiser, "Creation of a Substantial Financial Relation with a Company," memorandum to Stanford biochemistry faculty, 4 December 1981, PB, box 18, folder: Faculty Minutes.

86. Ibid.

87. Interview with Stanford University president Donald Kennedy, "Gene Splicing."

88. Donald Kennedy, "Commercialization and University Research: Prospectus for a Conference," 27 August 1981, Donald Kenney Papers, Stanford University Archives, unprocessed.

89. Barbara J. Culliton, "Pajaro Dunes: The Search for Consensus," *Science* 216 (1982): 155–58.

90. Kennedy, "Commercialization and University Research," 1.

91. Ibid., 3–4.

92. Hughes, *Genentech*, chapter 6, "Wall Street Debut," 137–64.

93. Kennedy, "Commercialization and University Research," 2.

94. Ibid., 3–4.

95. Culliton, "Pajaro Dunes," 156.

96. Gilbert returned to Harvard three years later.

97. Kornberg, *Golden Helix*, 231–32.

98. For example, Roger L. Geiger, *Knowledge and Money: Research Universities and the Paradox of the Market Place* (Stanford: Stanford University Press, 2005); Mowery et al., *Ivory Tower and Industrial Innovation*; and Derek Bok, *Universities in the Marketplace: The Commercialization of Higher Education* (Princeton: Princeton University Press, 2003).

99. Henry Etzkowitz, *MIT and the Rise of Entrepreneurial Science* (London: Routledge, 2003).

100. For example, Kenney, *Biotechnology*; and Jennifer Washburn, *University, Inc.: The Corporate Corruption of American Higher Education* (New York: Basic Books, 2005).

101. For a discussion on the articulation of the traditional distinction between academic scientists and industrialist in the Cold War context, see Steven Shapin, "Who Is the Industrial Scientist?: Commentary from Academic Sociology and from the Shop-Floor in the United States, ca. 1900–ca. 1970," in Karl Grandin, Nina Wormbs, Anders Lundgren and Sven Widmalm, eds., *The Science-Industry Nexus: History, Policy, Implications* (Sagamore Beach: Science History Publications, 2004), 337–63.

Conclusion

1. Eric J. Vettel, *Biotech: The Countercultural Origins of an Industry* (Philadelphia: University of Pennsylvania Press, 2006); and Sally S. Hughes, "Making Dollars out of DNA: The First Major Patent in Biotechnology and the Commercialization of Molecular Biology, 1974–1980," *Isis* 92 (2001): 541–75.

2. For a contemporary discussion on the commercialization of biomedicine regarding the sell-out narrative, see Sheldon Krimsky, *Science in the Private Interest: Has the Lure of Profits Corrupted Biomedical Research?* (Lanham: Rowman & Littlefield, 2003).

3. For a similar line of investigations that emphasized broader economic, legal, and institutional changes that played a significant role in the commercialization of academic research, see Elizabeth P. Berman, *Creating the Market University: How Academic Science Became an Economic Engine* (Princeton: Princeton University Press, 2011); and Philip Mirowski, *Science-Mart: Privatizing America Science* (Cambridge: Harvard University Press, 2011).

4. Bruno Latour, *The Pasteurization of France* (Cambridge: Harvard University Press, 1988); and Sheila Jasanoff, "Ordering Knowledge, Ordering Society," in Sheila Jasanoff, ed., *States of Knowledge: The Co-Production of Science and Social Order* (London: Routledge, 2004), 13–45.

5. For an excellent analysis on global biotechnology, see Paul Rabinow, *French DNA: Trouble in Purgatory* (Chicago: University of Chicago Press, 1999). For discussions on the regulation of global biotechnology, see Sheila Jasanoff, *Designs on Nature: Science and Democracy in Europe and the United States* (Princeton: Princeton University Press, 2005). The international context out of which molecular biology emerged has begun to be explored. For example, see Pnina G. Abir-Am, "From Multidisciplinary Collaboration to Transnational Objectivity: International Space as Constitutive of Molecular Biology, 1930–1970," in E. Crawford, T. Shinn, and S. Sörlin, eds., *Denationalising Science: The Context of International Scientific Practice* (Dordrecht: Kluwer Academic Publishers, 1992), 153–86; María J. Santesmases, "Severo Ochoa and the Biomedical Sciences in Spain under France, 1959–1975," *Isis* 91 (2000): 706–34; Hisao Uchida, "Building a Science in Japan: The Formative Decades of Molecular Biology," *Journal of the History of Biology* 26 (1993): 499–517; Angela N. H. Creager, "Tracing the Politics of Changing Postwar Research Practices: The Export of 'American' Radioisotopes to European Biologists," *Studies in History and Philosophy of Biological and Biomedical Sciences* 33 (2002): 367–88; Soraya de Chadarevian and Bruno J. Strasser, "Molecular Biology in Postwar Europe: Towards a 'Glocal' Picture," *Studies in History and Philosophy of Biological and Biomedical Sciences* 33 (2002): 361–65; Jean-Paul Gaudillière, *Inventer la biomédecine: La France, l'Amérique et la production des savoirs du vivant, 1945–1965* (Paris: Éditions la Découverte, 2002); Gaudillière, "Paris-New York Roundtrip: Transatlantic Crossings and the Reconstruction of the Biological Sciences in Post-War France," *Studies in History and Philosophy of Biological and Biomedical Sciences* 33 (2002): 389–417; and Aihwa Ong and Nancy N. Chen, eds., *Asian Biotech: Ethics and Communities of Fate* (Durham: Duke University Press, 2010).

Works Cited

Abelson, Philip H. "Academic Science and Industry." *Science* 183 (1974): 1251.

Abir-Am, Pnina G. "From Multidisciplinary Collaboration to Transnational Objectivity: International Space as Constitutive of Molecular Biology, 1930–1970." In *Denationalising Science: The Context of International Scientific Practice*, edited by E. Crawford, T. Shinn and S. Sörlin, 153–86. Dordrecht: Kluwer Academic Publishers, 1992.

———. "The Politics of Macromolecules: Molecular Biologists, Biochemists, and Rhetoric." *Osiris* 7 (1992): 164–91.

———. "Commemorative Practices in Science: Historical Perspectives on the Politics of Collective Memory." *Osiris* 14 (1999): 1–33.

———. "The First American and French Commemorations in Molecular Biology: from Collective Memory to Comparative History." *Osiris* 14 (1999): 324–70.

Ancker-Johnson, Betsy, and David B. Change. *U.S. Technology Policy: A Draft Study*. Office of the Assistant Secretary for Science and Technology, US Department of Commerce, National Technical Information Service, PB-263 806, March 1977.

Anderson, Warwick. "The Possession of Kuru: Medical Science and Biocolonial Exchange." *Comparative Studies in Society and History* 42 (2000): 713–44.

Apple, Rima D. "Patenting University Research: Harry Steenbock and the Wisconsin Alumni Research Foundation." *Isis* 80 (1989): 375–94.

Appel, Toby A. *Shaping Biology: The National Science Foundation and American Biological Research, 1945–1975*. Baltimore: Johns Hopkins University Press, 2000.

Arai, Ken-ichi. "Interview: Biological Research in Japan and Asia." *EMBO Reports* 2 (2001): 549–51.

Arrow, Kenneth J. "Economic Welfare and the Allocation of Resources for Invention." Rand Corporation, P-1856-RC, 15 December 1959, 1–23. Reprinted in *Science Bought and Sold: Essays in the Economics of Science*, edited by Philip Mirowski and Esther-Mirjam Sent, 165–80. Chicago: University of Chicago Press, 2002.

Asner, Glen R. "U.S. Department of Defense, and the Golden Age of Industrial Research." In *The Science-Industry Nexus: History, Policy, Implications*, edited by Karl Grandin, Nina Wormbs, and Sven Widmalm, 3–30. Sagamore Beach: Science History Publications, 2004.

Baldwin, Robert L. "Boundary Spreading in Sedimentation-Velocity Experiments." *Biochemical Journal* 65 (1957): 490–512.

Baumol, William J., Sue Anne Batey Blackman, and Edward N. Wolff. *Productivity and American Leadership: The Long View*. Cambridge: MIT Press, 1989.

Beermann, Wolfgang. "Chromomerenkonstanz und spezifische Modifikationen der Chromosomenstruktur in der Entwicklung und Organdifferenzierung von *Chironomus tentans*." *Chromosoma* 5 (1952): 139–98.

Bender, Welcome, Pierre Spierer, and David Hogness. "Chromosomal Walking and Jumping to Isolate DNA from the *Ace* and *rosy* Loci and the Bithorax Complex in *Drosophila melanogaster*." *Journal of Molecular Biology* 168 (1983): 17–33.

Berg, Douglas E., David A. Jackson, and Janet E. Mertz. "Isolation of a λdv Plasmid Carrying Bacterial *gal* Operon." *Journal of Virology* 14 (1974): 1063–69.

Berg, Paul. "Viral Genome in Transformed Cells: A Discussion on Animal Viruses as Genetic Modifiers of the Cell." *Proceedings of the Royal Society of London: Series B. Biological Sciences* 177 (1971): 65–76.

———. "Suppression: A Subversion of Genetic Decoding." *The Harvey Lectures, Series 67*, 247–72. New York: Academic Press, 1974.

———. "Nobel Lecture 1980: Dissections and Reconstructions of Genes and Chromosomes." In *Nobel Lectures, Chemistry 1971–1980*, edited by Tore Frängsmyr, 385–402. Singapore: World Scientific Publishing, 1993.

———, PhD. "A Stanford Professor's Career in Biochemistry, Science Politics, and the Biotechnology Industry." An oral history conducted in 1997 by Sally Smith Hughes, PhD. Regional Oral History Office, Bancroft Library, University of California, Berkeley, 2000.

Berg, Paul, and E. James Ofengand. "An Enzymatic Mechanism for Linking Amino Acids to RNA." *Proceedings of the National Academy of Sciences, USA* 44 (1958): 78–86.

Berg, Paul, and Maxine Singer. *George Beadle, an Uncommon Farmer: The Emergence of Genetics in the 20th Century*. Cold Spring Harbor: Cold Spring Harbor Laboratory Press, 2003.

Berman, Elizabeth P. "Why Did Universities Start Patenting? Institution-Building and the Road to the Bayh-Dole Act." *Social Studies of Science* 38 (2008): 835–71.

———. *Creating the Market University: How Academic Science Became an Economic Engine*. Princeton: Princeton University Press, 2011.

Beurton, Peter, Raphael Falk, and Hans-Jörg Rheinberger, eds. *The Concept of the Gene in Development and Evolution: Historical and Epistemological Perspectives*. Cambridge: Cambridge University Press, 2000.

Birnstiel, Max L., H. Wallace, J. L. Hirlin, and M. Fischberg. "Localization of the Ribosomal DNA Complements in the Nucleolar Organizer Region of *Xenopus laevis*." *National Cancer Institute Monograph* 23 (1966): 431–44.

Boffey, Philip M. "Federal Research Funds: Science Gets Caught in a Budget Squeeze." *Science* 158 (1967): 1286–88.

Bok, Derek. *Universities in the Marketplace: The Commercialization of Higher Education*. Princeton: Princeton University Press, 2003.

Bonner, James. *The Molecular Biology of Development*. Oxford: Oxford University Press, 1965.

Bonner, James, and Paul Ts'o, eds. *The Nucleohistones*. San Francisco: Holden-Day, 1964.

Boyer, Herbert W. "DNA Restriction and Modification Mechanisms in Bacteria." *Annual Review of Microbiology* 25 (1971): 153–76

Boyle, James. *Shamans, Software, and Spleens: Law and the Construction of the Information Society*. Cambridge: Harvard University Press, 1996.

———. *The Public Domain: Enclosing the Commons of the Mind*. New Haven: Yale University Press, 2008.

Brenner, Sydney. "Theories of Gene Regulation." *British Medical Journal* 21 (1965): 244–48.

Britten, Roy J., and Eric H. Davidson. "Gene Regulation for Higher Cells: A Theory." *Science* 165 (1969): 349–57.

Britten, Roy J., and David Kohne. "Repeated Sequences in DNA." *Science* 161 (1968): 529–40.

Broad, William J. "Patent Bill Returns Bright Idea to Inventor." *Science* 205 (1979): 473–76.

Brock, Thomas D. *The Emergence of Bacterial Genetics*. Cold Spring Harbor: Cold Spring Harbor Laboratory Press, 1990.

Brown, Donald D. "Some Genes were Isolated and Their Structure Studied before the Recombinant DNA Era." *BioEssays* 16 (1994): 139–43.

Bud, Robert. *The Uses of Life: A History of Biotechnology*. Cambridge: Cambridge University Press, 1993.

Burian, Richard M., Jean Gayon, and Doris T. Zallen. "Boris Ephrussi and the Synthesis of Genetics and Embryology." In *Developmental Biology: A Comprehensive Synthesis*, edited by Scott F. Gilbert, 7:207–27. New York: Plenum Press, 1991.

Cairns, John, Gunther S. Stent, and James D. Watson, eds. *Phage and the Origins of Molecular Biology*. Plainview: Cold Spring Harbor Laboratory Press, 1992 [1966].

Callan, Harold G. "The Organization of Genetic Units in Chromosomes." *Journal of Cell Science* 2 (1967): 1–7.

Cambrosio, Alberto, and Peter Keating. "Between Fact and Technique: The Beginnings of Hybridoma Technology." *Journal of the History of Biology* 25 (1992): 175–230.

Campbell, Allan M. "Episomes." *Advances in Genetics Incorporating Molecular Genetic Medicine* 11 (1962): 101–45.

———. "Bacteriophage λ as a Model System." *Bioessays* 5 (1986): 277–80.

Carbon, John, Paul Berg, and Charles Yanofsky. "Studies of Missense Suppression of Tryptophan Synthetase A-Protein Mutant A36." *Proceedings of the National Academy of Sciences, USA* 56 (1966): 764–71.

Carroll, Sean. *Endless Forms Most Beautiful: The New Science of Evo Devo and the Making of the Animal Kingdom.* New York: W. W. Norton, 2005.

Casjens, Sherwood, and Colin Manoil. "1991 Thomas Hunt Morgan Medal: Dale Kaiser." In "*Genetics Society of America: 1991 Records, Proceedings and Reports,*" supplement, *Genetics* 128 (1991): s12–s13.

Cassier, Maurice, and Christiane Sinding. "'Patenting in the Public Interest': Administration of Insulin Patents by the University of Toronto." *History and Technology* 24 (2008): 153–71.

Chamberlin, Michael, and Paul Berg, "Studies on DNA-directed RNA Polymerase: Formation of DNA-RNA Complexes with Single-Stranded φX174 DNA as Template." *Cold Spring Harbor Symposium on Quantitative Biology* 23 (1963): 67–75.

Champoux, James J., and David S. Hogness. "The Topography of λ DNA: Polyriboguanylic Acid Binding Sites and Base Composition." *Journal of Molecular Biology* 71 (1972): 383–405.

Chang, Annie C., and Stanley N. Cohen. "Genome Construction between Bacterial Species *In Vitro*: Replication and Expression of *Staphylococcus* Plasmid Genes in *Escherichia coli.*" *Proceedings of the National Academy of Sciences, USA* 71 (1974): 1030–34.

Chang, Annie C., Jack H. Nunberg, Randal J. Kaufman, Henry A. Erlich, Robert T. Schimke, and Stanley N. Cohen. "Phenotypic Expression in *E. coli* of a DNA Sequence Coding for Mouse Dihydrofolate Reductase." *Nature* 275 (1978): 617–24.

Cheit, Earl Frank. *The New Depression in Higher Education: A Study of Financial Conditions at 41 Colleges and Universities.* Carnegie Commission on Higher Education, and Ford Foundation. New York: McGraw-Hill, 1971.

Clever, Ulrich. "Puffing in Giant Chromosomes of Diptera and the Mechanism of its Control." In *The Nucleohistones*, edited by James Bonner and Paul Ts'o, 317–34. San Francisco: Holden-Day, 1964.

Coggeshall, Porter E., and Prudence W Brown. Committee on National Needs for Biomedical and Behavioral Research Personnel, Institute of Medicine, *The Career Achievements of NIH Predoctoral Trainees and Fellows.* Washington, DC: National Academy Press, 1984.

Cohen, Stanley N. "The Manipulation of Genes." *Scientific American* 233 (1975): 24–33.

———. "Recombinant DNA: Fact and Fiction." *Science* 195 (1977): 654–57.

———. "The Transplantation and Manipulation of Genes in Microorganisms." In *The Harvey Lectures*, 173–204. New York: Academic Press 1980.

———. "The Stanford DNA Cloning Patent." In *From Genetic Engineering to Biotechnology: The Critical Transition*, edited by W. J. Whelan and Sandra Black, 213–16. New York: Wiley, 1982.

———. "Shaw Prize Lecture." Peking University, 2004.

———, MD. "Science, Biotechnology, and Recombinant DNA: A Personal History." An oral history conducted in 1995 by Sally Smith Hughes, PhD. Regional Oral History Office, Bancroft Library, University of California, Berkeley, 2009.

Cohen, Stanley N., and Herbert W. Boyer. "Process for Producing Biologically Functional Molecular Chimeras." US Patent 4,237,224. Issued December 2, 1980. Filed November 4, 1974.

———. "Biologically Functional Molecular Chimeras, product patent (prokaryotic)." US Patent 4,468,464. Issued August 28, 1984. Filed initially November 4, 1974. Abandoned May 17, 1976. Refiled November 9, 1978.

———. "Biologically Functional Molecular Chimeras, product patent (eukaryotic)." US Patent 4,740,470. Issued April 26, 1988. Filed initially November 4, 1974. Abandoned May 17, 1976. Refiled April 20, 1984.

Cohen, Stanley N., and Annie C. Y. Chang. "Recircularization and Autonomous Replication of a Sheared R-Factor DNA Segment in *Escherichia coli* Transformants." *Proceedings of the National Academy of Sciences, USA* 70 (1973): 1293–97.

———. "Revised Interpretation of the Origins of the pSC101 Plasmid." *Journal of Bacteriology* 132 (1977): 734–37.

Cohen, Stanley N., Annie C. Chang, Herbert W. Boyer, and Robert B. Helling. "Construction of Biologically Functional Bacterial Plasmids *In Vitro*." *Proceedings of the National Academy of Sciences, USA* 70 (1973): 3240–44.

Cohen, Stanley N., Annie C. Chang, and Leslie Hsu. "Nonchromosomal Antibiotic Resistance in Bacteria: Genetic Transformation of *Escherichia coli* by R-Factor DNA." *Proceedings of the National Academy of Sciences, USA* 69 (1972): 2110–14.

Cohn, Edwin J. "History of the Development of the Scientific Policies of the University Laboratory of Physical Chemistry Related to Medicine and Public Health, Harvard University: A Memorandum on the Unwisdom of Projects and Reports (January 1952)." In Edwin J. Cohn, *A Collection of Pamphlets: Published to Record an Experiment in Organization for Research, Training, and Development in Science Basic to Medicine and Public Health*, 1–24. Cambridge: Harvard University Printing Office, 1952.

Cohn, Melvin. "Contributions of Studies on the β-galactosidase of *Escherichia coli* to Our Understanding of Enzyme Synthesis." *Bacteriological Review* 21 (1957): 140–68.

———. "The Wisdom of Hindsight." *Annual Review of Immunology* 12 (1994): 1–62.

Creager, Angela N. H. "In the Fly Room." *Historical Studies in the Physical and Biological Sciences* 25 (1995): 357–60.

———. "Wendell Stanley's Dream of a Free-standing Biochemistry Department at the University of California, Berkeley." *Journal of the History of Biology* 29 (1996): 331–60.

———. "'What Blood Told Dr Cohn': World War II, Plasma Fractionation, and the Growth of Human Blood Research." *Studies in History and Philosophy of Biological and Biomedical Sciences* 30 (1999): 377–405.

———. "Biotechnology and Blood: Edwin Cohn's Plasma Fractionation Project, 1940–1953." In *Private Science: Biotechnology and the Rise of Molecular Sciences*, edited by Arnold Thackray, 39–62. Philadelphia: University of Pennsylvania Press, 1998.

———. *The Life of a Virus: Tobacco Mosaic Virus as an Experimental Model, 1930–1965*. Chicago: University of Chicago Press, 2002.

———. "Tracing the Politics of Changing Postwar Research Practices: The Export of 'American' Radioisotopes to European Biologists." *Studies in History and Philosophy of Biological and Biomedical Sciences* 33 (2002): 367–88.

———. "The Industrialization of Radioisotopes by the U.S. Atomic Energy Commission." In *The Science-Industry Nexus: History, Policy, Implications*, edited by Karl Grandin, Nina Wormbs, and Sven Widmalm, 143–67. Sagamore Beach: Science History Publications, 2004.

———. "Mapping Genes in Microorganisms." In *From Molecular Genetics to Genomics: Mapping Cultures of Twentieth Century Genetics*, edited by Jean-Paul Gaudillière and Hans-Jörg Rheinberger, 9–41. London: Routledge, 2004.

———. "Adaptation or Selection? Old Issues and New Stakes in the Postwar Debates over Bacterial Drug Resistance." *Studies in History and Philosophy of Biological and Biomedical Sciences* 38 (2007): 159–90.

———. "Christian B. Anfinsen." In *New Dictionary of Scientific Biography*, edited by Noretta Koertge, 1:76–82. New York: Charles Scribner's Sons/Thomson Gale, 2008.

———. "Mobilizing Biomedicine: Virus Research between Lay Health Organizations and the U.S. Federal Government, 1935–1955." In *Biomedicine in the Twentieth Century: Practices, Policies, and Politics*, edited by Caroline Hannaway, 171–201. Amsterdam: IOS Press, 2008.

Creager, Angela N. H., and Jean-Paul Gaudillière. "Meanings in Search of Experiments and Vice-Versa: The Invention of Allosteric Regulation in Paris and Berkeley, 1959–1968." *Historical Studies in the Physical and Biological Sciences* 27 (1996): 1–89.

———. "Experimental Arrangements and Technologies of Visualization: Cancer as a Viral Epidemic, 1930–1960." *Heredity and Infection: The History of Disease Transmission*, edited by In Jean-Paul Gaudillière and Ilana Löwy, 203–41. London: Routledge, 2001.

Creager, Angela N. H., Elizabeth Lunbeck, and M. Norton Wise, eds. *Science without Laws: Model Systems, Cases, Exemplary Narratives.* Durham: Duke University Press, 2008.

Creager, Angela N. H., and Gregory J. Morgan. "After the Double Helix: Rosalind Franklin's Research on Tobacco Mosaic Virus." *Isis* 99 (2008): 239–72.

Crick, Francis. "Molecular Biology and Medical Research." *Journal of the Mount Sinai Hospital* 36 (1969): 178–88.

———. "General Model for the Chromosomes of Higher Organisms." *Nature* 234 (1971): 25–27.

Crotty, Shane. *Ahead of the Curve: David Baltimore's Life in Science.* Berkeley: University of California Press, 2001.

Cuzin, François, Marguerite Vogt, Marianne Dieckmann, and Paul Berg. "Induction of Virus Multiplication in 3T3 Cells Transformed by a Thermosensitive Mutant of Polyoma Virus. II. Formation of Oligometric Polyoma DNA Molecules." *Journal of Molecular Biology* 47 (1970): 317–33.

Danna, Kathleen, and Daniel Nathans. "Specific Cleavage of Simian Virus 40 DNA by Restriction Endonuclease of *Hemophilus influenzae.*" *Proceedings of the National Academy of Sciences, USA* 68 (1971): 2913–17.

Darling, George B. "Can We Pay for Our Medical Schools?" *Atlantic Monthly* (June 1950): 38–42.

Daston, Lorraine. "The Moral Economy of Science." *Osiris* 10 (1995): 3–24.

Davis, Natalie Zemon. *The Gift in Sixteenth-Century France.* Madison: University of Wisconsin Press, 2000.

de Chadarevian, Soraya. "Sequence, Conformation, Information: Biochemists and Molecular Biologists in the 1950s." *Journal of the History of Biology* 29 (1996): 361–86.

———. "Of Worms and Programmes: *Caenorhabditis elegans* and the Study of Development." *Studies in History and Philosophy of Biological and Biomedical Sciences* 29 (1998): 81–105.

———. *Designs for Life: Molecular Biology after World War II.* Cambridge: Cambridge University Press, 2002.

———. "The Making of an Entrepreneurial Science: Biotechnology in Britain, 1975–1995." *Isis* 102 (2011): 601–33.

de Chadarevian, Soraya, and Jean-Paul Gaudillière. "The Tools of the Discipline: Biochemists and Molecular Biologists." *Journal of the History of Biology* 29 (1996): 327–30.

de Chadarevian, Soraya, and Harmke Kamminga, eds. *Molecularizing Biology and Medicine: New Practices and Alliances, 1910s–1970s.* Amsterdam: Harwood Academic Publishers, 1998.

de Chadarevian, Soraya, and Bruno J. Strasser. "Molecular Biology in Postwar Europe: Towards a 'Glocal' Picture." *Studies in History and Philosophy of Biological and Biomedical Sciences* 33 (2002): 361–65.

Demsetz, Harold. "Toward a Theory of Property Rights." *American Economic Review* 57 (1967): 347–59.

Denison, Edward F. *Accounting for Slower Economic Growth: The United States in the 1970's.* Washington, DC: Brookings Institution, 1979.

Dickson, David. *The New Politics of Science.* New York: Pantheon Books, 1984.

Dulbecco, Renato. "Production of Plaques in Monolayer Tissue Cultures by Single Particles of an Animal Virus." *Proceedings of the National Academy of Sciences, USA* 38 (1952): 747–52.

———. "Basic Mechanisms in the Biology of Animal Viruses: Concluding Address." *Cold Spring Harbor Symposia on Quantitative Biology*, Vol. XXVII: Basic Mechanisms in Animal Virus Biology, 519–25. Cold Spring Harbor: Cold Spring Harbor Laboratory, 1962.

———. "Transformation of Cells in vitro by Viruses." *Science* 142 (1963): 932–36.

Dupree, A. Hunter. *Science and the Federal Government: A History of Policies and Activities.* Baltimore: Johns Hopkins University Press, 1957 [1986].

Echols, Harrison, and Carol Gross. *Operators and Promoters: The Story of Molecular Biology and Its Creators.* Berkeley: University of California Press, 2001.

Edgerton, David. "'The Linear Model Did Not Exist: Reflections on the History and Historiography of Science and Research in Industry in the Twentieth Century." In *The Science-Industry Nexus: History, Policy, Implications*, edited by Karl Grandin, Nina Wormbs, and Sven Widmalm, 31–57. Sagamore Beach: Science History Publications, 2004.

Egan, J. Barry, and David S. Hogness. "The Topography of Lambda DNA: Isolation of Ordered Fragments and the Physical Mapping of Point Mutations." *Journal of Molecular Biology* 71 (1972): 363–81.

Eisenberg, Rebecca S. "Public Research and Private Development: Patents and Technology Transfer in Government-Sponsored Research." *Virginia Law Review* 82 (1996): 1663–727.

Eskridge, Nancy K. "Dole Blasts HEW for 'Stonewalling' Patent Applications." *BioScience* 28 (1978): 605–6.

Etzkowitz, Henry. *MIT and the Rise of Entrepreneurial Science.* London: Routledge, 2002.

Falkow, Stanley. "I'll Have the Chopped Liver Please, or How I Learned to Love the Clone." *American Society for Microbiology News* 67:11 (2001): 555–59.

Feldman, Maryann, Alessandra Colaianni, and Kang Liu. "Commercializing Cohen-Boyer 1980–1997." DRUID Working Paper No 05–21, DRUID, Copenhagen Business School, Department of Industrial Economics and Strategy/Aalborg University, Department of Business Studies, 2005.

Finnegan, David J., Gerald M. Rubin, and Michael W. Young, and David S. Hogness, "Repeated Gene Families in *Drosophila melanogaster*." *Cold Spring Harbor Symposia on Quantitative Biology* 42:2 (1978): 1053–63.

Finnegan, Henderson, Farabow, Garrett & Dunner, LLP. *Opinion Regarding Valid-

ity, Enforceability and Infringement Issues Presented by the Cohen and Boyer Patents. Prepared for Leland Stanford Junior University. Confidential Opinion of Counsel, August 9, 1985.

Forman, Paul. "Behind Quantum Electronics: National Security as Basis for Physical Research in the United States, 1940–1960." *Historical Studies in the Physical and Biological Sciences* 18 (1987): 149–229.

Fortun, Michael. "The Human Genome Project and the Acceleration of Biotechnology." In *Private Science: Biotechnology and the Rise of the Molecular Sciences,* edited by Arnold Thackray, 182–201. Philadelphia: University of Pennsylvania Press, 1998.

Fox, Daniel M. "The Politics of NIH Extramural Program, 1937–1950." *Journal of the History of Medicine and Allied Sciences* 42 (1987): 447–66.

Fredrickson, Donald S. *The Recombinant DNA Controversy: A Memoir — Science, Politics, and the Public Interest 1974–1981.* Washington, DC: ASM Press, 2001.

Friedmann, Theodore, and Richard Roblin. "Gene Therapy for Human Genetic Disease?" *Science* 175 (1972): 949–55.

Fujimura, Joan H. *Crafting Science: A Sociohistory of the Quest for the Genetics of Cancer.* Cambridge: Harvard University Press, 1996.

Galison, Peter, Bruce Hevly, and Rebecca Lowen. "Controlling the Monster: Stanford and the Growth of Physics Research, 1935–1962." In *Big Science: The Growth of Large-Scale Research,* edited by Peter Galison and Bruce Hevly, 46–77. Stanford: Stanford University Press, 1992.

Garb, Solomon. *Cure for Cancer: A National Goal.* New York: Springer Publishing, 1968.

Gaudillière, Jean-Paul. "The Molecularization of Cancer Etiology in the Postwar United States: Instruments, Politics and Management." In *Molecularizing Biology and Medicine: New Practices and Alliances, 1910s–1970s,* edited by Soraya de Chadarevian and Harmke Kamminga, 139–70. Amsterdam: Harwood Academic Publishers, 1998.

———. *Inventer la biomédecine: La France, l'Amérique et la production des savoirs du vivant, 1945–1965.* Paris: Éditions la Découverte, 2002.

———. "Paris-New York Roundtrip: Transatlantic Crossings and the Reconstruction of the Biological Sciences in Post-War France." *Studies in History and Philosophy of Biological and Biomedical Sciences* 33 (2002): 389–417.

———. "How Pharmaceuticals became Patentable: The Production and Appropriation of Drugs in the Twentieth Century." *History and Technology* 24 (2008): 99–106.

Gaudillière, Jean-Paul, and Ilana Löwy, eds. *The Invisible Industrialist: Manufactures and the Production of Scientific Knowledge.* London: Routledge, 1998.

Gaudillière, Jean-Paul, and Hans-Jörg Rheinberger, eds. *From Molecular Genetics to Genomics: Mapping Cultures of Twentieth Century Genetics.* London: Routledge, 2004.

Geiger, Roger L. *Research and Relevant Knowledge: American Research Universities since World War II.* New York: Oxford University Press, 1993.

———. *Knowledge and Money: Research Universities and the Paradox of the Marketplace.* Stanford: Stanford University Press, 2004.

Gilbert, Scott F., ed. *Developmental Biology: A Comprehensive Synthesis.* Vol. 7. New York: Plenum Press, 1991.

———. "Entrance of Molecular Biology into Embryology." In *The Philosophy and History of Molecular Biology: New Perspectives,* edited by Sahotra Sarkar, 101–23. Dordrecht: Kluwer Academic, 1996.

Gillmor, C. Stewart. *Fred Terman at Stanford: Building a Discipline, a University, and Silicon Valley.* Stanford: Stanford University Press, 2004.

Goulian, Mehran, Arthur Kornberg, and Robert L. Sinsheimer. "Enzymatic Synthesis of DNA: XXVI. Synthesis of Infectious Phage φX174 DNA." *Proceedings of the National Academy of Sciences, USA* 58 (1967): 2321–28.

Grandin, Karl, Nina Wormbs, and Sven Widmalm, eds. *The Science-Industry Nexus: History, Policy, Implications.* Sagamore Beach: Science History Publications, 2004.

Greenberg, Daniel S. *Science, Money, and Politics: Political Triumph and Ethical Erosion.* Chicago: University of Chicago Press, 2001.

Greene, Jeremy A. *Prescribing by Numbers: Drugs and the Definition of Disease.* Baltimore: Johns Hopkins University Press, 2007.

Grote, Mathias. "Hybridizing Bacteria, Crossing Methods, Cross-checking Arguments: The Transition from Episomes to Plasmids (1961–1969)." *History and Philosophy of the Life Sciences* 30 (2008): 407–30.

Grunstein, Michael, and David Hogness. "Colony Hybridization: A Method for the Isolation of Cloned DNAs that Contain a Specific Gene." *Cell* 72 (1975): 3961–65.

Gurdon, John. "The Developmental Capacity of Nuclei Taken from Intestinal Epithelium Cells of Feeding Tadpoles." *Journal of Embryology and Experimental Morphology* 10 (1962): 622–40.

Haber, Edgar, Daniel E. Koshland, David Morris, Arthur Kornberg, David Hogness, I. Robert Lehman, Robert Baldwin, Paul Berg, A. Dale Kaiser, Philip Sharp, J. Michael Bishop, Robert J. Glaser, and Robert S. Schwartz. "Review of Kornberg's *The Golden Helix.*" *New England Journal of Medicine* 334:15 (1996): 994–95.

Hall, Stephen S. *Invisible Frontiers: The Race to Synthesize a Human Gene.* 1st ed. New York: Atlantic Monthly Press, 1987.

Halluin, Albert P. "Patenting the Results of Genetic Engineering Research: An Overview." In *Patenting of Life Forms,* edited by David W. Plant, Neils J. eimers and Norton D. Zinder, 67–126. Cold Spring Harbor: Cold Spring Harbor Laboratory, 1982.

Harbridge House, Inc. *Government Patent Policy Study: Final Report.* Prepared for the Federal Council for Science and Technology Committee on Government Patent Policy. Washington, DC: US Government Printing Office, 1968.

Haraway, Donna J. *Modest_Witness@Second_Millennium. FemaleMan©_Meets_ OncoMouse™: Feminism and Technoscience*. New York: Routledge, 1997.

Hardin, Garrett. "The Tragedy of the Commons." *Science* 162 (1968): 1243–48.

Hedgpeth, Joe, Howard M. Goodman, and Herbert W. Boyer. "DNA Nucleotide Sequence Restricted by the RI Endonuclease." *Proceedings of the National Academy of Sciences, USA* 69 (1972): 3448–52.

Heller, Michael A. "The Tragedy of the Anticommons: Property in the Transition from Marx to Markets." *Harvard Law Review* 111 (1998): 621–88.

Hellman, Alfred, M. N. Oxman, and Robert Pollack, eds. *Biohazards in Biological Research*. Cold Spring Harbor: Cold Spring Harbor Laboratory, 1973.

Hershey, Alfred D., ed. *The Bacteriophage Lambda*. Cold Spring Harbor: Cold Spring Harbor Laboratory, 1971.

Hershey, Alfred D., L. Ingraham, and E. Burgi. "Cohesion of DNA Molecules Isolated from Phage Lambda." *Proceedings of the National Academy of Sciences, USA* 49 (1963): 748–55.

Hogness, David S. "Induced Enzyme Synthesis." *Reviews of Modern Physics* 31 (1959): 256–68.

———. "The Structure and Function of the DNA from Bacteriophage Lambda." *Journal of General Physiology* 49 (1966): 29–57.

———. "The Structure and Function of Animal Chromosomes." National Science Foundation grant proposal, 1970. Document courtesy of David S. Hogness.

———. "The Arrangement and Function of DNA Sequences in Animal Chromosomes." National Institutes of Health grant proposal, 1972. Document courtesy of David S. Hogness.

Hogness, David S., Melvin Cohn, and Jacques Monod. "Studies on the Induced Synthesis of β-galactosidase in *Escherichia coli*." *Biochimica et Biophysica Acta* 16 (1955): 99–116.

Hollinger, David A. "Science as a Weapon in *Kulturkampfe* in the United States during and after World War II." *Isis* 86 (1995): 440–54.

Holmes, Frederic L. *Between Biology and Medicine: The Formation of Intermediary Metabolism*. Berkeley: University of California Press, 1992.

———. "Seymour Benzer and the Definition of the Gene." In *The Concept of the Gene in Development and Evolution: Historical and Epistemological Perspectives*, edited by Peter J. Beurton, Raphael Falk, and Hans-Jörg Rheinberger, 115–55. Cambridge: Cambridge University Press, 2000.

———. *Meselson, Stahl, and the Replication of DNA: A History of "The Most Beautiful Experiment in Biology."* New Haven: Yale University Press, 2001.

Hounshell, David A. "The Medium is the Message, or How Context Matters: The RAND Corporation Builds an Economics of Innovation, 1946–1962." In *Systems, Experts, and Computers: The System Approach in Management and En-*

gineering, World War II and After, edited by Agatha C. Hughes and Thomas P. Hughes, 255–310. Cambridge: MIT Press, 2000.

Hounshell, David A., and John K. Smith. *Science and Corporate Strategy*. Cambridge: Cambridge University Press, 1988.

Hughes, Sally S. "Making Dollars out of DNA: The First Major Patent in Biotechnology and the Commercialization of Molecular Biology, 1974–1980." *Isis* 92 (2001): 541–75.

———. *Genentech: The Beginnings of Biotech*. Chicago: University of Chicago Press, 2011.

Inglis, John, Joseph Sambrook, and Jan A. Witkowski, eds. *Inspiring Science: Jim Watson and the Age of DNA*. Cold Spring Harbor: Cold Spring Harbor Laboratory Press, 2003.

Jackson, David A., Robert H. Symons, and Paul Berg. "Biochemical Method for Inserting New Genetic Information into DNA of Simian Virus 40: Circular SV40 DNA Molecules Containing Lambda Phage Genes and Galactose Operon of *Escherichia coli*." *Proceedings of the National Academy of Sciences, USA* 69 (1972): 2904–9.

Jackson, Myles W. "Intellectual Property and Molecular Biology: Biomedicine, Commerce and the CCR5 Gene." Francis Bacon Lecture, Caltech, May 5, 2011.

Jacob, François. *Of Flies, Mice, and Men*. Cambridge: Harvard University Press, 2002.

Jacob, François, and Jacques Monod. "Genetic Regulatory Mechanisms in the Synthesis of Proteins." *Journal of Molecular Biology* 3 (1961): 318–56.

Jacob, François, and Elie L. Wollman. *Sexuality and the Genetics of Bacteria*. New York: Academic Press, 1961.

Jasanoff, Sheila. *Science at the Bar: Law, Science, and Technology in America*. Cambridge: Harvard University Press, 1995.

———. "Ordering Knowledge, Ordering Society." In *States of Knowledge: The Co-Production of Science and Social Order*, edited by Sheila Jasanoff, 13–45. London: Routledge, 2004.

———. *Designs on Nature: Science and Democracy in Europe and the United States*. Princeton: Princeton University Press, 2005.

Johns, Adrian. "Intellectual Property and the Nature of Science." *Cultural Studies* 20:2/3 (2006): 145–64.

———. *Piracy: The Intellectual Property Wars from Gutenberg to Gates*. Chicago: University of Chicago Press, 2009.

Johnson, William S. *A Fifty-Year Love Affair with Organic Chemistry*. Washington, DC: American Chemical Society, 1998.

Jong, Simcha. "How Organizational Structures in Science Shape Spin-off Firms: The Biochemistry Departments of Berkeley, Stanford, and UCSF and the Birth of the Biotech Industry." *Industrial and Corporate Change* 15 (2006): 251–83.

Jordan, Kathleen, and Michael Lynch. "The Sociology of a Genetic Engineering

Technique: Ritual and Rationality in the Performance of the 'Plasmid Prep.'"
In *The Right Tools for the Job: At Work in Twentieth-century Life Sciences*, edited by Adele Clarke and Joan H. Fujimura, 77–114. Princeton: Princeton University Press, 1992.

Kaiser, A. Dale. "The Production of Phage Chromosome Fragments and Their Capacity for Genetic Transfer." *Journal of Molecular Biology* 4 (1962): 275–87.

———. "Description of Work Leading up to Recombinant DNA." Unpublished
manuscript, August 1980. Document courtesy of A. Dale Kaiser.

Kaiser, A. Dale, and Robert L. Baldwin. "A Relation between Dinucleotide and
Base Frequencies in Bacterial DNAs." *Journal of Molecular Biology* 4 (1962):
418–19.

Kaiser, A. Dale, and David S. Hogness. "Transformation of *Escherichia coli* with
Deoxyribonucleic Acid Isolated from Bacteriophage-λdg." *Journal of Molecular Biology* 2 (1960): 392–415.

Kaiser, A. Dale, and R. B. Inman. "Cohesion and the Biological Activity of Lambda
DNA." *Journal of Molecular Biology* 13 (1965): 78–91.

Kaiser, A. Dale, and François Jacob. "Recombination between Related Temperate
Bacteriophages and the Genetic Control of Immunity and Prophage Localization." *Virology* 4 (1957): 509–21.

Kaiser, A. Dale, and Ray Wu. "Structure and Function of DNA Cohesive Ends."
Cold Spring Harbor Symposia on Quantitative Biology 33 (1968): 729–34.

Kay, Lily E. "Selling Pure Science in Wartime: The Biochemical Genetics of G.W.
Beadle." *Journal of the History of Biology* 22 (1989): 73–101.

———. *The Molecular Vision of Life: Caltech, the Rockefeller Foundation, and the
Rise of the New Biology*. Chicago: University of Chicago Press, 1993.

———. "Problematizing Basic Research in Molecular Biology." In *Private Science:
Biotechnology and the Rise of Molecular Sciences*, edited by Arnold Thackray,
20–38. Philadelphia: University of Pennsylvania Press, 1998.

———. *Who Wrote the Book of Life? A History of the Genetic Code*. Stanford:
Stanford University Press, 2000.

Keating, Peter, and Alberto Cambrosio. *Biomedical Platforms: Realigning the Normal and the Pathological in Late-Twentieth-Century Medicine*. Cambridge: MIT
Press, 2003.

Keller, Evelyn Fox. "Physics and the Emergence of Molecular Biology: A History
of Cognitive and Political Synergy." *Journal of the History of Biology* 23 (1990):
389–409.

———. "*Drosophila* Embryos as Transitional Objects: The Work of Donald Poulson and Christiane Nüsslein-Volhard." *Historical Studies in the Physical and
Biological Sciences* 26 (1996): 313–46.

———. "Developmental Biology as a Feminist Cause?" *Osiris* 12 (1997): 16–28.

———. "Decoding the Genetic Program: Or, Some Circular Logic in the Logic of
Circularity." In *The Concept of the Gene in Development and Evolution: His-*

torical and Epistemological Perspectives, edited by Peter Beurton, Raphael Falk, and Hans-Jörg Rheinberger, 159–77. Cambridge: Cambridge University Press, 2000.

Kelly, Thomas J., Jr., and Hamilton O. Smith. "A Restriction Enzyme from *Hemophilus influenzae*. II." *Journal of Molecular Biology* 51 (1970): 393–409.

Kennedy, Donald. "Interview: Gene Splicing: Patents, Problems, and 'Enormous Potential Benefit.'" *SIPI Scope* 9 (1981): 1–9.

Kenney, Martin. *Biotechnology: The University-Industrial Complex*. New Haven: Yale University Press, 1986.

Kerr, Clark. *The Uses of the University*. Cambridge: Harvard University Press, 1963.

Kevles, Daniel J. "The National Science Foundation and the Debate over Postwar Research Policy, 1942–1945: A Political Interpretation of Science—the Endless Frontier." *Isis* 68 (1977): 5–26.

———. *The Physicists: The History of a Scientific Community in Modern America*. 1st ed. New York: Knopf, 1977.

———. "Renato Dulbecco and the New Animal Virology: Medicine, Methods, and Molecules." *Journal of the History of Biology* 26 (1993): 409–42.

———. "Ananda Chakrabarty Wins a Patent: Biotechnology, Law and Society, 1972–1980." *Historical Studies in the Physical and Biological Sciences* 25 (1994): 111–35.

———. "Pursuing the Unpopular: A History of Courage, Viruses, and Cancer." In *Hidden Histories of Science*, edited by Robert B. Silvers, 69–114. New York: New York Review of Books, 1995.

———. "Principles, Property Rights, and Profits: Historical Reflections on University/Industry Tensions." *Accountability in Research* 8 (2001): 12–26.

Kitch, Edmund W. "The Nature and Function of the Patent System." *Journal of Law and Economics* 20 (1977): 265–90.

Kleinman, Daniel L. "Layers of Interests, Layers of Influence: Business and the Genesis of the National-Science-Foundation." *Science Technology and Human Values* 19 (1994): 259–82.

Kloppenburg, Jack Ralph. *First the Seed: the Political Economy of Plant Biotechnology, 1492–2000*. Cambridge: Cambridge University Press, 1988.

Knox, Richard A. "Stanford Medical School Suffers Fiscal Ideological Crises." *Science* 203 (1979): 148–52.

Kohler, Robert E. "The Management of Science: The Experience of Warren Weaver and the Rockefeller Foundation Programme in Molecular Biology." *Minerva* 14 (1976): 279–306.

———. "Walter Fletcher, F. G. Hopkins, and the Dunn Institute of Biochemistry: A Case Study in the Patronage of Science." *Isis* 69 (1978): 331–55.

———. *From Medical Chemistry to Biochemistry: The Making of a Biomedical Discipline*. Cambridge: Cambridge University Press, 1982.

————. *Partners in Science: Foundations and Natural Scientists, 1900–1945*. Chicago: University of Chicago Press, 1991.

————. "Systems of Production: *Drosophila, Neurospora,* and Biochemical Genetics." *Historical Studies in the Physical and Biological Sciences* 22 (1991): 87–130.

————. *Lords of the Fly:* Drosophila *Genetics and the Experimental Life*. Chicago: University of Chicago Press, 1994.

————. "Moral Economy, Material Culture and Community in *Drosophila* Genetics." In *The Science Studies Reader*, edited by Mario Biagioli, 243–56. London: Routledge, 1999.

Kornberg, Arthur. "Biologic Synthesis of Deoxyribonucleic Acid." *Science* 131 (1960): 1503–8.

————. "The Biologic Synthesis of Deoxyribonucleic Acid: Nobel Lecture, December 1959." In *Nobel Lectures, Physiology or Medicine 1942–1962*, 665–80. Amsterdam: Elsevier Publishing Company, 1964.

————. *For the Love of Enzymes: The Odyssey of a Biochemist*. Cambridge: Harvard University Press, 1989.

————. "Editorial: Science is Great, but Scientists are Still People." *Science* 257 (1992): 859.

————. *The Golden Helix: Inside Biotech Ventures*. Sausalito: University Science Books, 1995.

————, MD. "Biochemistry at Stanford, Biotechnology at DNAX." An oral history conducted in 1997 by Sally Smith Hughes, PhD. Regional Oral History Office. Bancroft Library, University of California, Berkeley, 1998.

Kriegstein, Henry J., and David S. Hogness. "Mechanism of DNA Replication in Drosophila Chromosomes: Structure of Replication Forks and Evidence for Bidirectionality." *Proceedings of the National Academy of Sciences, USA* 71 (1974): 135–39.

Krimsky, Sheldon. *Genetic Alchemy: The Social History of the Recombinant DNA Controversy*. Cambridge: MIT Press, 1982.

————. *Biotechnics and Society: The Rise of Industrial Genetics*. New York: Praeger, 1991.

————. *Science in the Private Interest: Has the Lure of Profits Corrupted Biomedical Research?* Lanham: Rowman & Littlefield, 2003.

Kutcher, Gerald J. *Contested Medicine: Cancer Research and the Military*. Chicago: University of Chicago Press, 2009.

Landecker, Hannah. "Sending Cells Around: How to Exchange Biological Matter." Paper presented in The Moment of Conversion: Exchange Networks in Modern Biomedical Science, Department of Anthropology, History, and Social Medicine, UC San Francisco, April 4–5, 2003.

————. *Culturing Life: How Cells Became Technologies*. Cambridge: Harvard University Press, 2007.

Latker, Norman J. *Patent Council, Department of Health, Education, and Welfare,*

US Congressional Hearing, Committee on Science and Technology, Government Patent Policy: The Ownership of Inventions Resulting from Federally Funded Research and Development, 94th Congress, 2nd session. Washington, DC: US Government Printing Office, 1976.

Latker, Norman J., and Ronald J. Wylie. "Utilization of Government-Owned Health and Welfare Inventions." *Journal of the Patent Office Society* 47 (1965): 868–79.

Latour, Bruno. *The Pasteurization of France*. Cambridge: Harvard University Press, 1988.

———. *Science in Action: How to Follow Scientists and Engineers through Society*. Cambridge: Harvard University Press, 1988.

Lear, John. *Recombinant DNA: The Untold Story*. New York: Crown Publishers, 1978.

Lederberg, Esther M. "Lysogenicity in *E. coli* K-12." *Genetics* 36 (1951): 560–69.

Lederberg, Joshua. "Dr. Lederberg Speaks Out on Biological Warfare Hazards." *Congressional Record* 112, Part 23 (September 11, 1970): 31395–96.

———. News release. *Stanford University News Service*. May 20, 1974.

Lee, Peter. *Medical Schools and the Changing Times: Nine Case Reports on Experimentation in Medical Education, 1950–60*. Evanston: Association of American Medical Colleges, 1962.

Lehman, I. Robert, Maurice J. Bessman, Ernest Simms, and Arthur Kornberg. "Enzymatic Synthesis of Deoxyribonucleic Acid. I. Preparation of Substrates and Partial Purification of an Enzyme from *Escherichia coli*." *Journal of Biological Chemistry* 233 (1958): 163–70.

Lenoir, Timothy. "Biochemistry at Stanford: A Case Study in the Formation of an Entrepreneurial Culture." Unpublished manuscript, April 21, 2002. Microsoft Word file.

Leslie, Stuart W. *The Cold War and American Science: The Military-Industrial-Academic Complex at MIT and Stanford*. New York: Columbia University Press, 1993.

Leslie, Stuart W., and R. H. Kargon, "Selling Silicon Valley: Frederick Terman's Model for Regional Advantage." *Business History Review* 70:4 (1996): 435–72.

Lewis, Edward B. "Genes and Developmental Pathways." *American Zoologist* 3 (1963): 33–56.

Lobban, Peter E. "The Generation of Transducing Phage *In Vitro*." Third essay for PhD examination, Stanford University, November 6, 1969. Document courtesy of Peter E. Lobban.

———. *An Enzymatic Method for End-to-End Joining of DNA Molecules*. PhD dissertation, Stanford University, 1972.

Lobban, Peter E., and A. D. Kaiser. "Enzymatic End-to-End Joining of DNA Molecules." *Journal of Molecular Biology* 78 (1973): 453–71.

Lock, Margaret. *Twice Dead: Organ Transplants and the Reinvention of Death*. Berkeley: University of California Press, 2002.

Lowen, Rebecca S. *Creating the Cold War University: The Transformation of Stanford*. Berkeley: University of California Press, 1997.

Ludmerer, Kenneth M. *Time to Heal: American Medical Education from the Turn of the Century to the Era of Managed Care*. Oxford: Oxford University Press, 1999.

Lwoff, André. "Lysogeny." *Bacteriological Reviews* 17 (1953): 269–337.

Machlup, Fritz. *An Economic Review of the Patent System. Study #15 of the Subcommittee on Patents, Trademarks, and Copyrights of the US Senate Committee on the Judiciary*, 85th Congress, 2nd session. Washington, DC: US Government Printing Office, 1958.

———. *The Production and Distribution of Knowledge in the United States*. Princeton: Princeton University Press, 1962.

Mackenzie, Michael, Alberto Cambrosio, and Peter Keating. "The Commercial Application of a Scientific Discovery: The Case of the Hybridoma Technique." *Research Policy* 17 (1988): 155–70.

Mandel, Morton, and Akiko Higa. "Calcium-Dependent Bacteriophage DNA Infection." *Journal of Molecular Biology* 53 (1970): 159–62.

Maniatis, Tom, E. F. Fritsch, and Joseph Sambrook. *Molecular Cloning: A Laboratory Manual*. Cold Spring Harbor: Cold Spring Harbor Laboratory Press, 1982.

Matsubara, Kenichi, and A. Dale Kaiser. "Lambda dv: An Autonomously Replicating DNA Fragment." *Cold Spring Harbor Symposia on Quantitative Biology* 33 (1968): 769–75.

McElheny, Victor K. "Animal Gene Shifted to Bacteria: Aid Seen to Medicine and Farm." *New York Times*, May 20, 1974.

Mendelsohn, Everett. "The Politics of Pessimism: Science and Technology Circa 1968." In *Technology, Pessimism, and Postmodernism*, edited by Yaron Ezrahi, Everett Mendelsohn, and Howard Segal, 151–73. Dordrecht: Kluwer Academic Publishers, 1994.

Merril, Carl R., Mark R. Geier, and John C. Petricciani. "Bacterial Virus Gene Expression in Human Cells." *Nature* 233 (1971): 398–400.

Merton, Robert K. "The Normative Structure of Science." In *The Sociology of Science: Theoretical and Empirical Investigations*, edited by Norman W. Storer, 267–78. Chicago: University of Chicago Press, 1973.

Mertz, Janet E., and Ronald W. Davis. "Cleavage of DNA by RI Restriction Endonuclease Generates Cohesive Ends." *Proceedings of the National Academy of Sciences, USA* 69 (1972): 3370–74.

———. Interview, March 9, 1977. Oral History Programs, Recombinant DNA Controversy Collection, MC 100, MIT Institute Archive.

Mertz, Janet E., Ronald W. Davis, and Paul Berg. "Characterization of the Cleavage Site of the R1 Restriction Enzyme." Abstract presented at the Fourth Annual Tumor Virus Meeting (arranged by Philip A. Sharp), August 16–19, 1972, Cold Spring Harbor Laboratory, Cold Spring Harbor, New York.

Mirowski, Philip. *Science-Mart: Privatizing America Science*. Cambridge: Harvard University Press, 2011.

Mirowski, Philip, and Esther-Mirjam Sent. "The Commercialization of Science, and the Response of STS." In *Handbook of Science, Technology and Society Studies*, edited by Edward J. Hackett, Olga Amsterdamska, Michael Lynch, and Judy Wajcman, 635–89. Cambridge: MIT Press, 2007.

Miyoshi, Masao. "Ivory Tower in Escrow." *Boundary 2* 27 (2000): 7–50.

Monod, Jacques, and François Jacob. "General Conclusions: Teleonomic Mechanisms in Cellular Metabolism, Growth, and Differentiation." *Cold Spring Harbor Symposia on Quantitative Biology* 26 (1961): 389–401.

Monod, Jacques, Ernest Borek, and André Lwoff, eds. *Of Microbes and Life*. New York: Columbia University Press, 1971.

Morange, Michel. "From the Regulatory Vision of Cancer to the Oncogene Paradigm, 1975–1985." *Journal of the History of Biology* 30 (1997): 1–29.

———. "The Transformation of Molecular Biology on Contact with Higher Organisms, 1960–1980: From a Molecular Description to a Molecular Explanation." *History and Philosophy of the Life Sciences* 19 (1997): 369–93.

———. "The Developmental Gene Concept: History and Limits." In *The Concept of the Gene in Development and Evolution: Historical and Epistemological Perspectives*, edited by Peter Beurton, Raphael Falk, and Hans-Jörg Rheinberger, 193–213. Cambridge: Cambridge University Press, 2000.

———. "Francois Jacob's Lab in the Seventies: The T-Complex and the Mouse Developmental Genetic Program." *History and Philosophy of the Life Sciences* 22 (2000): 397–411.

———. *A History of Molecular Biology*. Cambridge: Harvard University Press, 2000.

Morrow, John F. "Recombinant DNA Techniques." In *Methods in Enzymology*, edited by Ray Wu, 68:3–24. New York: Academic Press, 1979.

Morrow, John F., and Paul Berg. "Cleavage of Simian Virus 40 DNA at a Unique Site by a Bacterial Restriction Enzyme." *Proceedings of the National Academy of Sciences, USA* 69 (1972): 3365–69.

Morrow, John F., Stanley N. Cohen, Annie C. Chang, Herbert W. Boyer, Howard M. Goodman, and Robert B. Helling. "Replication and Transcription of Eukaryotic DNA in *Escherichia coli*." *Proceedings of the National Academy of Sciences, USA* 71 (1974): 1743–47.

Mowery, David C., and Bhaven N. Sampat. "Patenting and Licensing University Inventions: Lessons from the History of the Research Corporation." *Industrial and Corporate Change* 10 (2001): 317–55.

———. "University Patents and Patent Policy Debates in the USA, 1925–1980." *Industrial and Corporate Change* 10 (2001): 781–814.

Mowery, David, Richard R. Nelson, Bhaven N. Sampt, and Arvids A. Ziedonis, eds. *Ivory Tower and Industrial Innovation: University-Industry Technology Trans-*

fer Before and After the Bayh-Dole Act in the United States. Stanford: Stanford University Press, 2004.

Nash, George H. *Conservative Intellectual Movement in America since 1945*. New York: Basic Books, 1976.

National Institutes of Health. *Recombinant DNA Research*. Vol 2. Documents Relating to "NIH Guidelines for Research Involving Recombinant DNA Molecules." Prepared by the Office of the Director, NIH, Department of Health, Education, and Welfare Publication No. (NIH) 78–1139, June 1976–November 1977. Washington, DC: US Government Print Office, 1978.

National Science Foundation. *Technology in Retrospect and Critical Events in Science*. Prepared by the Illinois Institute of Technology Research Institute under Contract NSF-C535, December 15, 1968.

Neumann-Held, Eva M., and Christoph Rehmann-Sutter, eds. *Genes in Development: Re-reading the Molecular Paradigm*. Durham: Duke University Press, 2006.

"News: Genetic Manipulation to be Patented?" *Nature* 261 (1976): 624.

Newswatch. "What Japanese Industry is Doing in Biotechnology," July 6, 1981.

Newsweek. "Gene Transplanters." June 17, 1974.

Nomura, Masayasu. "Switching from Prokaryotic Molecular Biology to Eukaryotic Molecular Biology." *Journal of Biological Chemistry* 284 (2009): 9625–35.

November, Joseph A. *Biomedical Computing: Digitizing Life in the United States*. Baltimore: Johns Hopkins University Press, 2012.

O'Mara, Margaret Pugh. *Cities of Knowledge: Cold War Science and the Search for the Next Silicon Valley*. Princeton: Princeton University Press, 2005.

Onaga, Lisa. "Ray Wu as Fifth Business: Deconstructing Collective Memory in the History of DNA Sequencing." *Studies in History and Philosophy of Biological and Biomedical Sciences* 46 (2014): 1–14.

Ong, Aihwa, and Nancy N. Chen, eds. *Asian Biotech: Ethics and Communities of Fate*. Durham: Duke University Press, 2010.

Overtveldt, Johan Van. *The Chicago School: How the University of Chicago Assembled the Thinkers Who Revolutionized Economics and Business*. Chicago: B2 Books, 2007.

Owens, Larry. "Patents, the 'Frontiers' of American Invention, and the Monopoly Committee of 1939: Anatomy of a Discourse." *Technology and Culture* 32 (1991): 1076–93.

Ozeki, Haruo, and Hideo Ikeda. "Transduction Mechanisms." *Annual Review of Genetics* 2 (1968): 245–78.

Park, Buhm Soon. "The Development of the Intramural Research Program at the National Institutes of Health after World War II." *Perspectives in Biology and Medicine* 46 (2003): 383–402.

Peyrieras, Nadine, and Michel Morange. "The Study of Lysogeny at the Pasteur Institute (1950–1960): An Epistemologically Open System." *Studies in History and Philosophy of Biological and Biomedical Sciences* 33 (2002): 419–30.

Pisano, Gary P. *Science Business: The Promise, the Reality, and the Future of Bio-tech*. Cambridge: Harvard Business School Press, 2006.

Podolsky, Scott H., and Alfred I. Tauber. *The Generation of Diversity: Clonal Selection Theory and the Rise of Molecular Immunology*. Cambridge: Harvard University Press, 1997.

Posner, Richard A. *Economic Analysis of the Law*. Boston: Little Brown, 1973.

Rabinow, Paul. *Essays on the Anthropology of Reason*. Princeton: Princeton University Press, 1996.

———. *Making PCR: A Story of Biotechnology*. Chicago: University of Chicago Press, 1996.

———. *French DNA: Trouble in Purgatory*. Chicago: University of Chicago Press, 1999.

Rabinow, Paul, and Talia Dan-Cohen. *A Machine to Make a Future: Biotech Chronicles*. Princeton: Princeton University Press, 2005.

Rai, Arti K. "Regulating Scientific Research: Intellectual Property Rights and the Norms of Science." *Northwestern University Law Review* 94 (1999): 77–152.

Rasmussen, Nicholas. "The Forgotten Promise of Thiamin: Merck, Caltech Biologists, and Plant Hormones in a 1930s Biotechnology Project." *Journal of the History of Biology* 32 (1999): 245–61.

Reimers, Niels. "Mechanisms for Technology Transfer: Marketing University Technology." In *Technology Transfer: University Opportunities and Responsibilities, A Report on the Proceedings of a National Conference on the Management of University Technology Resources*, 100–108. Case Western Reserve University, October 15 and 16, 1974.

———. "Tiger by the Tail," *Chemtech* 17 (1987): 464–71.

———. "Stanford's Office of Technology Licensing and the Cohen/Boyer Cloning Patents." An oral history conducted in 1997 by Sally Smith Hughes, PhD. Regional Oral History Office, Bancroft Library, University of California, Berkeley, 1998.

———. "A Personal History of the Stanford University Office of Technology Licensing." Unpublished manuscript, 2003. Document courtesy of Stanford Office of Technology Licensing.

Reingold, Nathan. "Science and Government in the United States since 1945." *History of Science* 32 (1994): 361–86.

Rettig, Richard A. *Cancer Crusade: The Story of the National Cancer Act of 1971*. Princeton: Princeton University Press, 1977.

Rheinberger, Hans-Jörg. "Beyond Nature and Culture: A Note on Medicine in the Age of Molecular Biology." *Science in Context* 8 (1995): 249–63.

———. *Toward a History of Epistemic Things: Synthesizing Proteins in the Test Tube*. Stanford: Stanford University Press, 1997.

———. "Gene Concepts: Fragments from the Perspective of Molecular Biology." In *The Concept of the Gene in Development and Evolution: Historical and Epis-*

temological Perspectives, edited by Peter J. Beurton, Raphael Falk, and Hans-Jörg Rheinberger, 219–39. Cambridge: Cambridge University Press, 2000.

Riley, Paddy. "Clark Kerr: From the Industrial to the Knowledge Economy." In *American Capitalism: Social Thought and Political Economy in the Twentieth Century*, edited by Nelson Lichtenstein, 71–87. Philadelphia: University of Pennsylvania Press, 2006.

Robbins-Roth, Cynthia. *From Alchemy to IPO: The Business of Biotechnology*. Cambridge: Perseus Publishing, 2000.

Rosenberg, Charles E. *No Other Gods: On Science and American Social Thought*. Revised and expanded edition. Baltimore: Johns Hopkins University Press, 1997.

Rowland, Bertram. "Bertram Rowland and the Cohen/Boyer Cloning Patent." Available at http://www.law.gwu.edu/Academics/FocusAreas/IP/Pages/Cloning.aspx.

Rubin, Gerald M., and Edward B. Lewis, "A Brief History of *Drosophila*'s Contributions to Genome Research." *Science* 287 (2000): 2216–18.

Sambrook, Joseph, Heiner Westphal, P. R. Srinivas, and Renato Dulbecco. "Integrated State of Viral DNA in SV40-Transformed Cells." *Proceedings of the National Academy of Sciences, USA* 60 (1968): 1288–95.

Sandelin, Jon. "A History of the Association of University Technology Managers." Unpublished manuscript, October 2005. Document courtesy of Jon Sandelin.

Santesmases, María J. "Severo Ochoa and the Biomedical Sciences in Spain under Franco, 1959–1975." *Isis* 91 (2000): 706–34.

Sarkar, Sahotra, ed. *The Philosophy and History of Molecular Biology: New Perspectives*. Dordrecht: Kluwer Academic, 1996.

———. "From Genes as Determinants to DNA as Resource: Historical Notes on Development and Genetics." In *Genes in Development: Re-reading the Molecular Paradigm*, edited by Eva M. Neumann-Held and Christoph Rehmann-Sutter, 77–95. Durham: Duke University Press, 2006.

Saxenian, AnnaLee. *Regional Advantage: Culture and Competition in Silicon Valley and Route 128*. Cambridge: Harvard University Press, 1994.

Schwartz, Robert S. "Review of *The Golden Helix: Inside Biotech Ventures*." *New England Journal of Medicine* 333:19 (1995): 1292–93.

Selya, Rena. *Salvador Luria's Unfinished Experiment: The Public Life of a Biologist in a Cold War Democracy*. PhD dissertation, Harvard University, 2002.

Sgaramella, Vittorio, J. H. van de Sande, and H. Gobind Khorana. "Studies on Polynucleotides, C: A Novel Joining Reaction Catalyzed by the T4-Polynucleotide Ligase." *Proceedings of the National Academy of Sciences, USA* 69 (1972): 1468–75.

Shapin, Steven. "The House of Experiment in Seventeenth-Century England." *Isis* 79 (1988): 373–404.

———. "Who Is the Industrial Scientist?: Commentary from Academic Sociology and from the Shop-Floor in the United States, ca. 1900–ca. 1970." In *The Science-Industry Nexus: History, Policy, Implications*, edited by Karl Grandin, Nina Wormbs, Anders Lundgren and Sven Widmalm, 337–63. Sagamore Beach: Science History Publications, 2004.

———. *The Scientific Life: A Moral History of a Late Modern Vocation*. Chicago: University of Chicago Press, 2008.

Shapin, Steven, and Simon Schaffer. *Leviathan and the Air-Pump: Hobbes, Boyle, and the Experimental Life*. Princeton: Princeton University Press, 1989.

Sherwin, Chalmers W., and Raymond S. Isenson. "Project Hindsight: A Defense Department Study of the Utility of Research." *Science* 156 (1967): 1571–77.

Singer, Maxine. "Leon Heppel and the Early Days of RNA Biochemistry." *Journal of Biological Chemistry* 278 (2003): 47351–56.

Singer, Maxine, and Paul Berg. *Genes and Genomes: A Changing Perspective*. Mill Valley: University Science Books, 1991.

Singer, Maxine, and Dieter Söll. "Guidelines for DNA Hybrid Molecules." *Science* 181 (1973): 1114.

Sinsheimer, Robert L. *Interview with Robert L. Sinsheimer*. Pasadena: California Institute of Technology Archives, 1992.

Slater, Leo B. "Industry and Academy: The Synthesis of Steroids." *Historical Studies in the Physical and Biological Sciences* 30 (2000): 443–80.

———. *War and Disease Biomedical Research on Malaria in the Twentieth Century*. New Brunswick: Rutgers University Press, 2009.

Smith, Hamilton O. and Kent W. Wilcox. "A Restriction Enzyme from *Hemophilus influenzae*. I: Purification and General Properties." *Journal of Molecular Biology* 51 (1970): 379–91.

Smith, Mark A. *The Right Talk: How Conservatives Transformed the Great Society into the Economic Society*. Princeton: Princeton University Press, 2007.

Spath, Susan B. *C. B. van Niel and the Culture of Microbiology, 1920–1965*. PhD dissertation, University of California, Berkeley, 1999.

Spiegelman, Sol. "Differentiation as the Controlled Production of Unique Enzymatic Patterns." In *Growth in Relation to Differentiation and Morphogenesis*, edited by J. F. Danielli and K. Brown, 286–325. Cambridge: Cambridge University Press, 1948.

"Stanford Shuts Open Door to Cohen-Boyer Patent File as Application Hangs Fire." *Biotechnology Newswatch* 2 (1982): 1.

Stanier, Roger Y., and C. B. van Niel. "The Concept of a Bacterium." *Archiv für Mikrobiologie* 42 (1962): 17–35.

Stent, Gunter. "The Operon: On Its Third Anniversary." *Science* 144 (1964): 816–20.

———. *The Coming of the Golden Age: a View of the End of Progress*. Garden City: Natural History Press, 1969.

Stokes, Donald. *Pasteur's Quadrant: Basic Science and Technological Innovation*. Washington, DC: Brookings Institution, 1997.

Strack, Hans B., and A. Dale Kaiser. "On Structure of Ends of Lambda DNA." *Journal of Molecular Biology* 12 (1965): 36–49.

Strasser, Bruno J. "Who Cares about the Double Helix?" *Nature* 422 (2003): 803–4.

———. "Collecting and Experimenting: The Moral Economies of Biological Databases, 1960s–1980s." *Preprints of the Max-Planck Institute for the History of Science* 310 (2006): 105–23.

———. "World in One Dimension: Linus Pauling, Francis Crick and the Central Dogma of Molecular Biology." *History and Philosophy of the Life Sciences* 28 (2006): 491–512.

———. "The Experimenter's Museum: GenBank, Natural History, and the Moral Economies of Biomedicine." *Isis* 102 (2011): 60–96.

Strickland, Stephen P. *Politics, Science, and Dread Disease: A Short History of United States Medical Research Policy.* Cambridge: Harvard University Press, 1972.

Technology Transfer: University Opportunities and Responsibilities, A Report on the Proceedings of a National Conference on the Management of University Technology Resources. Cleveland: Case Western Reserve University, October 15 and 16, 1974.

Teitelman, Robert. *Gene Dreams: Wall Street, Academia, and the Rise of Biotechnology.* New York: Basic Books, 1989.

Teles, Steven M. *The Rise of the Conservative Legal Movement: The Battle for Control of the Law.* Princeton: Princeton University Press, 2008.

Temussi, P. A. "Automatic Comparison of the Sequence of Calf Thymus Histones." *Journal of Theoretical Biology* 50 (1975): 25–33.

Thackray, Arnold, ed. *Private Science: Biotechnology and the Rise of Molecular Sciences.* Philadelphia: University of Pennsylvania Press, 1998.

Thomas, C. A., Jr. "The Genetic Organization of Chromosomes." *Annual Review of Genetics* 5 (1971): 237–56.

Thompson, Charis. *Making Parenthood: The Ontological Choreography of Reproductive Technology.* Cambridge: MIT Press, 2005.

Thompson, Edward P. "The Moral Economy of the English Crowd in the Eighteenth Century." *Past and Present* 50 (1971): 76–136.

Time. "Pure Knowledge Vs. Pure Profit." September 28, 1981.

Uchida, Hisao. "Building a Science in Japan: The Formative Decades of Molecular Biology." *Journal of the History of Biology* 26 (1993): 499–517.

US Congressional Hearing. *Committee on Science and Technology, Government Patent Policy: The Ownership of Inventions Resulting from Federally Funded Research and Development.* Washington, DC: US Government Printing Office, 1976.

US General Accounting Office. *Problem Areas Affecting Usefulness of Results of Government-Sponsored Research in Medical Chemistry: A Report to the Congress.* Washington, DC: US Government Printing Office, 1968.

Vagelos, P. Roy, and Louis Galambos. *Medicine, Science, and Merck*. Cambridge: Cambridge University Press, 2004.

Varmus, Harold. "The Pastorian: A Legacy of Louis Pasteur." In *Advances in Cancer Research*, edited by Gorge F. Woude and George Klein, 69:1–16. New York: Academic Press, 1996.

Vettel, Eric J. "The Protean Nature of Stanford University's Biological Sciences, 1946–1972." *Historical Studies in the Physical and Biological Sciences* 35 (2004): 95–113.

———. *Biotech: The Countercultural Origins of an Industry*. Philadelphia: University of Pennsylvania Press, 2006.

Vogt, Marguerite, and Renato Dulbecco. "Virus-Cell Interaction with a Tumor-Producing Virus." *Proceedings of the National Academy of Sciences, USA* 46 (1960): 365–70.

Waddington, Conrad H. "Gene Regulation in Higher Cells." *Science* 166 (1969): 639–40.

Warhol, Andy. *The Philosophy of Andy Warhol: From A to B and Back Again*. New York: Harcourt Brace Jovanovich, 1975.

Washburn, Jennifer. *University, Inc.: The Corporate Corruption of American Higher Education*. New York: Basic Books, 2005.

Watkins, J. F., and Renato Dulbecco. "Production of SV40 Virus in Heterokaryons of Transformed and Susceptible Cells." *Proceedings of the National Academy of Sciences, USA* 58 (1967): 1396–403.

Watson, James D. *Molecular Biology of the Gene*. New York: W. A. Benjamin, 1965.

———. *The Double Helix: A Personal Account of the Discovery of the Structure of DNA*. New York: Atheneum, 1968.

Watson, James D., and John Tooze. *The DNA Story: A Documentary History of Gene Cloning*. San Francisco: W. H. Freeman and Co., 1981.

Weber, Marcel. "Representing Genes: Classical Mapping Techniques and the Growth of Genetical Knowledge." *Studies in History and Philosophy of Biological and Biomedical Sciences* 29 (1998): 295–315.

———. "Walking on the Chromosome: *Drosophila* and the Molecularization of Development." In *From Molecular Genetics to Genomics: Mapping Cultures of Twentieth Century Genetics,* edited by Jean-Paul Gaudillière and Hans-Jörg Rheinberger, 63–78. London: Routledge, 2004.

———. "Redesigning the Fruit Fly: The Molecularization of *Drosophila*." In *Science without Laws: Model Systems, Cases, Exemplary Narratives*, edited by Angela N. H. Creager, Elizabeth Lunbeck, and M. Norton Wise, 23–45. Durham: Duke University Press, 2008.

Weber, Marcel. *Philosophy of Experimental Biology*. Cambridge: Cambridge University Press, 2004.

Weigle, Jean J., and Max Delbrück. "Mutual Exclusion between an Infecting Phage and a Carried Phage." *Journal of Bacteriology* 62 (1951): 301–18.

Weiner, Charles. "Patenting and Academic Research: Historical Case Studies." In *Owning Scientific and Technical Information: Value and Ethical Issues*, edited by Vivian Weil and John W. Snapper, 87–109. New Brunswick: Rutgers University Press, 1989.

Weiner, Jonathan. *Time, Love, Memory: A Great Biologist and His Quest for the Origins of Behavior*. New York: Knopf, 1999.

Wellerstein, Alex. "Patenting the Bomb: Nuclear Weapons, Intellectual Property, and Technological Control." *Isis* 99 (2008): 57–87.

Wensink, Pieter C., David Finnegan, John E. Donelson, and David S. Hogness. "A System for Mapping DNA Sequences in the Chromosome of *Drosophila melanogaster*." *Cell* 3 (1974): 315–25.

Westphal, Heiner and Renato Dulbecco. "Viral DNA in Polyoma- and SV40-Transformed Cell Lines." *Proceedings of the National Academy of Sciences, USA* 59 (1968): 1158–65.

Wilcox, G., J. Abelson, C. F. Fox, eds. *Proceedings of the 1977 ICN/UCLA Symposium: Eucaryotic Genetic Systems*. New York: Academic Press, 1977.

Williams, Bruce. "The Economic Impact of Science and Technology in Historical Perspective." *Minerva* 20 (1982): 301–12.

Wilson, Edward O. *Naturalist*. Washington, DC: Island Press/Shearwater Books, 1994.

Wisnioski, Matthew. *Engineers for Change: Competing Visions of Technology in 1960s America*. Cambridge: MIT Press, 2012.

Wollman, Elie L. "Bacterial Conjugation." In *Phage and the Origins of Molecular Biology*, edited by John Cairns, Gunther S. Stent and James D. Watson, 216–25. Plainview: Cold Spring Harbor Laboratory Press, 1966/1992.

Wood, William and Paul Berg, "The Effect of Enzymatically Synthesized Ribonucleic Acid on Amino Acid Incorporation by a Soluble Protein-Ribosome System from *Escherichia coli*." *Proceedings of the National Academy of Sciences, USA* 48 (1962): 94–104.

Wright, Susan. "Recombinant DNA Technology and Its Social Transformation, 1972–1982." *Osiris* 2 (1986): 303–60.

———. *Molecular Politics: Developing American and British Regulatory Policy for Genetic Engineering, 1972–1982*. Chicago: University of Chicago Press, 1994.

Yamamoto, Keith. "Faculty Members as Corporate Officers: Does Cost Outweigh Benefit?" In *From Genetic Experimentation to Biotechnology: The Critical Transition*, edited by William Whelan and Sandra Black, 195–201. New York: John Wiley & Sons, 1982.

Yanofsky, Charles. "What Will Be the Fate of Research on Prokaryotes?" *Cell* 65 (1991): 199–200.

Yanofsky, Charles, B. C. Carlton, D. R. Helinski, J. R. Guest, and U. Henning. "On Colinearity of Gene Structure and Protein Structure." *Proceedings of the National Academy of Sciences, USA* 51 (1964): 266–72.

Yi, Doogab. "Cancer, Viruses, and Mass Migration: Paul Berg's Venture into Eu-karyotic Biology and the Advent of Recombinant DNA Research and Tech-nology, 1967–1974." *Journal of the History of Biology* 41 (2008): 589–636.

————. "The Scientific Commons in the Marketplace: The Industrialization of Biomedical Materials at the New England Enzyme Center, 1963–1980." *History and Technology* 25 (2009): 69–87.

————. "Who Owns What? Private Ownership and the Public Interest in Recom-binant DNA Technology in the 1970s." *Isis* 102 (2011): 446–74.

Young, Michael W., and David S. Hogness. "A New Approach for Identifying and Mapping Structural Genes in *Drosophila melanogaster*." In *Proceedings of the 1977 ICN/UCLA Symposium: Eukaryotic Genetic Systems*, edited by G. Wilcox, J. Abelson, C. F. Fox, 315–31. New York: Academic Press, 1977.

Index

gene cloning experiment with Chang, 107–9; letter to Reimer about possible withdrawal of recombinant DNA patent application, 156–57; Miles Symposium and, 156–57; offer to write for *Scientific American* on gene cloning, 131; official patent filing of recombinant DNA invention, 131; as outsider of Stanford's Biochemistry Department, 115, 122–28; participation in early recombinant DNA research network, 84–5; patent and its relation to university and private industry, 153; plasmid as cloning vector, 101–2; plasmid research, 100–101; prior arts dispute with Mertz, 85, 95–99; pSC101 as strategic tool, 122–23; reactions to public criticisms on recombinant DNA patent application, 156–57; on recombinant DNA risk controversy, 105; relationship between patenting and priority, 130; relation to Stanford's Biochemistry Department via Hurwitz, 99; renege on giving up personal royalties, 136, 267n81; temporary withheld of pSC101, 123–26. *See also* Berg, Paul; Boyer, Herbert W.; Mertz, Janet E.; recombinant DNA technology
Cohn, Melvin 18, 31, 38
coinventors of recombinant DNA technology, 2, 128, 130, 152
countercultural movement 7, 54, 63, 80
Creager, Angela: boundary between public and private knowledge in biotechnology, 224n17; fine-structure mapping, 71; invention of bacterial sexuality, 63; mobile genes, 76; moral economy of science and competition, 230n48; plasmid research, 226n27; role of voluntary health activists, 225n19
Crick, Francis: exchange over secrecy in science with Berg, 99; model for chromosomes of higher organisms, 255n16; relationship between molecular biology and medicine, 54–55, 62, 240n7; structure of DNA, 33
Cuzin, François, 69

Darling, George B., 27–28
Davis, Ronald: collaboration with Janet Mertz, 94; prior arts dispute regarding

recombinant DNA patent application, 85, 94–99, 132–34; systematic study on eukaryotic chromosomes, 124. *See also* Mertz, Janet E.
de Chadarevian, Soraya, 225n19, 230n45, 232n10
Delbrück, Max, 29, 33, 58
Department of Commerce, 157; attempt to institute uniform government patent policy, 161–62; Ancker-Johnson as assistant secretary for science and technology of, 157; intervention with recombinant DNA patent applications, 162; suspension of order to accelerate recombinant DNA patent applications, 164
Department of Defense (DOD), 9–10; licensing policy of, 145; Project Hindsight of, 142; support for university research, 228n36
Department of Health, Education, and Welfare (DHEW), 143; ability to repel patent title waver, 157; Califano as secretary of, 164; Dole's criticism of, 165, 260n6; patent council of, 165; patent policy of, 144; private industry's concerns on title policy of, 144–45; support for university research, 228n36; title policy of, 145–46
Diamond v. Chakrabarty, 167, 172
DNA Department, 15, 19; early research of, 33–41
DNAX, 174; as academic research institution, 195; acquisition by Schering-Plough, 202–3; alternative to commercial biotechnology, 187–95; business flexibility of, 192; business plan of, 196–99; criticisms of startup biotech venture identified by, 190–91; discussion on potential joint venture with Japanese companies, 199–202; equity incentive of, 194–95; founding of, 174; global enterprise, 199–203; immunogenetic engineering project of, 195–99; organizational structure of, 191–92; relation with academia, 192–93; Research Institute of Molecular and Cellular Biology of, 191; scientific advisory board of, 191; Zaffaroni and, 189–90
Dole, Bob (Robert), criticism of DHEW's patent policy, 165, 260n6